Sustainable Civil Infrastructures

About this Series

Sustainable Infrastructure impacts our well-being and day-to-day lives. The infrastructures we are building today will shape our lives tomorrow. The complex and diverse nature of the impacts due to weather extremes on transportation and civil infrastructures can be seen in our roadways, bridges, and buildings. Extreme summer temperatures, droughts, flash floods, and rising numbers of freeze-thaw cycles pose challenges for civil infrastructure and can endanger public safety. We constantly hear how civil infrastructures need constant attention, preservation, and upgrading. Such improvements and developments would obviously benefit from our desired book series that provide sustainable engineering materials and designs. The economic impact is huge and much research has been conducted worldwide. The future holds many opportunities, not only for researchers in a given country, but also for the worldwide field engineers who apply and implement these technologies. We believe that no approach can succeed if it does not unite the efforts of various engineering disciplines from all over the world under one umbrella to offer a beacon of modern solutions to the global infrastructure. Experts from the various engineering disciplines around the globe will participate in this series, including: Geotechnical, Geological, Geoscience, Petroleum, Structural, Transportation, Bridge, Infrastructure, Energy, Architectural, Chemical and Materials, and other related Engineering disciplines.

More information about this series at http://www.springer.com/series/15140

Wen-Chieh Cheng · Junsheng Yang
Jinfeng Wang
Editors

Tunneling in Soft Ground, Ground Conditioning and Modification Techniques

Proceedings of the 5th GeoChina International Conference 2018 – Civil Infrastructures Confronting Severe Weathers and Climate Changes: From Failure to Sustainability, held on July 23 to 25, 2018 in HangZhou, China

Springer

Editors
Wen-Chieh Cheng
School of Civil Engineering
Xi'an University of Architecture and Technology
Xi'an, China

Jinfeng Wang
Zhejiang University
HangZhou, China

Junsheng Yang
Central South University
Changsha, China

ISSN 2366-3405 ISSN 2366-3413 (electronic)
Sustainable Civil Infrastructures
ISBN 978-3-319-95782-1 ISBN 978-3-319-95783-8 (eBook)
https://doi.org/10.1007/978-3-319-95783-8

Library of Congress Control Number: 2018948639

This Springer imprint is published by the registered company Springer Nature Switzerland AG
The registered company address is: Gewerbestrasse 11, 6330 Cham, Switzerland

Contents

Introduction

Given that urbanization is promptly proceeded by not only coastal cities in China but also others in western region, scale and number of their infrastructures are deemed to be critical for a self-review during urbanization. Subway system, one of the essential infrastructures, can basically be utilized to represent the pace of urbanization. Shield tunneling has been used for subway tunnel construction. However, tunneling in soft ground may lead to great potential of triggering geo-hazards and causing damages to nearby properties. This volume presents a series of high-quality papers discussing expansive soil stabilization, tunnel bore characterization, local voids effect on cracking performance of an existing tunnel lining, insulation layer design for high altitude cold region tunnel, explosive consumption of rock fragmentation by blasting in tunneling, tunneling-induced ground settlement prediction, canopy filling influence on shallow tunnels, etc. The studies can be shared with researchers and engineers as well as experts from various engineering disciplines around the globe and utilized for prevention of geo-hazards and mitigation of damages, while implementing similar underground works in the future. This volume is part of the proceedings of the 5th GeoChina International Conference on Civil Infrastructures Confronting Severe Weathers and Climate Changes: From Failure to Sustainability, held in HangZhou, China 2018. Hangzhou is renowned for its historic relics and natural beauty.

Predicting Subgrade Resilient Modulus for Use in the MEPDG Using Common Soil Indices

Mena I. Souliman[1(✉)], Christopher J. Strunk[1], and Lubinda F. Walubita[2]

[1] The University of Texas at Tyler, Tyler, TX, USA
msouliman@uttyler.edu
[2] The Texas A&M University System, College Station, TX, USA

Abstract. In roadway pavement design, one must consider the performance of the pavement structure and mixture design. The tendency for pavement to be more flexible in nature gives increased importance to the properties and condition of the underlying soil subgrade. Traditional foundation design normally considers the static strength response of the soil from stationary loads. Roadway pavement design, however, must consider the dynamic strength response created by moving traffic loads. This dynamic strength of soil, or "stiffness," is known as its resilient modulus (M_R). The resilient modulus is determined by costly and time-consuming laboratory tests. For these reasons, it would be greatly beneficial to find an effective, yet simpler method to determine the values of M_R. The goal of this study was to utilize multiple regression analyses to determine the relationships between common soil index properties, M_R, and use these relationships to create an equation to model. The methodology used in the creation of these equations will be described as well as a statistical evaluation of their performance. Subgrade soil property data for 253 different soils found in northeast Texas was mined from Arizona State University's "National Catalog of Natural Subgrade Properties Needed for the ME-PDG Input." The common soil index properties studied include five sieve analysis gradations, liquid limit, and plasticity index. These properties were analyzed against M_R.

1 Introduction

In 2008 the American Association of State Highway and Transportation Officials (AASHTO) produced the most recent version of the Mechanistic-Empirical Pavement Design Guide (MEPDG) in order to describe pavement design methodology. The guidance has been validated with extensive field performance data for the latest in pavement design. The MEPDG design approach identifies foundation analysis as a key step of the process. Within this analysis, M_R determination and other hydraulically related factors are important characteristics. In fact, resilient modulus determination is required for all unbound paving layers and the foundation (AASHTO 2008).

The resilient modulus is recognized as an important property in subgrade performance. It provides a measure of the stiffness of the subgrade soil. M_R values are

W.-C. Cheng et al. (eds.), *Tunneling in Soft Ground, Ground Conditioning and Modification Techniques*, Sustainable Civil Infrastructures,
https://doi.org/10.1007/978-3-319-95783-8_1

determined through complex laboratory tests performed on undisturbed samples by the application of a confining pressure and a repeated tri-axial loading to represent the loading effects of traffic (Coleri 2007). These tests are time-consuming and expensive.

2 Objective

The goal of this study was to predict the values of the resilient for use in the MEPDG with subgrade soil index properties retrieved from soil families compiled by a national soil property database and located within the northeast portion of the state of Texas. Prediction models were created using a mechanistic-empirical method by utilizing multiple linear regression statistical analysis in Microsoft Excel. Statistical relationships between soil index properties were analyzed against resilient modulus in these models. The produced models were evaluated for validity using statistical verification methods. The result of this research shall provide initial empirical formulas that can be used by researchers and engineers for the determination of M_R for soils found in northeast Texas. This methodology shall be applicable for use in any localized region of the United States for the creation of similar predictive models.

3 Literature Review

Resilient modulus of subgrade material is defined as the resilient response after a number of repeated loads. It is calculated by dividing the resilient strain by the deviatoric stress. After a large number of tests, the amount of permanent deformation of the material approaches zero (Huang 1993).

The importance of resilient modulus in flexible pavement design is not to be underestimated. In the MEPDG resilient modulus is a required input for designing all unbound pavement layers and the foundation (AASHTO 2008). This sentiment is reflected among all other researchers involved in pavement design. One researcher makes note that, “In any mechanistically based design/analysis procedure for flexible pavement, the resilient modulus of pavement materials is the prime input material property necessary for determining deflection is layered systems, resilient stress and strains, and for analyzing the performance of the system.” (Davies 2004).

Laboratory methods of determining resilient modulus is time-consuming and expensive. Other researchers have attempted to predict resilient modulus values using simpler methods than full laboratory testing.

Ravindra (2004) developed models to estimate the resilient modulus of base and subgrade soils from in situ test devices. Two types of cohesive soils and three types of granular soils commonly used in Louisiana were considered. Three types of in situ devices Geo-gauge, Light Falling Weight Deflectometer, Dynamic Cone Penetrometer and a laboratory repeated triaxial tests were conducted. The resilient modulus predicted for both type of soils was in agreement with the lab measured resilient modulus.

Nguyen and Mohajerani (2016) proposed a simplification of the normal methods of determining resilient modulus for pavement design. Their methods showed that acceptable M_R values could be obtained by using fewer tri-axial loading tests than standard testing procedures and a select set of model equations.

Coleri (2007) used a statistical approach in his Master of Science thesis for developing a relationship between resilient modulus and soil index properties of unbound materials. Data was collected by performing tests on soil samples representative of those found in Turkey. The final developed model predicting M_R held a relatively high R^2 value of 0.8120, showing good fit, and is as follows (Coleri 2007):

$$M_R = 27.4079\left(\frac{(LL + 28.5) \times (DDstnd + 19.5)}{Wcc}\right)^{1.79706} + \left(\frac{PP200}{100}\right)^{0.166677} \quad (1)$$

where,

LL = Liquid Limit
DDstnd = Maximum Dry-Density (Mg/m^3)
Wcc = Compaction Water Content (%)
PP200 = Percent Passing No. 200 Sieve (%).

A 1994 study compiled by K.P. George in conjunction with the Mississippi Department of Transportation, the United States Department of Transportation, and the Federal Highway Administration noted that establishing reliable correlation equations from in situ soil test data with resilient modulus values would be desirable. The statistical study suggests that the top Mississippi soil index properties influencing M_R include moisture content, degree of saturation, material passing #200 sieve, plasticity index, and density. For the purpose of determining if any M_R predicting model was suitable for soil types found in Mississippi, the strongest results were found using a LTTP-FHWA study program equation. The governing equation takes the form (George 1994):

$$M_R = k_1 P_a \left(\frac{\theta}{p_a}\right)\left(\left(\frac{\tau_{oct}}{p_a}\right) + 1\right)^{k_s} \quad (2)$$

where,

k_1, k_2, and k_3 = Material-Specific Regression Constants
P_a = Atmospheric Pressure
θ = Bulk Stress
τ_{oct} = Octahedral Shear Stress.

Previous studies have detailed that the resilient modulus is an important property required for input in the MEPDG. Testing procedures take considerable time and money. Research has been conducted in evaluating alternative methodologies for determining M_R with some success.

4 Methodology

4.1 Database Creation

The National Database of Subgrade Soil-Water Characteristic Curves and Selected Soil Properties for Use in the MEPDG (hereby known as, database) produced in 2010 by Arizona State University was utilized as the source for data in this study since it was created for direct use in the MEPDG. The database can be used for research in the creation of M_R models (Zapata 2010) (Figs. 1 and 2).

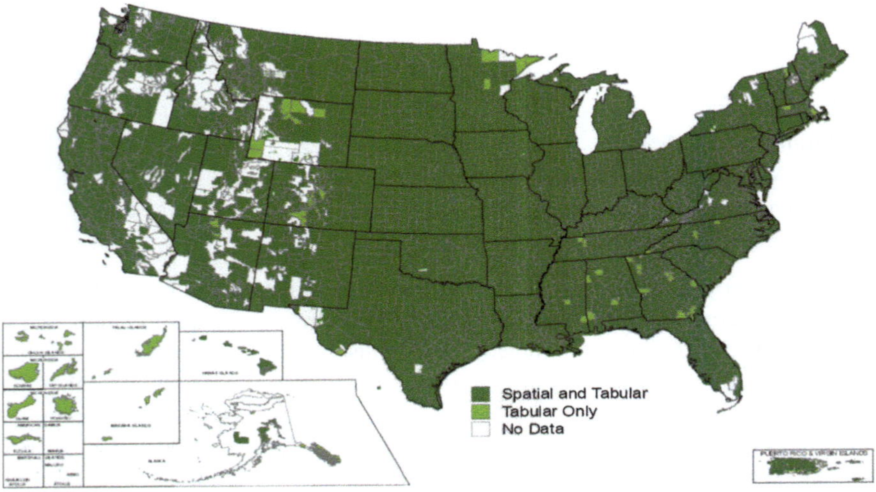

Fig. 1. Soil survey data available (Zapata 2010)

The subgrade soil data collected for this study was compiled by selecting the localized area of northeast Texas. Specifically, soil map regions 5, 6, 7, 11, 12, and 13 of the state as defined by the database and shown in Fig. 3. Subgrade soil index properties, M_R values were retrieved from these regions in the database and compiled in Excel. The soil index properties included in this study include particle sizes (percent passing sieve number) P(4), P(10), P(40), P(200), and P(0.002), Liquid Limit (LL), and Plasticity Index (PI). Soil units with missing information were excluded from this study. A total of 253 subgrade soil types found in northeast Texas were collected and used for this study.

4.1.1 Data Analysis

The data was first randomized using a simple RAND() function in Excel, assigning a number to each soil type. The data was sorted from smallest to largest randomly generated number. The first 80% of the data (202 datasets) was selected for model generation, with the remaining 20% of the data (51 datasets) separated for model validation discussed later in this study.

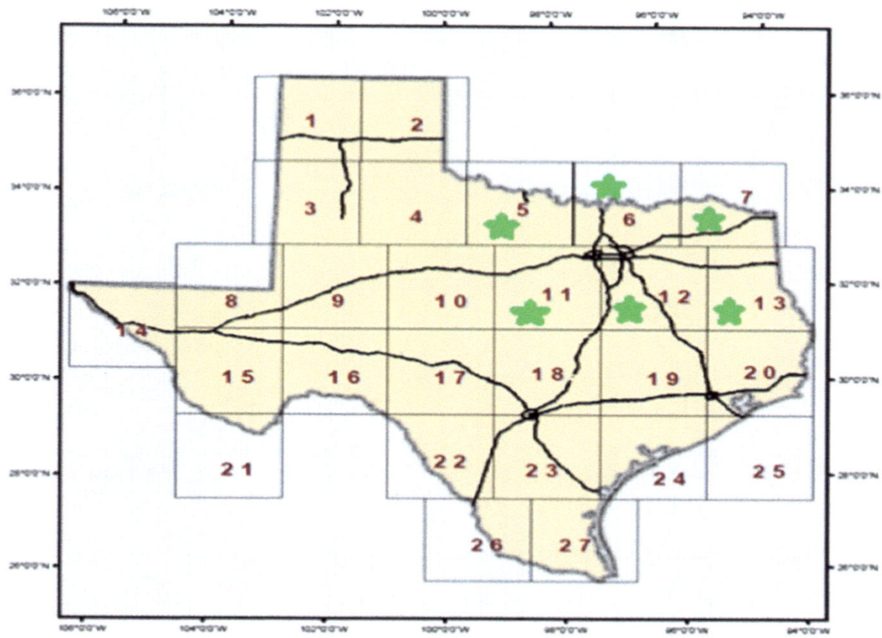

Fig. 2. Database map regions in the state of Texas. Regions 5, 6, 7, 11, 12, and 13 were chosen for this study

Relationships between soil index properties (independent variables) and M_R was determined by plotting each independent variable against each dependent variable. Within each plot, trend lines were created to observe the statistical correlations and mathematical relationships identified.

The relationships between M_R and each soil index property are shown in Fig. 4. Strong relationships were identified between M_R and independent variables P(0.002), LL, and PI. A moderately strong relationship occurs between M_R and P(200). Poor correlations exist with the remaining soil index properties.

4.1.2 Model Creation

Having identified the mathematical correlations between the dependent and independent variables, the independent variables were transformed by their relationship with the dependent variables. Multiple linear regression analysis was performed in Excel to create statistical models that will predict the dependent variable based on the independent variable and its coefficient. After the first analysis was performed, the independent variables were analyzed to determine which, if any, were statistically significant in the model. Independent variables with a P-value greater than 0.05 were eliminated from the model. The independent variables with a P-value less than 0.05

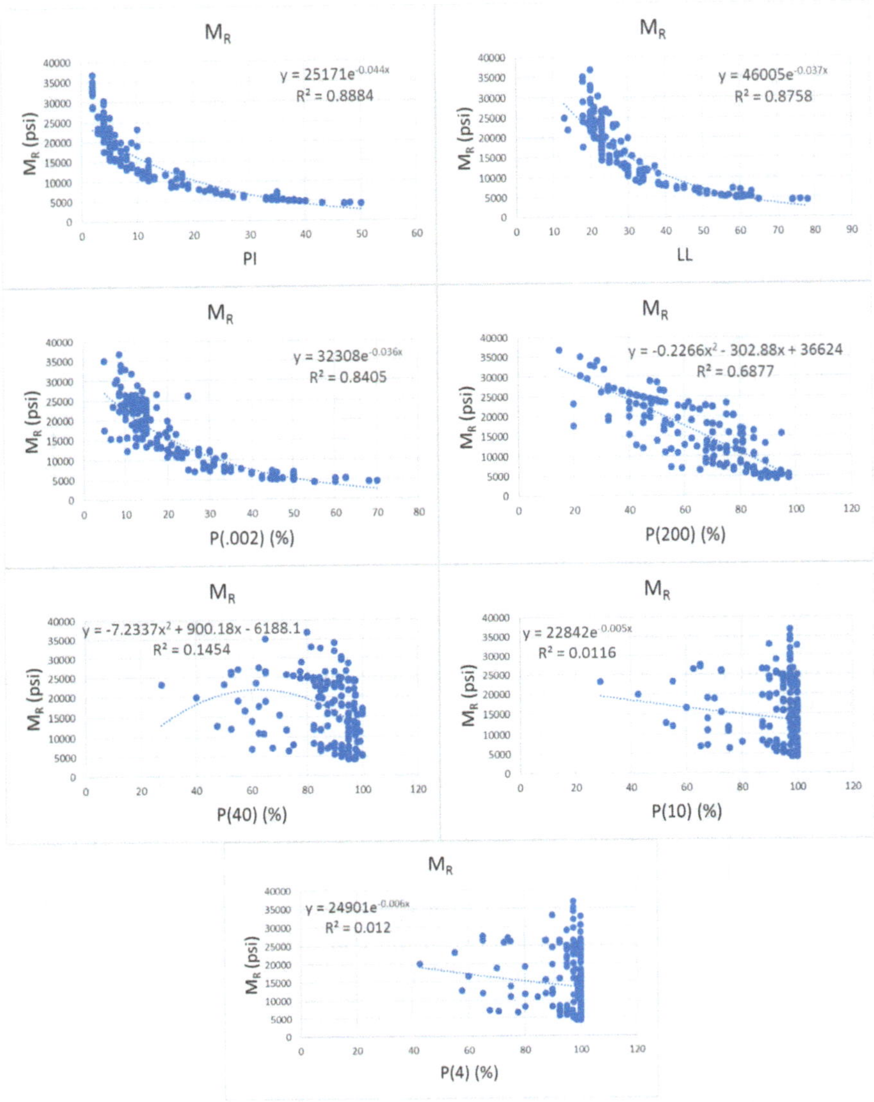

Fig. 3. Soil property relationships with resilient modulus

were deemed statistically significant and were selected for additional multiple linear regression analysis runs. This process was repeated until all independent variables possessed a P-value of less than 0.05.

Results from each multiple linear regression run are available upon request. The final outputs for each dependent variable is shown in Table 1.

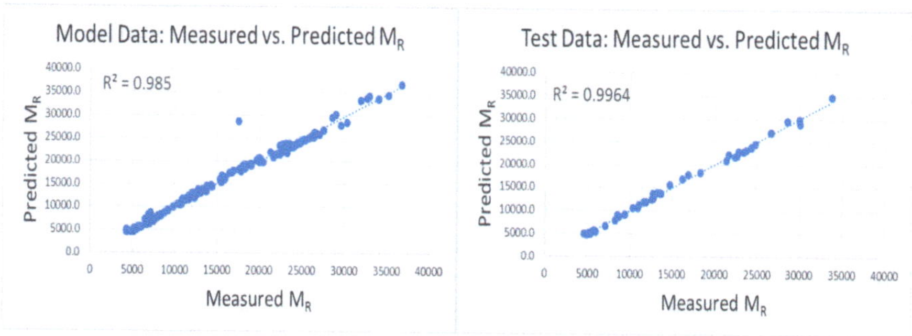

Fig. 4. Graphical summary of M_R model performance

Table 1. Final multiple regression analysis for M_R

Summary output					
Regression Statistics					
Multiple R	0.9924675				
R square	0.9849917				
Adjusted R square	0.9846089				
Standard error	1012.0469				
Observations	202				
ANOVA					
	df	SS	MS	F	Significance F
Regression	5	1.32E+10	2.60E+09	2572.7	1.39E−176
Residual	196	2.01E+08	1,024,239		
Total	201	1.34E+10			
	Coefficients	Standard error	t Stat	P-value	
Intercept	45,555.7	822.16	55.41	1.08E−121	
P(200)	−293.07	24.96	−11.74	1.87E−24	
P(200)^2	1.28	0.21	6.18	3.59E−09	
ln(P(0.002))	891.23	321.79	2.77	6.15E−03	
PI	219.01	20.68	10.59	5.17E−21	
ln(PI)	−10,018	273.35	−36.65	1.17E−89	

5 Results

5.1 Proposed Model

From the results of the multiple linear regression analysis, new model was created for the dependent variables M_R. The model for M_R showed the highest correlation between the dependent and independent variables. The R^2 value of 0.98 for the model is very high. The proposed equation is as follows:

$$M_R = 45{,}555.69 - 293.07 \times P(200) + 1.275 \times P(200)^2 + 891.23 \times \ln(P(0.002)) + 219.01 \times PI - 10{,}018.347 \times \ln(PI) \quad (3)$$

Model Performance and Validation

The performance of the proposed model was tested against model soil index property data that was used in the creation of the model as well as additional test data that was set aside for model validation purposes. The measured value of the dependent variable was compared against the predicted value in three ways. These validation tests included average percent error, a Student's t-Test, and a graphical representation of the two values with trend line regression values and standard error divided by the standard deviation (S_e/S_y) ratios that describe how spread out the data is.

Performance and Validation of M_R Model:

- Model Data (%) Error: 3.90%
- Test Data (%) Error: 2.74%

From Fig. 4 we can conclude that, the M_R model is statistically valid and shows very good performance for both model and test data. The percent error is very low, the P-values for the one-tail Student's t-Test are greater than 0.05 indicating no statistical difference between datasets, the trend line R^2 values are both very high, and the S_e/S_y ratio is low which indicate a low spread in the results. The model performs well enough that one should be able to input the soil index properties for any subgrade soil in northeast Texas and expect to receive an accurate assessment of the resilient modulus.

6 Conclusion

This study attempted to create a model that would predict resilient modulus by using multiple linear regression analysis to fit common subgrade soil index properties that can easily be identified with quick and inexpensive field tests. Northeast Texas subgrade soil data collected from the national database intended for use with the MEPDG was used in the creation of these models. The goal was to be able to use these models to aid engineers in the design of flexible pavement systems.

The results of the study provided new models for predicting M_R value. The M_R model performed well and could be used for analysis of northeast Texas soil. Further research in the application of this methodology in different regions of the United States would be useful in determining its validity in all areas. The statistical model presented in this study should be analyzed with additional soil properties to determine any additional correlations that could improve the performance of the models.

Acknowledgements. The contents presented in this paper reflect the views of the authors who are responsible for the facts and accuracy of the data presented herein and do not necessarily reflect the official views or policies of any agency or institute.

References

American Association of State Highway and Transportation Officials (AASHTO): Mechanistic-empirical pavement design guide (2008)

Coleri, E.: Relationship between resilient modulus and soil index properties of unbound materials. Unpublished Master's Thesis, Middle East Technical University, Ankara, Turkey (2007)

Davies, B.O.A.: A model for the prediction of subgrade soil resilient modulus for flexible-pavement design: influence of moisture content and climate change. Unpublished Master's Thesis, The University of Toledo, Toledo, OH (2004)

George, K.P.: Prediction of resilient modulus from soil index properties. Department of Civil Engineering, The University of Mississippi, Oxford, MS (2004)

Huang, Y.H.: Pavement Analysis and Design. Prentice Hall, Englewood Cliffs, New Jersey (1993)

Ravindra, G.: Development of resilient modulus prediction models for base and subgrade pavement layers from in-situ devices test results. Master's Thesis, Louisiana State University (2004)

Nguyen, B.T., Mohajerani, A.: Resilient modulus of fine-grained soil and a simple testing and calculation method for determining an average resilient modulus value for pavement design. Transp. Geotech. **7**, 59–70 (2016)

Zapata, C.E.: Research Results Digest 347: A National Database of Subgrade Soil-Water Characteristic Curves and Selected Soil Properties for Use with the MEPDG. National Cooperative Highway Research Program, Transportation Research Board of the National Academies. ISBN: 978-0-309-43012-8 (2010)

Evaluating the Application of Microbial Induced Calcite Precipitation Technique to Stabilize Expansive Soils

Bhaskar Chittoori(✉) and Sikha Neupane

Department of Civil Engineering, Boise State University, Boise, USA
bhaskarchittoori@boisestate.edu,
sikhaneupane@u.boisestate.edu

Abstract. Expansive soils, also known as swell-shrink soils have been a problem for civil infrastructures including roads and foundations from ancient times. The use of chemical additives such as cement and lime to stabilize expansive soils is a common practice among geotechnical engineers, especially for lightly loaded structures. However, several occurrences of subgrade failures have been observed after stabilizing with chemical additives. Hence, engineers are in search of sustainable stabilization alternatives. Microbial Induced Calcite Precipitation (MICP) is gaining attention as an environmentally friendly soil improvement technique. Several researchers have successfully tested its feasibility in mitigating liquefaction-induced problems in sandy soils. In this research, the authors are evaluating its effectiveness in stabilizing expansive soils. For this purpose two natural expansive soils with high and low plasticity properties were subjected to MICP treatments. The soil samples were first augmented with bacterium Sporosarcina Pasteurii and then treated with Calcium Chloride and Urea. Variables such as microbial concentrations and curing times were studied in this research. Geotechnical testing including Atterberg limits and unconfined compression strength were performed to evaluate the efficacy of MICP treatments. Preliminary results indicate that there is a reduction in plasticity and swelling characteristics of the soils and increase in the unconfined compression strength.

1 Introduction and Background

The highly plastic expansive soils cause swelling and shrinking (volume change) with changes in moisture content. Due to these volumetric changes structures built on expansive soils tend to undergo moderate to severe cracking problems (Mitchell 1986; Nelson and Miller 1992). In particular, lightly loaded structures such as one or two story residential and industrial structures and pavements have experienced severe damage (Petry and Little 2002) often associated with substantive repair and mitigation costs. The use of chemical additives such as cement and lime to stabilize these problematic soils is a common practice among geotechnical engineers, especially for lightly loaded structures. However, several occurrences of subgrade failure have been observed after stabilization with chemical additives which indicates a technology gap of sustainable stabilization of expansive soils. Soil stabilization via Microbial Induced

W.-C. Cheng et al. (eds.), *Tunneling in Soft Ground, Ground Conditioning and Modification Techniques*, Sustainable Civil Infrastructures,
https://doi.org/10.1007/978-3-319-95783-8_2

Calcite Precipitation (MICP) is one of the several applications of bio-remediated processes that could fill this gap. This technique employs microbes as a primary contributor for soil stabilization. Successful implementation of MICP will have its application in a wide variety of civil engineering fields such as, stability for retaining walls, embankments and dams; controlling soil erosion; stabilizing cohesionless soils; increasing bearing capacity of shallow and deep foundations; and reducing liquefaction potential of soils (Kucharski et al. 2005; Ivanov and Chu 2008; Kavazamjian and Karatas 2008; Montoya et al. 2014).

Microbes are often responsible for the chemical cementation of soil in nature due to the precipitation of cementing materials into the voids of soils and rocks (Ivanov and Chu 2008). Microbes can precipitate cementing materials such as calcium, magnesium, iron, manganese, and aluminum, which are crystallized to form carbonates, silicates, phosphates, sulfides and hydroxides (DeJong et al. 2006). The prime role of microbes in precipitation of minerals is their ability to create an alkaline environment through various physiological activities (Douglas and Beveridge 1998). Calcium carbonate (calcite) precipitation is observed to be a general mineral precipitation process in the microbial world under the ambient environment (Bang et al. 2001).

Calcite mineralization can occur as a by-product of microbial metabolic activity such as photosynthesis, urea hydrolysis, sulfate reduction and iron reduction. During these different metabolic processes, the alkalinity or pH of the system increases, favoring the calcite precipitation (Knorre and Krumbein 2000). It is believed that bacteria are dominant soil inhabitants. There are 10^6 to 10^{12} bacterial cells in a gram of soil (Torsvik et al. 1990). Sporosarcina pasteurii (previously known as Bacillus pasteurii) species of Bacillus group, a common alkaliphilic soil bacterium has high urease enzyme activity (Dejong et al. 2006). S. pasteurii use urea as an energy source which hydrolyzes Urea ($CO(NH_2)_2$) into ammonia (NH_3) and carbonic acid (H_2CO_3) · NH_3 and H_2CO_3 equilibrate in water to form bicarbonate (HCO_3^-), ammonium (NH_4^+) and hydroxide (OH^-) ions. It is during this stage the pH of system increases and shifts the HCO_3^- equilibrium to form carbonate ion (CO_3^{2-}). The CO_3^{2-} produced will precipitate calcite ($CaCO_3$) in the presence of Ca^{2+} (Dejong et al. 2006). The calcite precipitation is influenced mainly by four factors: calcium ion concentration, dissolved inorganic carbon (DIC) concentration, pH and availability of nucleation sites (Hammes et al. 2003). This precipitation between particles helps in reducing the permeability, compressibility and increasing soil strength (DeJong et al. 2010).

In this research, two expansive soils were treated using MICP technique to study the efficiency of this technique in stabilizing expansive soils. Two methods of application were investigated using unconfined compressive strength and one-dimensional swell test as performance indicators. Variables such as microbial concentrations and number of treatment cycles were also evaluated. This research is an initial step in understanding the applicability of MICP to expansive soils.

2 Application Methods

MICP can be achieved in two ways: (a) *Bio-stimulation*—This method involves the modification of the environmental condition by stimulating the indigenous bacteria present in the soil, which is typically achieved by introducing nutrients into the soil. (b) *Bio-augmentation*—This method involves the introduction of the required microbes along with nutrients needed to stimulate the microbes into the soil. In this research, two approaches to bio-augmentation method were studied; Application Method-1 (AM-1) and Application Method-2 (AM-2).

In AM-1, the microbes were mixed in the soil sample along with substrates and the mixed sample was used for further testing. This approach is similar to conventional expansive soil treatment methods using lime or cement. The mixed sample was then compacted at the maximum dry unit weight (MDUW) and optimum moisture content (OMC). The compacted sample was then cured for seven days before being tested for unconfined compressive strength (UCS) and 1-Dimensional (1-D) swell tests.

In AM-2, soil samples were prepared as in the case of the AM-1 method. However, in this case, the prepared samples were placed in a specially designed nutrient delivery system instead of being cured at constant temperature and humidity. Using this system, substrate solutions were passed through the soil samples, and the effluent was collected. For each microbial concentration soils, samples were subjected to one, three and seven pore volumes of effluent. One pore volume (PV) here represents the volume of voids present in the soil sample compacted at MDUW and OMC. A collection of effluent is termed as 'treatment cycle' in this research. After collecting respective pore volumes, samples were then tested for UCS and 1-D swell tests.

2.1 Nutrient Delivery System

In order to provide nutrients to the bacteria mixed into the soil, substrate solutions consisting of Urea and $CaCl_2$, need to be passed through the soil sample. As the permeability of expansive soils is very low ($<10^{-6}$ cm/s) gravity feeding was not a practical option in view of the time needed to complete each treatment cycle. Hence, for this purpose a nutrient solution delivery system was developed as shown in Fig. 1. In this system, the soil sample is housed in a chamber made of schedule 40 PVC tube of 9.4 cm diameter. This chamber is fastened between two PVC plates with dimensions, 15.2 cm × 15.2 cm. This chamber can hold pressures up to 138 kPa. This chamber has two inlets and two outlets as shown in Fig. 1. One inlet is connected to the reservoir containing substrate solution while the other inlet is connected to a pressure-regulated container. The reservoir was used to fill the chamber with substrate solution while the pressure-regulated container which also contains the substrate solutions was used to drive the substrate into the soil sample. Similarly, one outlet is used to drain the chamber while the other outlet is used to collect the effluent through the soil sample.

In this system, the soil sample is surrounded by substrate solution which gets pushed through the holes present in the top cap. This arrangement allows the use of single pressure chamber to both apply confinement as well as inlet pressure. The soil sample having dimensions of 7.6 cm (diameter) × 15.2 cm (height) is placed between

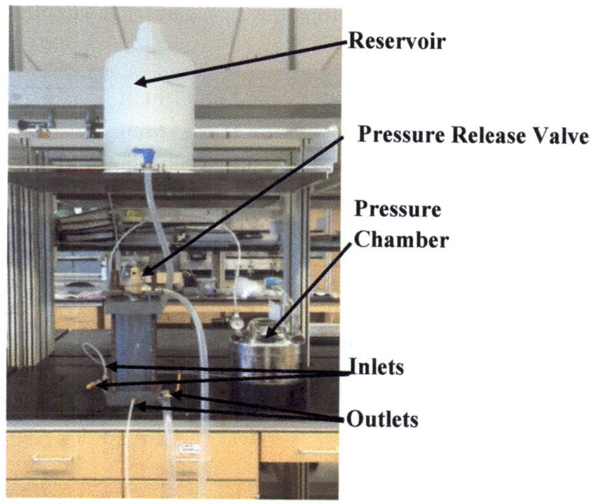

Fig. 1. Nutrient delivery system used in this research

the top cap and the base pedestal and is wrapped around by latex membrane in order to protect the sample from surficial erosion. The top cap and base pedestal are facilitated with grooves to hold O-rings. The O-rings hold the membrane in place and also prevent water from entering inside the sample. The top cap and bottom pedestal have holes in them through which substrate solution passes through and gets in and out of the soil sample.

3 Materials Used in This Research

3.1 Soil Characteristics

Two types of soils along state highway US-95 between Mileposts 16.0 to 18.0 near Marsing, Idaho were selected for this research. The plasticity characteristics of these soils ranged from low to high Plasticity Index (PI). These soils were designated as S1 (low to medium PI) and S2 (high PI). According to the Unified Soil Classification System, both of these soils were classified as high compressible clays identified with the notation CH. Characterization tests such as gradation, Atterberg limits, and compaction characteristics test were conducted on both control soils as per American Standard Testing Methods (ASTM) ASTM D6913, ASTM D4318, and ASTM D698, respectively. In addition to these tests, engineering tests such as Unconfined Compressive Strength (UCS) and 1-D Swell were also performed on the control soils as per ASTM D2166 and ASTM D4546 respectively. These results are presented in Table 1.

Table 1. Engineering properties of natural soil samples

Soil notation		S1	S2
Atterberg limits	Liquid limit	54	115
	Plastic limit	39	53
	Plasticity index	15	62
Maximum dry density (kN/m^3)		13.6	12.0
Optimum moisture content (%)		30	34
% finer than 0.075 mm		70	74
Unified soil classification system		CH	CH
Unconfined compressive strength (kPa)	Saturated	24.5	28.6
	Unsaturated	58.8	239.5
1-D swell strain (%)		2.83	8.85

3.2 Microbial Characteristics

The bacterial strain used in this research was ureolytic bacteria, Sporosarcina pasteurii (formerly known as Bacillus pasteurii). The growth media used to grow the microorganisms was primarily Laurel Broth (LB). The microbial concentration for the AM-1 method was maintained at 10^8 microbial colonies per gram of soil. In the case of the AM-2 method, two microbial concentrations were studied, 10^8 and 10^{10} microbial colonies per gram. Commercially available urea and calcium chloride were used in this research as substrates. The concentration of urea and calcium chloride was 333 and 250 mM respectively. The concentration of substrate was established from the previous researches conducted on sand through MICP technique.

In order to maintain the consistency of microbial concentration throughout the research, colony formation unit (CFU) method was adopted to determine the concentration of microbes in a given solution. For this purpose, S. Pasteurii was cultured in Laurel Broth (LB), incubated for 48 h at room temperature. After 48 h of inoculation, the optical density (OD) of the cultured microbes was measured. OD is the method of measuring the concentration of microbes in a sample by measuring the turbidity of the sample at certain wavelength usually 600 nm (Madigan et al. 2012). These cultured microbes were then serially diluted in various ratios such as 1:200, 1:40,000, 1:8,000,000. After serial dilution, 100 μL of the diluted media was taken and then plated in an LB plate. LB plate was prepared by mixing 10 g of LB and 6 g of agar in 400 ml of distilled water. The media after autoclaving was poured into the petri dish. The media solidifies after few hours due to the presence of agar. After 48 h of plating, the number of colonies was counted. The CFU/ml for each serial dilution is given as per Eq. (1).

$$\text{CFU/ml} = \frac{\text{Number of colonies per ml plated}}{\text{Total dilution factor}} \tag{1}$$

4 Micp Evaluation

In order to evaluate the effectiveness of MICP in stabilizing expansive soils, UCS and 1-D swell tests were chosen as performance indicators. For AM-1 treated samples, these tests were conducted after seven days of curing while for AM-2 soil samples the tests were performed after 1 PV, 3 PV and 7 PV of treatments. The treated samples were of same dimensions as UCS tests hence this test was performed on the treated samples at the end of the testing period with any sample alteration. In the case of 1-D swell tests, the samples were trimmed to a diameter of 6.35 cm and thickness of 2.54 cm with the help of the oedometer ring. Samples in the oedometer ring were oven dried in order to let the samples swell from a very dry state. A similar procedure was performed on control soils as well. As explained earlier, 1-D swell tests were performed according to the ASTM-D4546, method A where the samples are allowed to swell to a maximum value before bringing them back to their initial volume. In this paper, only the swell strain data is discussed and not the swell pressure data. The results of these tests are discussed in the following sections.

4.1 Application Method-1

Soil samples treated using AM-1 protocol were tested for UCS and 1-D swell test after seven days of curing. The UCS values for seven days cured samples are shown in Fig. 2a for both the soil samples. It can be observed from this figure that the UCS value increased from 58.8 to 88.0 kPa for S1 soil sample with an increase of 49.5%, while UCS value decreased by 39.4% for an S2 soil sample. The reduction in case of S2 soil could be due to the high plasticity nature of this soil and inadequate substrate present in the sample. As microbes require moisture to survive, seven days curing may have made microbes dormant and inactive. As soil samples S2 have high fines content (74%, passing through sieve no #200), this may have made the mobility of microbes less possible. Pore size distribution and the proportion of pore filled with water plays an important role in the contact between microbes and soil particles (Chenu and Stotzky 2002).

The 1-D swell strain data for seven days cured samples are shown in Fig. 2b. It can be observed that the swell strain values decreased by 11% from 2.83 to 2.52% for S1 soil and by 44.1% from 8.85 to 4.95% for S2 soil. From Fig. 2b, it is evident that bio-augmentation was effective for S1 sample with low plasticity in reducing 1-D swell percentage and increase in strength. Reduction in swelling was also observed for S2. However, the strength did not increase in case of S2 soil samples. One of the reasons for the reduction in swelling may be due to the cationic exchange in the clay particles due to the presence of calcium chloride in the substrate solution.

4.2 Application Method-2

Soil samples treated using AM-2 protocol were tested for UCS and 1-D swell test after collecting one pore volume (1 PV), three pore volumes (3 PV) and seven pore volumes (7 PV) of substrate effluent. The results obtained from both tests are discussed here. In this application method, two different microbial concentrations, M1 and M2 were

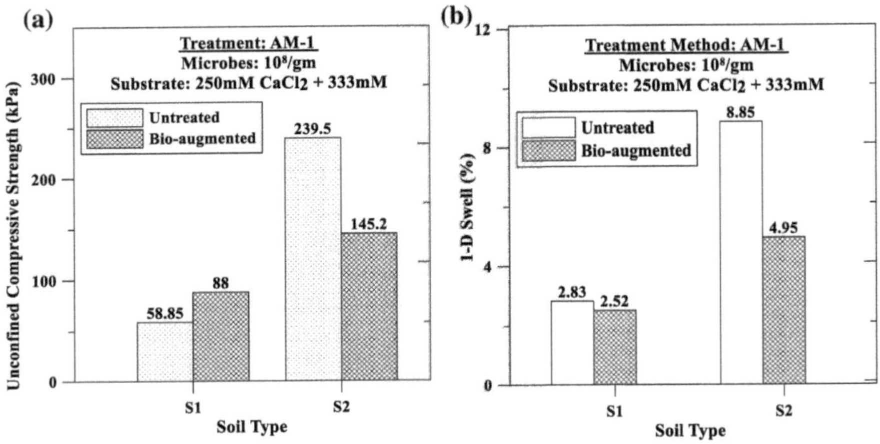

Fig. 2. Test data for AM-1 treated soil samples **a** UCS test **b** 1-D Swell test

studied. M1 and M2 represent 10^8 and 10^{10} microbial colonies per gram of dry soil, respectively. The UCS test results for both soil types and for both microbial concentrations are presented in Fig. 3. It can be observed from this Figure that for S1 soil treated with M1 concentration, the UCS value gradually increased from 25.8 to 54.2 kPa i.e. by 121% of untreated soil strength after 7 PV. However, the treatment did not have a similar effect on the strength of S2 soil. There was a slight increase in UCS value from 28.6 to 32.2 kPa after 7 PV treatment which is an increase in UCS value by 12.6%. UCS values also increased when soils were treated with M2 concentration. The UCS was observed to be 32.8 kPa for S1 soil samples after 7 PV. The increase in the percentage of UCS for S1 after 7 PV was observed to be 34.2% while little or no change in UCS value was observed in the case of S2.

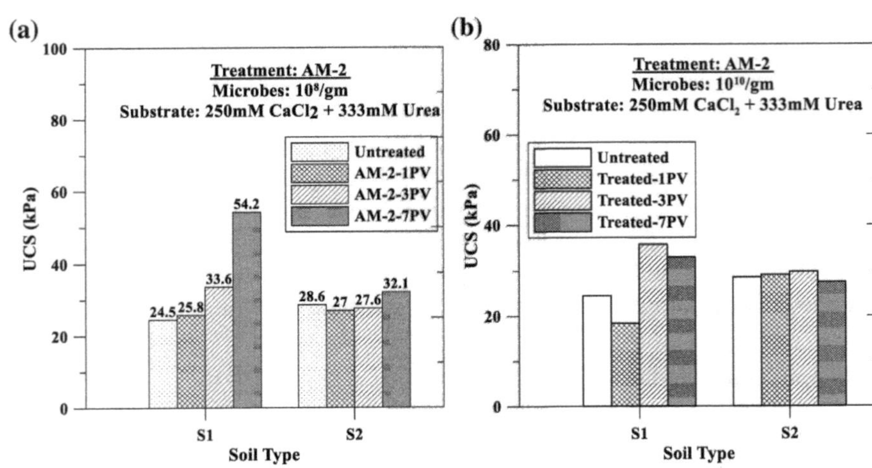

Fig. 3. UCS test results of AM-2 treated soil samples for **a** M1 concentration and **b** M2 concentration

In the case of AM-2, it is evident from Fig. 3a that with the increase in a number of pore volumes, the strength also increased. That is microbes in the soil require enough retention period to produce urease enzyme required to hydrolyze urea. Retention period helps bacteria to dwell into the liquid media (Burbank et al. 2013). It is also observed that the MICP technique whether bio-augmented or bio-stimulated is favored in low plasticity index soil such as S1. As in both the cases, the UCS value increased by 49 and 121% respectively as compared to the untreated S1. However very little or no increase in UCS value was noticed in the case of S2 soil, and this may be due to the absence of microbial activity.

From Fig. 3 it can also be observed that increase in microbial concentration did not increase the UCS value. Ramachandran et al. (2001) concluded that higher concentration of bacteria had no improvement in strength. They suggested that slower rates of calcite formation were more prominent in imparting higher strength than higher rates. Comparison between the 10^8 and 10^{10}/gm microbial concentration for each pore volume shows that the increase in microbial concentration did not increase the strength of these samples. The factors that influence the precipitation of calcite are mainly the concentrations of Ca^{2+} and CO_3^{2-}, pH of the system and the nucleation site. Bacterial cell surface acts as a nucleation site for the precipitation of the calcite. The solubility product (K_{sp}) of calcite is very low (3.3×10^{-9} mol L^{-1} at 25 °C), and for precipitation of calcite supersaturation of Ca^{2+} and CO_3^{2-} must exist. Since calcite has very low K_{sp}, supersaturation can be achieved by simply mixing Ca^{2+} and CO_3^{2-} together in moderate concentrations. However, when the reaction takes place rapidly, the crystals formed are very small and powder like with little or no cementation strength (Whiffin 2004). In order to have large crystal precipitation over an extended period of time with higher cementation strength, the supersaturating product concentration should remain low. The supersaturation of CO_3^{2-} is also influenced by the pH of the system. pH can be regulated by the dissociation of urea into NH_4^+. CO_3^{2-} concentration remains very low below pH 8. Thus, the size of crystal can be increased or decreased by decreasing or increasing the pH of the system (Whiffin 2004). The presence of higher microbes at the beginning of the might have contributed to higher rates of calcite formation thereby hindering strength development in the case of M2 concentration.

The 1-D swell test results of both samples with a microbial concentration of 10^8/gm are presented in Fig. 4a for all three pore volumes. It was observed that the swell strain reduced in the case of S1 samples after all three treatment cycles. Reduction in swell strain was also observed for S2 after 7 PV. Similar results were obtained when both soils were treated with a microbial concentration of 10^{10}/gm after 7 PV of treatments which is shown in Fig. 4b.

1-D swell strain reduced for S1 soil samples after all three pore volumes. It was also observed that after seven pore volumes, the swell reduction was possible i.e. higher the treatment cycles (or retention period) lower the swell strain. Reduction in swelling was also observed for S2 soil. However, the reduced values were still considered problematic swelling strains. One of the reasons for the reduction in swelling for S2 soil may be due to the cationic exchange in the clay particles due to the presence of calcium chloride as substrate. As there was no increase in strength, it was assumed that microbial activity was minimal in this soil.

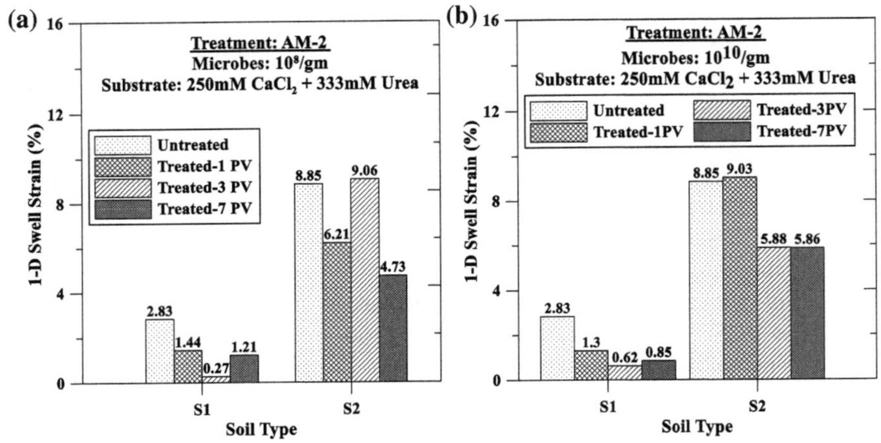

Fig. 4. 1-D Swell Strain data for AM-2 treated soil samples **a** M1 concentration and **b** M2 concentration

5 Summary

Two expansive soils with varying plasticity characteristics were tested to evaluate the effectiveness of MICP in stabilizing expansive soils. These soils were subjected to two methods of MICP treatments, and their performance was measured by monitoring swelling potential and unconfined compressive strength with various treatments. Variables such as soil type, bacterial population during augmentation, along with the number of treatment cycles were studied in this research. This research is the initial step to understand the applicability of MICP in expansive soils. The data presented in this research supports the applicability of MICP in expansive soils particularly in the case of low plasticity soils such as S1. However, changes in geotechnical properties of high plastic soils such as S2 soil samples' after MICP treatment needs further testing to understand the feasibility of MICP technique in high plastic soils.

References

Bang, S.S., Galimat, J.K., Ramakrishan, V.: Calcite precipitation induced by polyurethane-immobilized Bacillus pasteurii. Enzyme Microb. Technol. **28**, 404–409 (2001)

Burbank, M., Weaver, T.L., Williams, T., Williams, B., Crawford, R.: Geotechnical tests of sands following bioinduced calcite precipitation catalyzed by indigenous bacteria. ASCE J. Geotech. Genenviron. Eng. **139**, 928–936 (2013)

Chenu, C., Stotzky, G.: Interactions between microorganisms and soil particles: an overview. In: Interactions Between Soil Particles and Microorganisms: Impact on the Terrestrial Ecosystem. IUPAC. Wiley, Manchester, UK, pp. 1–40 (2002)

DeJong, J.T., Fritzges, M.B., Nusslein, K.: Microbial induced cementation to control sand response to undrained shear. ASCE J. Geotech. Geoenviron. Eng. **132**(11), 1381–1392 (2006)

DeJong, J.T., Mortensen, M.B., Martinez, B.C., Nelson, D.C.: Bio-mediated soil improvement. Ecol. Eng. **36**, 197–210 (2010)
Douglas, S., Beveridge, T.J.: Mineral formation by bacteria in natural communities. FEMS Microbiol. Ecol. **26**, 79–88 (1998)
Hammes, F., Seka, A., Van Hege, K., Van de Wiele, T., Vanderdeelen, J., Siciliano, S.D., Verstraete, W.: Calcium removal from industrial wastewater by bio-catalytic $CaCO_3$ percipitation. J. Chem. Technol. Biotechnol. **78**, 670–677 (2003)
Ivanov, I., Chu, J.: Applications of microorganisms to geotechnical engineering for bioclogging and biocementation of soil in situ. Rev. Environ. Sci. Biotechnol. (2008)
Kavazamjian, E., Karatas, I.: Microbiological improvements of the physical properties of soil. In: Proceedings of 6th International Conference on Case Histories in Geotechnical Engineering, Arlington, VA, 11–16 Aug 2008
Knorre, H., Krumbein, W.: Bacterial calcification. In: Microbial Sediments. Springer, Berlin, Heidelberg, pp. 25–31 (2000)
Kucharski, E.S., Cord-Ruwisch, R., Whiffin, V., Al-Thawadi, S.M.: Microbial biocementation. Application number US11/793,555. International application number PCT/AU2005/001927 (2005)
Madigan, M., Martinko, J., Stahl, D., Clark, D.: Microbial growth. In: Biology of Microorganism, 13th edn. Pearson Education, p. 159 (2012)
Montoya, B.M., DeJong, J.T., Boulanger, R.W.: Dynamic response of liquefiable sand improved by microbial-induced calcite precipitation. In: Bio- and Chemo-Mechanical Processes in Geotechnical Engineering, pp. 125–135 (2014)
Nelson, J.D., Miller, D.J.: Expanisve Soils: Problems and Practice in Foundation and Pavement Engineering. Wiley, New York (1992)
Petry, T.M., Little, D.N.: Review of stabilization of clays and expansive soils in pavements and lightly loaded structures-History, practice, and future. J. Mater. Civ. Eng. **14**, 447–460 (2002)
Ramachandran, S.K., Ramakrishnan, V., Bang, S.S.: Remediation of concrete using micro-organisms. ACI Mater. J., 3–9 (2001)
Torsvik, V., Goksoyr, J., Daae, F.L.: High diversity in DNA of soil bacteria. Appl. Environ. Microbiol. **56**(3), 782–787 (1990)
Whiffin, V.S.: Microbial $CaCO_3$ precipitation for the production biocement. Doctoral dissertation, Murdoch University (2004)

Transverse Cracks Cause Analysis on Rock-Anchored Beam During the Construction Process of Hydropower Station Underground Powerhouse

Fei Ai[1(✉)], Jian Liu[1], Nianshui Chen[2], and Haibin Xiao[3]

[1] State Key Laboratory of Geomechanics and Geotechnical Engineering, Institute of Rock and Soil Mechanics, Chinese Academy of Science, Wuhan 430071, China
fai@whrsm.ac.cn

[2] Northwest Engineering Corporation Limited, Power China, Xian 710065, China

[3] Huaneng Lancang River Hydropower Corporation Limited, Kunming 650214, China

Abstract. The different size of transverse cracks was appeared on the upper surface of rock-anchored beam in the process of excavation of underground powerhouse of hydropower station. Detailed cause analysis of these cracks was made based on the in situ geological condition, construction process, monitoring data, geophysical survey and numerical simulation. The results showed that the layered rock and faults are the premise of differential deformation of surrounding rock. Monitoring data and geophysical test results indicated that sidewall surficial surrounding rock was cracked by the blasting excavation which promoted the differential deformation of surrounding rock, easy to the generation of transverse crack on rock-anchor beam. Numerical analysis results revealed that differential vertical deformation of surrounding rock is the most important reason for the rock-anchored beam transverse cracks while the lithological differences as well as the existence of the faults are the main reason for the differences on the vertical deformation. The procedure of excavation and the nearby cavern excavation which resulted stress redistribution and differences on the deformation also leaded to another cause of rock-anchored beam transverse crack.

1 Introduction

Rock-anchored beam as a common structure of hydropower station underground plant, can effectively reduce the span of underground power plant, conducive to plant stability and reduce project cost. The long anchor is used to reinforce concrete structure in the rock wall. The stress conditions and carrying capacity are not only with the anchor parameters, beam structure but also the surrounding rock geological conditions, room scale, blasting excavation and other factors. Rigid body limit equilibrium method and force vector balance method are used for the stability of rock-anchored beam. The finite

W.-C. Cheng et al. (eds.), *Tunneling in Soft Ground, Ground Conditioning and Modification Techniques*, Sustainable Civil Infrastructures,
https://doi.org/10.1007/978-3-319-95783-8_3

element method is used to check and optimize the design of the rock-anchored beam, supplemented by structural model test and on-site detection and monitoring.

The rock-anchored beam is often cracked in the excavation period of the underground powerhouse. The cracks are usually the transverse cracks and the longitudinal cracks. Under the action of the engine load may cause rock-anchored beam structure damage. The problem of cracks in rock-anchored beam has become a technical problem in the construction of underground powerhouse. With the construction of a large number of underground power plants in the hydropower station, the research on the causes of cracks and the main influencing factors of rock-anchored beam has made great progress. Chen proposed that the fault on the rock of rock-anchored beam length direction would make the stress and strain quite different which cause cracks on the rock-anchored beam. Wang found that the differential deformation of the surrounding rock is the main reason of the cracks on the rock-anchored beam and the main reason of the differential deformation of the surrounding rock is because of faults and fractures. The research of Xu indicated the local stress concentration leads to the generation of uneven deformation which leads the cracks on the rock-anchored beam. At present, the research on the crack of rock-anchored beam is still limited to the mechanism of individual cracks in rock-anchored beam from geology, monitoring and numerical simulation. There is lack of deep research on the distribution law and mechanism of cracks on the rock-anchored beam.

In this paper, the distribution of transverse cracks in rock-anchored beam and its mechanism are analyzed synthetically from geological conditions, construction process, monitoring data, geophysical test and numerical simulation. It provides a reference for the similar projects.

2 Brief Introductions of Underground Plant

The underground powerhouse of the hydropower station is located in the right bank of the Lancang River. It is mainly composed of the main powerhouse, the main transformer room and the tailrace surge chamber. Three main caverns are arranged in parallel. The direction of the axis is NE50° with spacing 41.5 and 35.5 m. The main powerhouse outline size is 189 m × 26.7 m × 70.25 m (length × width × height), the main transformer room size is 143.7 m × 18.0 m × 34.3 m and tailrace surge chamber size is 116.50 m × 20.0 m × 65.9 m.

The minimum depth of the underground powerhouse is about 129 m, the maximum depth is about 331 m and lateral depth is of 124–154 m. The maximum in situ stress value of the rock mass in the main powerhouse area is about 10.0 MPa, and the in situ stress is composed of self-gravity stress and tectonic stress. The vertical stress ratio between the measured vertical stress component and the overburden mass is between 0.94 and 1.55.

The surrounding rock of the underground powerhouse is mainly distributed between sandstone and slate, and the formation of rock is NW350°–NE5° NE (SE)∠80°–87° and 50°–65° from the main powerhouse axis. The rock mass has obvious layered or thin lamellar structure, medium intensity, and the surrounding rock class is dominated by class III, a few of class II or IV.

According to geological conditions and duration requirements, the underground powerhouse was divided into X layers' excavation, the specific program shown in Fig. 1. After the excavation of the layer III completed, the rock-anchored beam was constructed. The surrounding rock support scheme was based on the system bolts, the system anchor cables, the hanging wire microfiber concrete and the use of the random anchor cables. The rock bolts used on the rock-anchored beam are in the form of an upper two rows of tensioned bolts and a lower one row of compression bolts. The tensioned bolts are Φ32 @ 75 cm, rod length 12.0 m, into rock 10.0 m, inclination angle $\beta 1 = 20°$, $\beta 2 = 15°$. The compression bolt is Φ32 mm @ 75 cm, rod length 9.0 m, into rock 7.7 m.

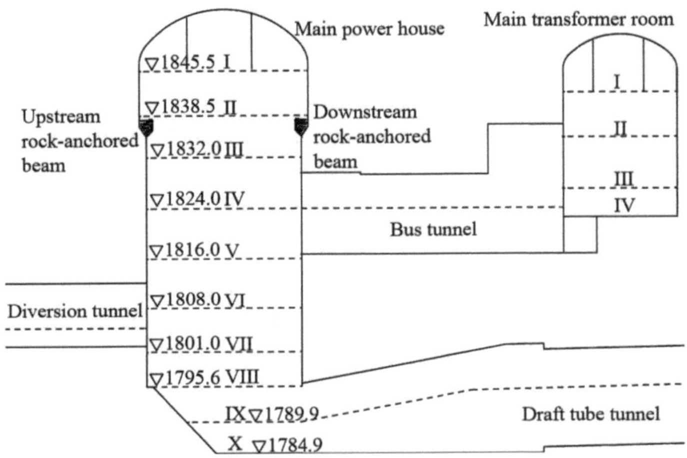

Fig. 1. Construction sequence of main powerhouse

3 Rock-Anchored Beam Crack Introductions

The rock-anchored beam of main powerhouse was poured the first warehouse concrete on November 11, 2014 and finished on December 16, 2014. August 24, 2015, it found that there were different sizes and vertical transverse cracks on the upstream and downstream rock-anchored beam platform surface, as shown in Table 1 and Fig. 2.

Fourteen transverse cracks were found on the upstream rock-anchored beam, mainly distributed on the 0 + 106.00–0 + 155.00 with length 0.5–1.25 m, no penetration, small cracks width, belonged concrete surface cracks. There are thirteen transverse cracks on the downstream rock-anchored beam, mainly distributed on the 0 + 45.00–0 + 49.00. Three cracks in the area of 0 + 48.00–0 + 49.00 were basically penetrated the surface, and one of them was extended to the concrete spray layer of rock wall. The crack extension length of about 2.1 m, as shown in Fig. 3.

Table 1. Transverse crack statistics of rock-anchored beam

Area	Location	Length (m)	Description
Upstream	0 + 096.00	1.1	Top platform surface
	0 + 106.00	1.25	Top platform surface, extended to the track
		1.1	Top platform surface, extended to the track
		1.14	Top platform surface, extended to the track
		1.18	Top platform surface, extended to the track
		0.6	Top platform surface
	0 + 113.00	1.1	Top platform surface, extended to the track
		0.5	Top platform surface
	0 + 137.00	0.78	Three cracks at top platform surface
	0 + 155.00	1.26	Three cracks at top platform surface
Downstream	0 + 045–049	2.1	Ten cracks at top platform surface
	0 + 068.80	1.1	Top platform surface
		0.8	Top platform surface
	0 + 096.30	0.7	Top platform surface

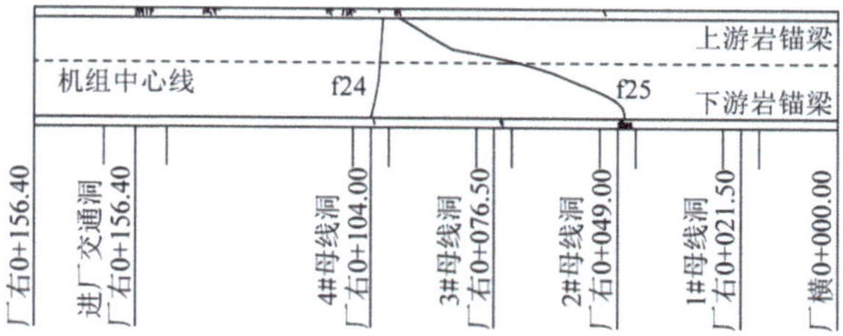

Fig. 2. Transverse crack pattern of rock-anchored beam

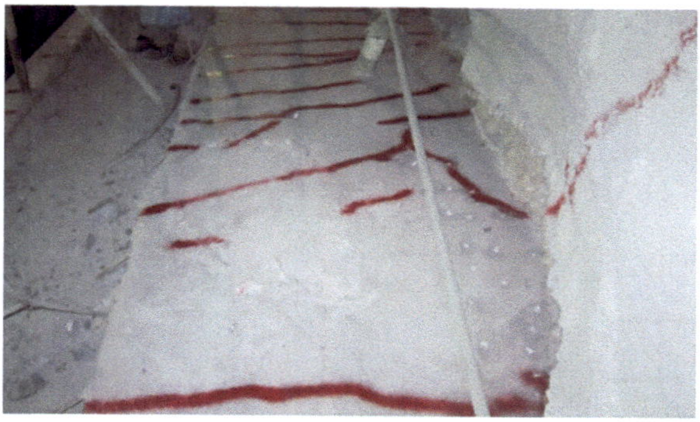

Fig. 3. Transverse cracks of downstream rock-anchored beam at 0 + 48.00

4 Analysis of Cracks in Rock-Anchored Beam

4.1 Geological Conditions Analysis

The surrounding rock of two sides of the wall revealed the lithology of sandstone and slate and sandy slate. The rock mass was fresh and slightly weathered, and the wall was dry. A set of steep fissures were developed parallel to the axis of main powerhouse.

The area of rock-anchored beam is dominated by III1–III2 surrounding rock, and the 0 + 100–0 + 174 m in the upstream wall was developed a group of steep fissures intersecting the small angle with the wall. Fault f24 and f25 are developed on the rock-anchored beam area, where fault f24 was SE145° NW∠76°, width 2–10 cm, filled with fractured rock, yellow muddy, calcareous cement and local overhead and fault f25 was NE82° NW∠69°, width 20–40 cm, filled with broken rock block. The location of fault f24 and f25 on the upstream and downstream rock-anchored beam is shown in Fig. 2, which can be found that there is existence of transverse cracks in the exposed areas of faults.

4.2 Analysis of Monitoring Data of Rock-Anchored Beam

The monitoring items on the rock-anchored beam include: crack monitoring, crack opening monitoring, anchoring stress monitoring. The monitoring arrangement are shown in Fig. 4.

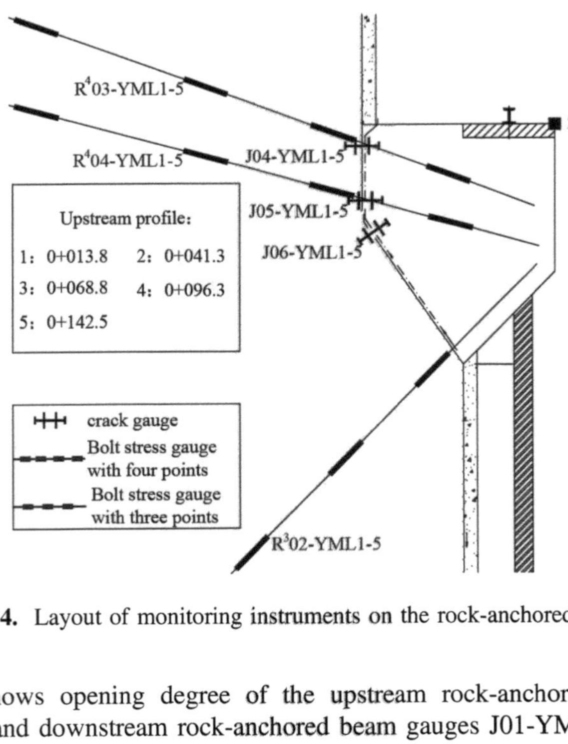

Fig. 4. Layout of monitoring instruments on the rock-anchored beam

Figure 5 shows opening degree of the upstream rock-anchored beam gauges J04-YML 1-5 and downstream rock-anchored beam gauges J01-YML1-5 at different

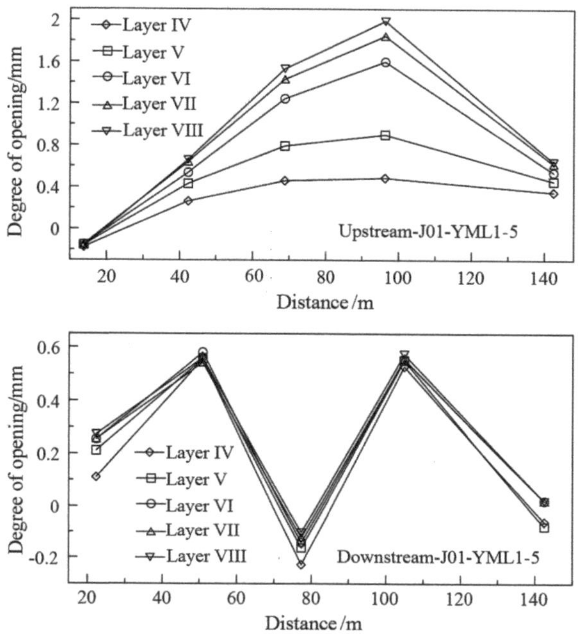

Fig. 5. Deformation curves of joints opening of rock-anchored beam along the axis of main excavation layers. The results show that the maximum opening degree of the upstream rock-anchored beam is close to 2 mm, which is located in the 0 + 96.3 area. The maximum opening degree of the downstream rock-anchored beam is close to 0.6 mm, which is located in the area of 0 + 50.0. In addition, the measured value of the upper rock-anchored beam is affected by the blasting excavation of the layers of IV–V. The opening degree of the layer IV of the J01-YML4 at the 0 + 96.3 is 0.4 mm, which increased to 1.6 mm at layer VI. Whereas the blast has less impact on the downstream rock-anchored beam joints. Compared with the opening degree results and the distribution of transverse cracks on the rock-anchored beam, it found that there are many areas with large opening, and the corresponding transverse cracks are also more.

Figure 6 shows the bolt stress data of the upstream rock-anchored beam four-point bolt stress gauge R403-YML1-5 and downstream rock-anchored beam four-point bolt stress meter R401-YML1-5 measured depth of 0.5 m at different excavation layers. The results show that the bolt stress of the upstream and downstream rock bolts is in good agreement with the curve. The maximum bolt stress of the upstream rock-anchored beam is close to 400 MPa, which is located in the 0 + 96.3 area. The bolt stress was 200 MPa in the layer IV, and increased to 370 MPa at the layer VI, while the bolt stress of the downstream rock bolt is less affected by the blasting excavation. Compared with the results of the bolt stress and the distribution of transverse cracks on the rock-anchored beam (see Fig. 2), it can be found that the area of the bolt stress is large and the corresponding transverse cracks are also more.

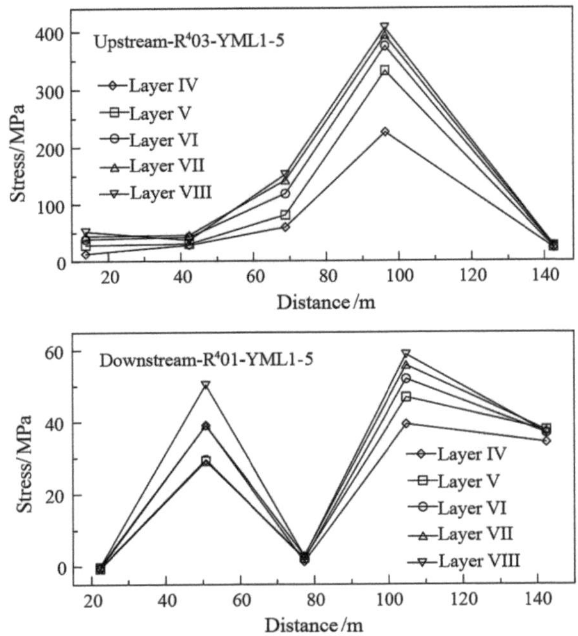

Fig. 6. Anchor stress curves of rock-anchored beams along the axis of main powerhouse

4.3 Monitoring Data of Surrounding Rock of Rock-Anchored Beam

The monitoring project of the surrounding rock of the rock-anchored beam includes deformation monitoring, bolt load monitoring, bolt stress monitoring and geophysical testing. The monitoring arrangement is shown in Fig. 7.

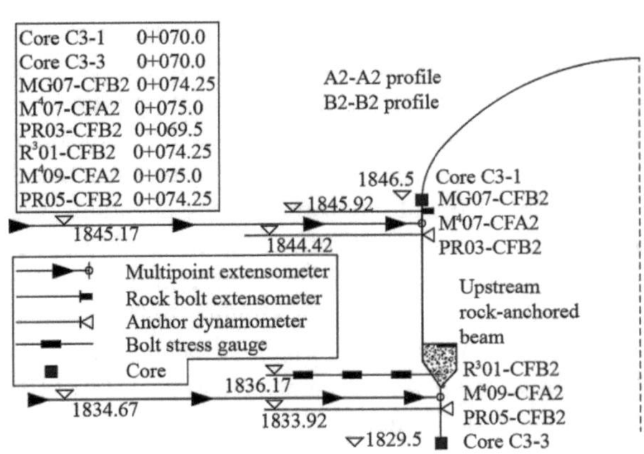

Fig. 7. Layout of monitoring instruments on the upstream sidewall of main powerhouse

Figure 8 shows the displacement-time curve of multi-point extensometer M407-CFA2 on the 0 + 075.0 upstream side wall 1845.17 elevation. Figure 9 shows the curve of bolt load with time of the bolt dynamometer MG07-CFB2 on the 0 + 074.25 upstream side wall 1845.92 elevation. Figure 10 shows the curve of anchorage cable load with time of the anchorage cable dynamometer PR03-CFB2 on the 0 + 069.5 upstream side wall 1844.42 elevation. Figure 11 shows the curve of anchor stress with time of the anchor stress gauge R301-CFB2 on the 0 + 074.25 upstream side wall 1836.17 elevation. Figure 12 shows the panoramic image of boreholes on upstream side wall of main powerhouse on the 0 + 070.00 upstream side wall 1846.5 elevation.

Based on the comprehensive analysis of Figs. 8, 9, 10, 11 and 12, it can be found that the influence of the excavation of the layers IV and V, the displacement of the upstream wall, the bolt stress, the bolt and anchor stress are significantly increased. After the excavation of layer IV, the curves gradually converge. In addition, the deformation of the surrounding rock in the 0 + 075.0 area occurs mainly in the 0–7 m depth (see Fig. 8). There is a significant correlation between the sharp increase of bolt stress and the differential deformation of shallow rock (<7 m depth). The bolt stress gauge R301-CFB2on the 0 + 074.25 upstream wall 1836.17 elevation is mainly impacted by the blasting excavation of layer IV and V. The stress of depth of 7.0 m is out of stress range which close to 500 MPa (see Fig. 11). Geological data show that the lithology of the area is sandy slate, and the panoramic image of the adjacent area

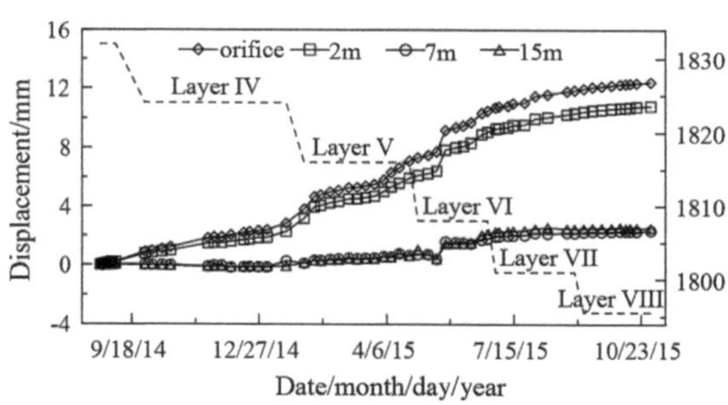

Fig. 8. Displacement curves of multi-point extensometer M407-CFA2

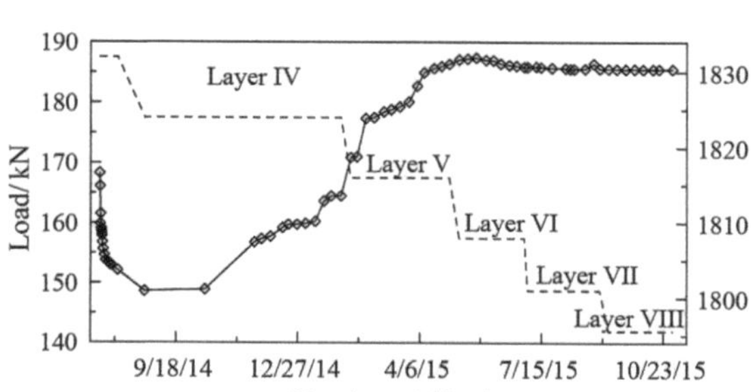

Fig. 9. Curve of bolt load with time of the bolt dynamometer MG07-CFB2

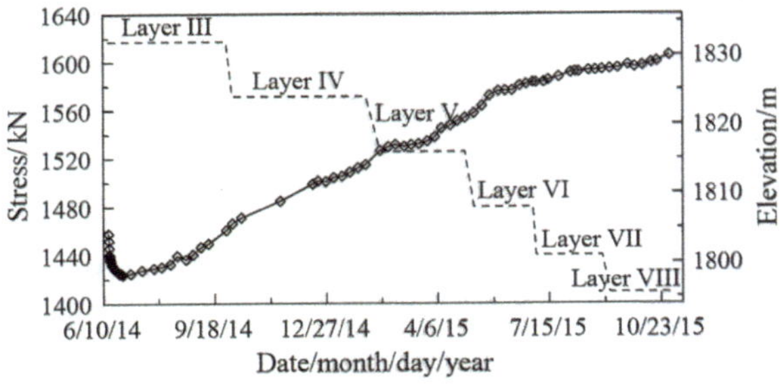

Fig. 10. Curve of anchorage cable load with time of the anchorage cable dynamometer PR03-CFB2

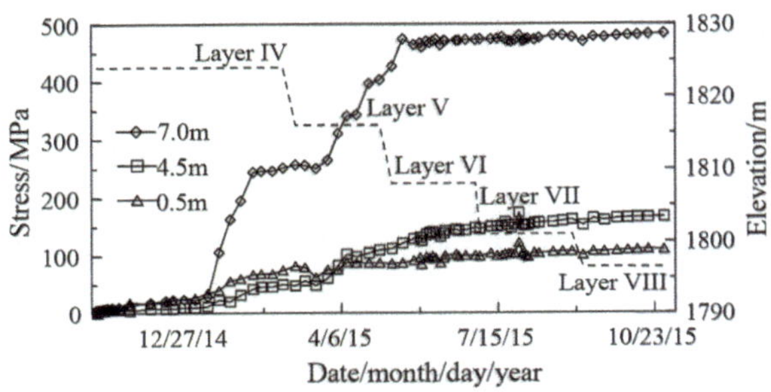

Fig. 11. Curve of anchor stress with time of the anchor stress gauge R301-CFB2

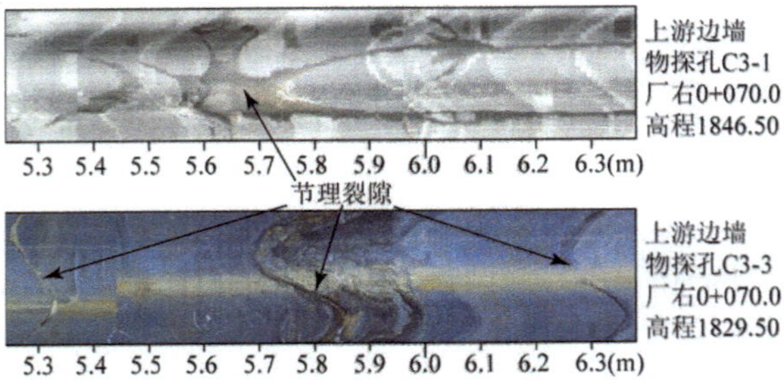

Fig. 12. Panoramic image of boreholes on upstream side wall of main powerhouse

(Fig. 12) shows that there are a lot of joint fissures in the shallow area of the wall of the surrounding rock (5–8 m), which is affected by the blasting excavation. The interlayer joint fissures produce shear deformation, leading to shallow cracking of the strata. Therefore, the values of the multi-point extensometer and bolt stress gauge showed a significant increase in this region. Shallow cracking promotes the differential deformation of surrounding rock, which is easy to produce transverse cracks of rock-anchored beam.

4.4 Numerical Simulation Analysis

4.4.1 Calculation Area and Calculation Model

According to the geologic map of 1:1000 planar terrain in the underground plant area, the numerical model is mainly considered for the influence of the terrain, stratigraphic lithology and fault on the geostress field. The calculated coordinates of the model are: the origin of the coordinate is located at the center of the main powerhouse, the x-axis positive is the downstream of main powerhouse, the y-axis positive is the left of main powerhouse, the z-axis positive is vertical upward. The simulation range of the model is to calculate the coordinate origin as the center, x direction from −450 m to 550 m, y direction from −575 m to 425 m, z direction from the elevation 1400 m simulation to the top of the mountain. The three-dimensional finite element model of the cavern group and the surrounding rock in the underground powerhouse area is built, the total number of discrete units is 931,691, and the total number of nodes is 157,458.

The simulation area consists of six strata, from left to right: ⑩ P2a-Sb + Ss sandstone, ⑨ P2a-Sb slate, ⑧ P2a-Ss sandstone, ⑦ P2a-Sb + Ss sand slate, ⑥ P2a-Ss sandstone, ⑤ P2a-Sb + Ss sand slate, simulated faults including f24, f25. The model grid is shown in Fig. 13.

4.4.2 Calculate Parameters

The Mohr-Coulomb constitutive model is used for rock and rock-anchored beams. The interface between rock wall and rock-anchored beam is simulated by interface unit in FLAC3D software. Bolt and anchor cable are simulated by cable unit. The physical and mechanical parameters of rock mass are used to evaluate the results of the design institute and the site geophysical test, as shown in Table 2.

4.4.3 Numerical Calculation Results Analysis

The key points of the upstream and downstream rock-anchored beams are selected to analyze the displacement of the rock-anchored beam along the main powerhouse axis. The simulation results of the vertical displacement of the key point 2 of the upstream (see Fig. 4) and the key point 1 (centerline line symmetry with the key point 2) of the downstream rock-anchored beam are analyzed.

Figure 14 shows the vertical displacement of key point 2 and key point 1 along the direction of the axis. Overall, with the blasting excavation, the vertical displacement of the upstream and downstream rock bolts showed a tendency to increase layer by layer. The maximum vertical displacement of the upstream and downstream rock-anchored beams is close to 11 mm at the excavation of layer VIII. The maximum vertical displacement of the upstream rock-anchored beam occurs in the area of 0 + 100.00. The main reason is that the faults f24 and f25 are exposed in this area, which makes the mechanical properties of the surrounding rock deteriorate and the vertical displacement increases. The difference between the vertical deformation of the rock, resulting in the differential displacement of the rock-anchored beam, eventually leading to rock-anchored beam transverse cracks. The results of the numerical simulation are in agreement with the actual occurrence of multiple transverse cracks in the region. The difference in vertical deformation of the rock-anchored beam is the main possible cause of transverse cracks.

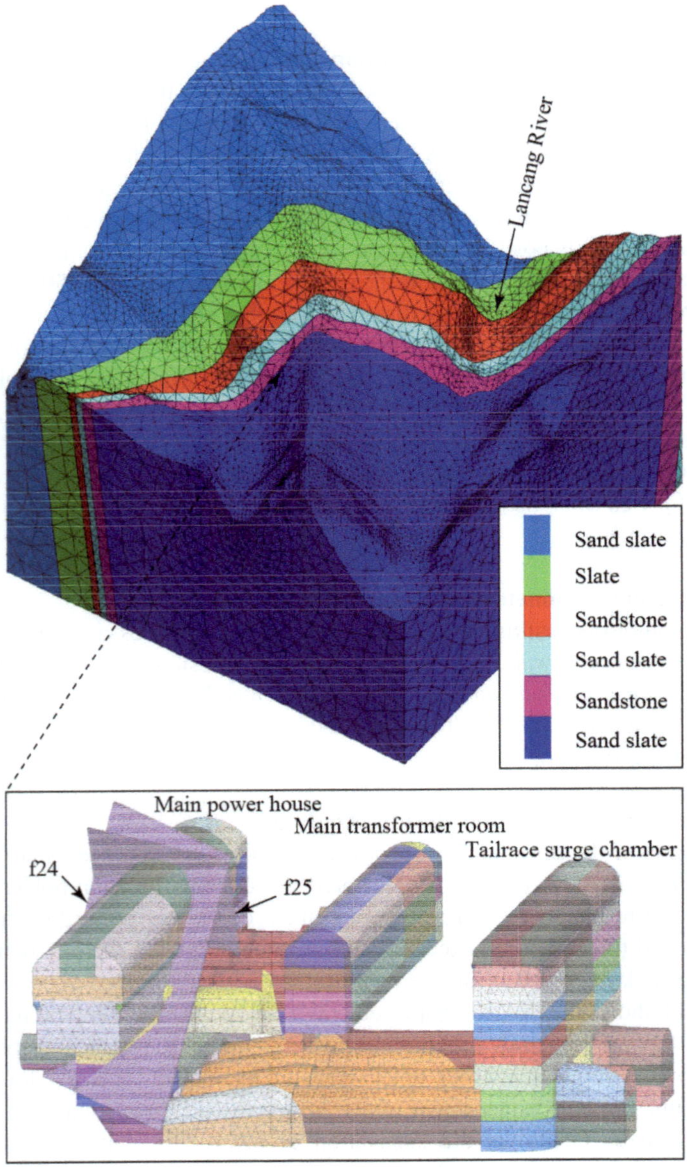

Fig. 13. Models of mountain and underground caverns

The maximum vertical displacement of the downstream rock-anchored beam occurs in the 0 + 060.00, and there are many vertical displacement fluctuations along the main powerhouse axis. For the area of 0 + 040.00–0 + 050.00 of downstream rock-anchored beam, the vertical displacement is found to change abruptly. When the excavation of the layer VIII, the vertical displacement at 0 + 040.00 is about 5 mm while vertical displacement is 9 mm at 0 + 040.00. The differential vertical deformation of this region

Table 2. Physical mechanical parameters of materials

Lithology	Elasticity modulus	Poisson's ratio	Density	Cohesion	Internal friction angle
	GPa	/	g/cm^2	MPa	°
Sandstone	12	0.265	2.74	1.5	47.73
Slate	9	0.3	2.77	1.1	43.53
Sand slate	11	0.28	2.75	1.35	46.4
Rock-anchored beam	24.5	0.17	2.58	1.8	60
Fault	3	0.31	2.7	0.5	33

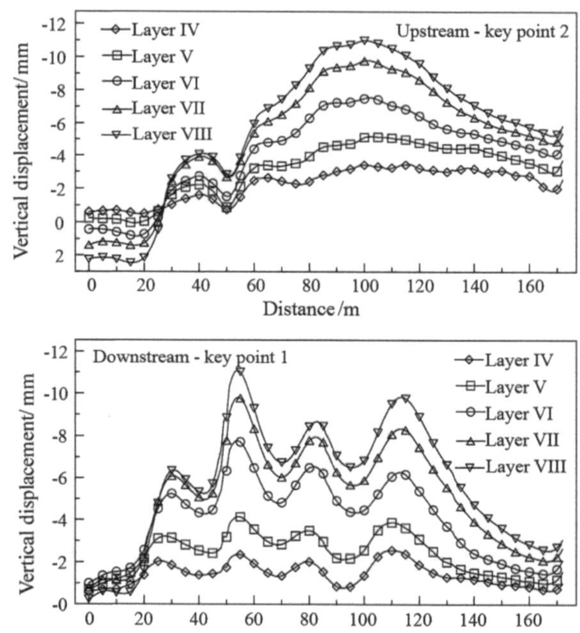

Fig. 14. Vertical displacement curves of key points 1 and 2 of rock-anchored beams along the axis of main powerhouse

directly lead to ten transverse cracks (see Fig. 3). On the one hand, fault f25 was exposed on the 0 + 050.00 of downstream rock-anchored beam. The degradation of mechanical properties of surrounding rock in fault exposed area leads to vertical deformation increases. On the other hand, 5.63 m blow of downstream rock-anchored beam is the roof arch of bus tunnel. The excavation of the bus tunnel made the rock between the rock-anchored beam and the bus tunnel deform to the direction of the bus tunnel. The stress redistribution and differential deformation caused by the excavation of the adjacent cavern are also another cause of the transverse cracks on the rock-anchored beam.

5 Conclusions

Based on the geological conditions, the latest monitoring data, the results of geophysical test and the numerical analysis, the transverse cracks on the upstream and downstream rock-anchored beams are analyzed. The results show that:

The lithology of the surrounding rock-anchored beams is sandstone and sandy slate, and many fractures and steep fissures are developed. The surrounding rock are class III1–III2. The layered and fractured surround rock condition is the premise of differential deformation of rocks. The explosion of faults f24 and f25 resulted in localized stress concentration, leading to a large difference in stress and strain in the direction of the length of rock-anchored beam, which is easy to produce transverse cracks of rock-anchored beam.

The monitoring data of rock-anchored beam show that there is a correspondence between the opening degree of the seam and the bolt stress and the crack distribution of the rock-anchored beam. Usually the area with large values of opening degree and bolt stress is prone to appear transverse crack. Due to the blasting excavation of the layers IV and V, a large number of joint fissures are generated in the shallow area (5–8 m) of the surrounding rock, resulting in a significant increase in the measured values of the multi-point extensometer and the bolt dynamometer in this area. Shallow cracking promotes the differential deformation of surrounding rock, which is easy to produce transverse cracks of rock-anchored beam.

The numerical results show that the difference in the vertical deformation of the rock-anchored beam is the main possible cause of the transverse cracks in the rock-anchored beam. The difference of the lithology and the existence of the fault are the main reasons leading to the difference of the vertical deformation. The excavation and the stress redistribution and differential deformation caused by adjacent cavern are also another cause of transverse cracks on the rock-anchored beam.

Acknowledgements. The financial supports from the National Natural Science Foundation of China (No. 41602320) are gratefully acknowledged. We also like to acknowledge the reviewers for their constructive remarks.

References

Cao, P., et al.: 2D-analysis for rock-bolt crane girder in underground powerhouse of Dongfeng plant and its verification. Chin. J. Rock Mech. Eng. **15**(4), 360–369 (1996)

Cen, B., Du, B.: Analysis and control of concrete cracking of rock-bolted crane girder. Railway Constr. Technol. **6**, 41–44 (1998)

Fu, J., et al.: Stability research on the crane beam on the palisades in the Three Gorges Project underground powerhouse. Chin. J. Rock Mech. Eng. **20**(S1), 1710–1713 (2001)

Fu, S., et al.: Study on key problems of finite element analysis of crane beam on rock wall. Chin. J. Rock Mech. Eng. **27**(10), 2124–2129 (2008)

Jin, F., et al.: Study on calculating method for rock-bolt crane girder design. Rock Soil Mech. **26** (S1), 175–179 (2005)

Li, S., et al.: Computational simulation and analysis of fracturing mechanism of crane girder bolted to rock wall of chamber. Chin. J. Rock Mech. Eng. **23**(10), 1730–1734 (2004)

Tang, J., et al.: Cause analysis of longitudinal cracks of rock-anchored beam and its developmental trend in underground powerhouses. Chin. J. Rock Mech. Eng. **28**(5), 1000–1009 (2009)

Tu, Z., Cui, W.: Study on in-situ test of rock-bolted crane girder in Xiaowan hydropower station. Rock Soil Mech. **28**(6), 1139–1144 (2007)

Wang, K., et al.: Study on site monitoring and stability of layered surrounding rocks of underground powerhouses of hydropower station. Chin. J. Rock Mech. Eng. **26**(10), 2148–2152 (2007a)

Wang, D., et al.: Cause analysis of cracking of rock-bolted crane girder in an underground power ground. Chin. J. Rock Mech. Eng. **26**(10), 2125–2129 (2007b)

Xiao, M., et al.: Three dimensional numerical analysis of crane beam with rock bolt under complex geological condition. Chin. J. Rock Mech. Eng. **22**(S1), 2294–2298 (2003)

Xu, Q., et al.: Application and verification of composite element method in the rock bolt crane girder of underground powerhouse. Chin. J. Rock Mech. Eng. **26**(S2), 4108–4114 (2007)

Xu, F., et al.: Formation mechanism and its countermeasures for cracks in rock-anchored beam during the construction process of underground caverns at the Houziyan Hydropower Station. J. Hydraul. Eng. **46**(S1), 242–247 (2015)

Yang, J., et al.: Three-dimensional nonlinear analysis of rock-anchored beam of underground house of hydropower station. J. Sichuan Univ. (Eng. Sci. Ed.) **35**(2), 31–35 (2003)

Yin, R., et al.: Analyses of deformation and stress caused by geostress on rock-anchored beam of underground powerhouse. Chin. J. Rock Mech. Eng. **32**(S2), 3681–3687 (2013)

Zeng, J., et al.: Numerical emulation analysis of rock bolt crane in underground powerhouse beam during construction and running periods. Chin. J. Rock Mech. Eng. **25**(S1), 3051–3056 (2006)

Zhao, Z., Zeng, Y.: The bearing mechanism model experiment research of rock-bolted crane girder. Chin. J. Geotech. Eng. **24**(2), 150–153 (2002)

Performance Evaluation of Jacking Force Models for Tunnel Bore Conditions Characterisation

Jason Wen-Chieh Cheng[1(✉)], James C. Ni[2], Jack Shui-Long Shen[3], Bruce Zhi-Feng Wang[4], and Arul Arulrajah[5]

[1] Institute of Tunnel and Underground Structure Engineering, School of Civil Engineering, Xi'an University of Architecture and Technology, Xi'an 710055, China
w-c.cheng@xauat.edu.cn

[2] Department of Civil Engineering, National Taipei University of Technology, Taipei 10608, Taiwan
ckni@ntut.edu.tw

[3] School of Naval Architecture, Ocean and Civil Engineering, Shanghai Jiao Tong University, Shanghai 200240, China
slshen@sjtu.edu.cn

[4] School of Highway, Chang'an University, Xi'an 710064, China
zhifeng.wang@chd.edu.cn

[5] Department of Civil and Construction Engineering, Swinburne University of Technology, Victoria 3122, Australia
aarulrajah@swin.edu.au

Abstract. Since the ability of many available jacking force models to characterise the tunnel bore conditions is limited. A simple approach to characterise the tunnel bore conditions is proposed and applied to a case study where four sewer pipelines of the Shulin district sewer network in Taipei County, Taiwan were constructed to verify its validity. Four jacking force models are benchmarked in this study. Based upon the given soil properties and pipe dimensions as well as the pipe buried depth, the calculated normal contact pressure (σ') from each model and the measured frictional stress (τ) in each baseline section are utilised for the back-analysis of the frictional coefficient (μ_{avg}). The μ_{avg} values outside the range of 0.1–0.3 recommended for lubricated drives can be ascribed to the increasing pipe friction resulting from the excessive pipe deviation or ground closure or the gravel formation not being long enough to establish lower face resistance. JMTA has indicated a further potential use in assessment of the interface performance during pipe-jacking works.

1 Introduction

Jacking force prediction highly correlates with the design and selection of a pipe-jacking system (Cheng et al. 2016). Despite many jacking force models available for guiding the provision of jacking capacity, their ability to characterise the conditions of tunnel bore is limited (Cheng et al. 2017). In this study, a pipe-jacking data from

W.-C. Cheng et al. (eds.), *Tunneling in Soft Ground, Ground Conditioning and Modification Techniques*, Sustainable Civil Infrastructures,
https://doi.org/10.1007/978-3-319-95783-8_4

four sewer pipelines of the Shulin district sewer network in Taipei County, Taiwan were analysed using the baseline technique. The specific objectives of this study are (i) to benchmark four well-known jacking force models and (ii) to propose an approach that is capable of characterising the tunnel bore conditions.

2 Jacking Force Models

The four well-established jacking force models that were benchmarked are: (i) JMTA (JMTA 2000), (ii) Ma Baosong (Ma 2008), (iii) Shimada and Matsui (Shimada and Matsui 1998), and (iv) Pellet-Beaucour and Kastner (Pellet-Beaucour and Kastner 2002). The review results are listed in Table 1. Since in the Pellet-Beaucour and Kastner model, the friction resistance, F_s, calculated is based upon the vertical soil stress on the pipe crown (σ_{EV}) relying on the active trap-door experiment (Terzaghi 1936). Table 2 lists the empirical parameters used in each standard.

3 A Simple Approach for Characterisation of Tunnel Bore Conditions

A simple approach to characterise the tunnel bore conditions is detailed as follows: (i) establish the baseline of jacking forces, (ii) calculate the normal contact pressure based upon the given soil properties, pipe dimensions, and pipe buried depth, (iii) extract the measured frictional stress in each baseline section, (iv) back-analyse the frictional coefficient (μ_{avg}) in each baseline section, and (v) determine the tunnel bore conditions by comparing the back-analysed μ_{avg} values to recommended values for lubricated drives.

3.1 Project Geological Conditions and Pipe-Jacking Works

Based on the geological investigation results from five 15-m deep boreholes, the soil succession at this worksite comprises a backfill layer underlaid by a silty sand layer, followed by a dense to very dense, poorly-graded to well graded sand and gravel layer. The piezometric level is at the depth of 4.8 m, as shown in Fig. 1. The slurry pipe-jacking method was utilised for carrying out the excavation of four pipe-jacking drives in the Shulin district in the Taipei County, Taiwan. During pipe-jacking works, bentonite slurry was utilised to maintain the stability of the cutting face and also to pump tunnelling spoils to decantation chambers which remove coarse materials from the tunnelling spoils throughout slurry circulation. Additionally, the tunnelling spoils collected were used to prepare samples for the particle size distribution (PSD) analysis. The PSD curves of the samples are shown in Fig. 2. A 1-m long, 1200-mm internal diameter reinforced concrete pipe was adopted in this project. The tunnel bore was excavated using the 1500-mm diameter cutting wheel, thereby leading to an overcut annulus of 30 mm. A special bentonite lubricant with 2% polymer was utilised for mitigating the impacts of loss of lubricant while excavating in gravel formation. The characteristics for the four drives are summarised in Table 3.

Table 1. Review of jacking force models

Model	JMTA	Ma Baosong	Shimada and Matsui	Pellet-Beaucour and Kastner
F_0	$F_0 = 10 \times 1.32\pi \times D_e \times SPT\text{-}N$	$F_0 = 1/4 \times \pi \times D_e^2 \times [K_0\Sigma(\gamma_i'h_i) + \gamma_w h_w]$	$F_0 = P_w \times A$ where P_w is slurry pressure and A is the area of tunnel face	F_0 = initial jacking or first load
F_s	$F_s = \pi \times D_e \times \tau \times L + \omega \times \mu \times L$	$F_s = 1.2 \times [\mu \times (2P_V + 2P_H + P_B)]$ where $P_V = K_p \times \gamma' \times h \times D_e \times L$, $P_H = \gamma' \times (h + D_e/2) \times D_e \times L \times \tan^2(45 - \phi'/2)$, and $P_B = \omega \times L$	$F_s = [p \times b \times \mu_1 + P_w \times \mu_2 \times (\pi D_2 - b)] \times L$ where b (contact width) = $1.6 \times (P_u \times k_d \times C_e)$ for which $P_u = (D_1^2 - D_2^2) \times \gamma_c \times 0.25 \times \pi$, $k_d = (D_1 \times D_2)/(D_1 - D_2)$, and $C_e = (1 - n_1^2)/E_1 + (1 - n_2^2)/E_2$	$F_s = \mu \times L \times D_e \times (\pi/2) \times [(\sigma_{EV} + \gamma D_e/2) + K_2(\sigma_{EV} + \gamma D_e/2)]$ where K_2 = thrust coefficient of soil arching
σ'	$\sigma' = \gamma' \times (D_e/2)/\tan\phi' + \omega/(\pi \times D_e)$	$\sigma' = K_p \times \gamma' \times h + \gamma' \times (h + D_e/2) \times \tan^2(45 - \phi'/2) + \omega/(2 \times D_e)$	$\sigma' = 2P_u \times [1 - (x^2/a^2)]^{0.5}/(\pi \times a)$ for which $a = b/2$ and x = distance to either side of centerline of the area of contact	$\sigma_{EV} = b \times (\gamma - 2 \times C/b)/(2 \times K \times \tan\delta) \times (1 - e^{-2 \times K \times \tan\delta \times (h/b)})$ where C = cohesion

Table 2. Empirical parameters used in each standard

Empirical parameters	Initial Terzaghi's definition (Terzaghi 1951)	GB 50332 (GB 50332-02 2002)	ATV A 161 (German ATV 1990)	ASTM F 1962 (ASTM 2011)	BS EN 1594 (BS 2009)
b (Silo width) (m)	$D_e \times [1 + 2\tan(45° - \phi/2)]$	$D_e \times [1 + \tan(45° - \phi/2)]$	$D_e \times 3^{0.5}$	$1.5 \times D_e$	$D_e \times [1 + 2\tan(45° - \phi/2)]$
δ (Angle of wall friction in shear plane) (°)	ϕ	30	$\phi/2$	$\phi/2$	ϕ
K (soil pressure coefficient)	1	$\tan^2(45° - \phi/2)$	0.5	$\tan^2(45° - \phi/2)$	$1 - \sin\phi$

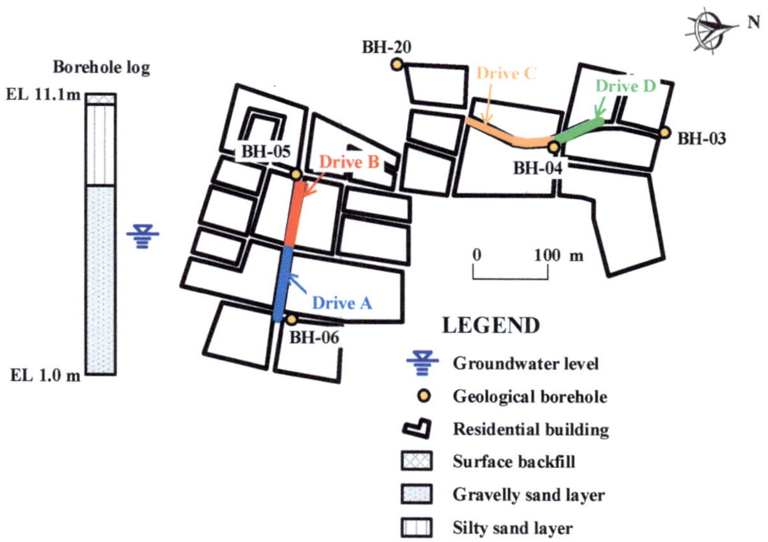

Fig. 1. Location of geological boreholes, typical soil log and design alignment

3.2 Calculated Normal Contact Pressure

Figures 3, 4, 5 and 6 show the variation in the jacking force and torque of cutting wheel at the four pipe-jacking drives. Pellet-Beaucour and Kastner (2002) declared that local variations of total jacking load are generally linked to the varying face resistance and that the minima of total jacking load correspond to very low face resistance. Prior to the analysis of pipe-jacking data, a line running through the minima of total jacking load, referred to as "baseline of jacking forces", was established to distinguish the friction and face resistance from the total jacking load.

Since the four drives were constructed with the h/D_e values varying from 6.7 to 7 (see Table 3), one single stratum has been assumed in the calculation of normal contact pressures (σ'). Considering the groundwater table of 4.5 m depth, the σ' value at the

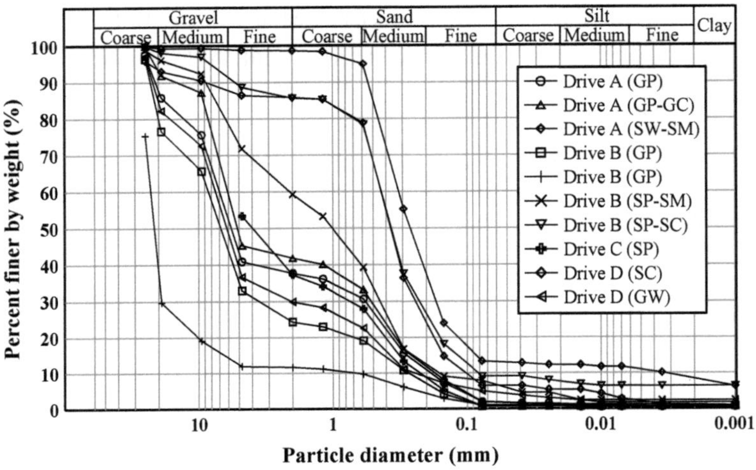

Fig. 2. Particle size distribution of tunnelling spoils for all pipe-jacking drives

Table 3. Characteristics of the four pipe-jacking drives

Drives	A	B	C	D
Geology against baseline section	1: Gravel; 2: fine soil governed gravel or sand deposit; 3: gravel	1, 2 and 4: Gravel; 3: fine soil governed gravel or sand deposit	1, 2 and 3: Gravel; 4: fine soil governed gravel or sand deposit	1: Gravel; 2, 3 and 4: fine soil governed gravel or sand deposit
Length (m)	73	126	75	102
Depth (m)	10.3	10.3	10.8	10.8
h/D_e	6.7	6.7	7	7
GW level (m)	4.5	4.5	4.5	4.5
D_e (m)	1.44	1.44	1.44	1.44
F_0 (kPa)	388	499	555	555
Pipe self-weight (kN/m)	12.7	12.7	12.7	12.7
Buoyancy (kN/m)	17	17	17	17
Theoretical overcut (L/m)	138	138	138	138
Avg. volume of lubricant (L/m)	378	381	552	534

Note 1, 2, 3, and 4 represent the numbering of baseline section

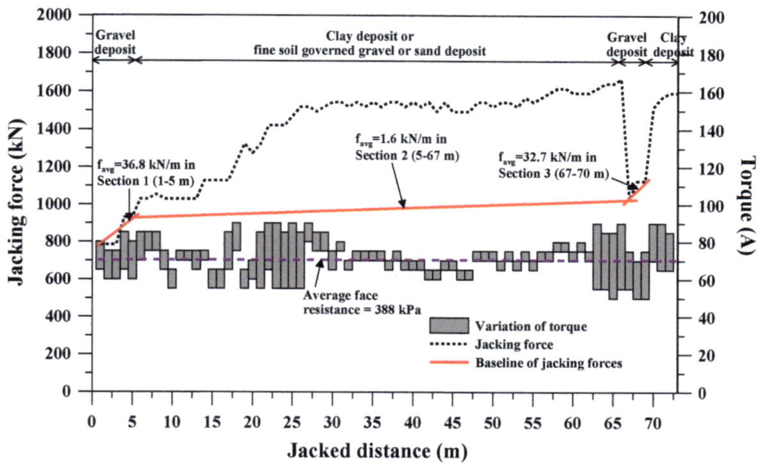

Fig. 3. Variation of jacking force in tunnelling at drive A and baseline of jacking forces

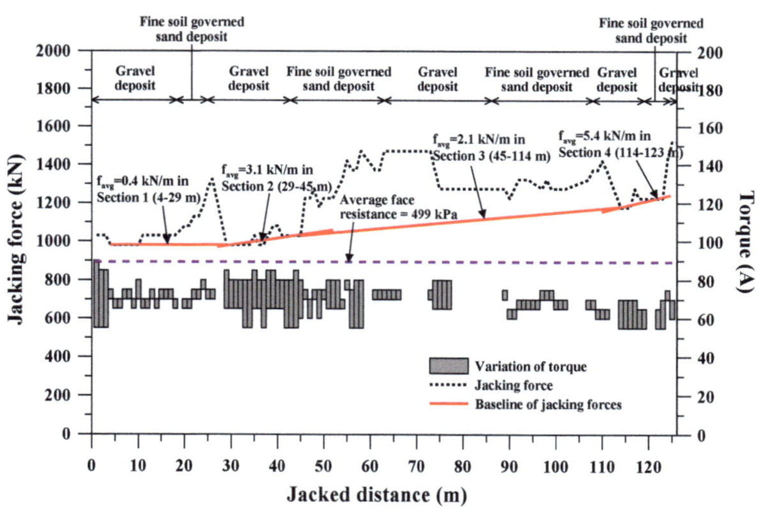

Fig. 4. Variation of jacking force in tunnelling at drive B and baseline of jacking forces

four drives can be yielded, respectively, through the contact pressure equation of the models by substituting the parameters listed in Table 4. The maximum and minimum of the σ' values in each model are depicted in Fig. 7, together with the upper bound value resulting from soil prism above the pipe. As can be seen, the σ' values varying from 13.8 to 14.8 kPa in JMTA were the minima amongst all the models, while in ASTM F 1962, the σ' values varying from 101.7 to 122.2 kPa were the maxima. A calculation sample given in Fig. 8 is for further clarification purposes. It can be seen from Fig. 8 that the σ' value in ASTM F 1962, ATV A 161, GB 50332, and BS EN 1594 increased a bit at small h/D_e and maintained at a constant value at large h/D_e. The

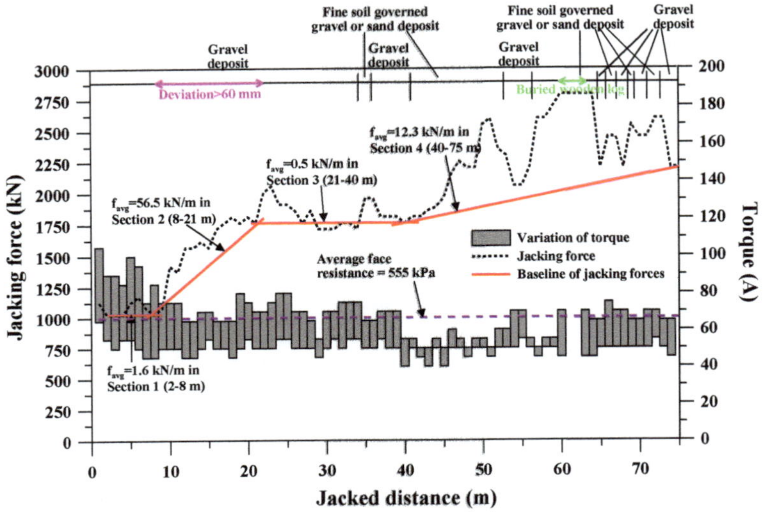

Fig. 5. Variation of jacking force in tunnelling at drive C and baseline of jacking forces

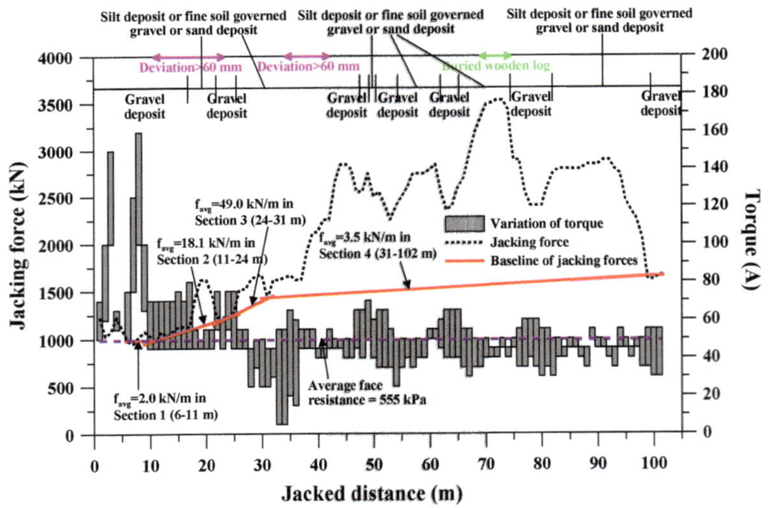

Fig. 6. Variation of jacking force in tunnelling at drive D and baseline of jacking forces

σ' value in Ma Baosong was in a linear relationship with the h/D_e. The σ' value in Terzaghi and JMTA as well as Shimada and Matsui maintained at constant values of 26.9 and 15.3 as well as 40.4, respectively, in all the h/D_e values. The two constant values from Terzaghi and JMTA, respectively, are due to the fact that the soil arching factor in Terzaghi and JMTA converged sooner than the other models (see Fig. 9) and the effect of soil arching quickly had a steady influence on the σ' value. It is known that Shimada and Matsui model assumed a stable tunnel bore and the calculation of σ' value

Table 4. Parameters adopted in the jacking force models

JMTA model	**ϕ' (°)**	**C' (kPa)**	**γ' (kN/m^3)**	**D_e (m)**
	34 (Gravelly soil); 26.3 (Clayey soil)	0	10.09 (Gravelly soil); 8.03 (Clayey soil)	1.44
Ma Baosong model	**K_p (kN/m^3)**	**h (Soil height above pipe) (m)**		**ω (kN/m)**
	0.3 (Gravelly soil); 0.5 (Clayey soil)	9.6 (Drives A and B); 10.1 (Drives C and D)		12.7
Shimada and Matsui model	**k_d $(D_1 \times D_2)/(D_1 - D_2)$**	**D_1 (m)**	**D_2 (m)**	**E_2 (kPa)**
	36	1.5	1.44	4e7
	E_1 (kPa)	**n_1**	**n_2**	**b (m)**
	22,373 (Gravelly soil); 8829 (Clayey soil)	0.35 (Gravelly soil); 0.49 (Clayey soil)	0.2	0.21 (Gravelly soil); 0.32 (Clayey soil)
	$C_e = (1 - n_1^2)/E_1 + (1 - n_2^2)/E_2$ (kPa^{-1})	**x (m)**	**$p_{max} = 2P_u \times [1 - (x^2/a^2)]^{0.5}/(\pi \times a)$**	
	4e−5 (Gravelly soil); 9e−5 (Clayey soil)	0	75.4 (Gravelly soil); 50.9 (Clayey soil)	
Pellet-Beaucour and Kastner model	**C' (kPa)**	**ϕ' (°)**	**γ' (kN/m^3)**	**h (m)**
	0	34 (Gravelly soil); 26.3 (Clayey soil)	10.09 (Gravelly soil); 8.03 (Clayey soil)	9.6 (Drives A and B); 10.1 (Drives C and D)

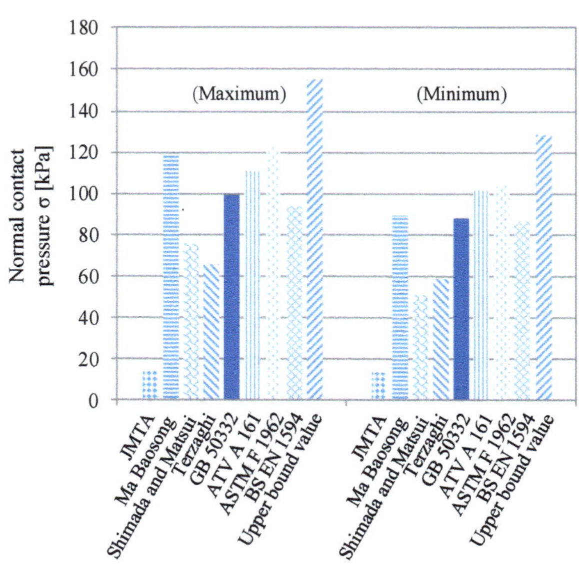

Fig. 7. Comparison of the calculated normal contact pressure between the jacking force models

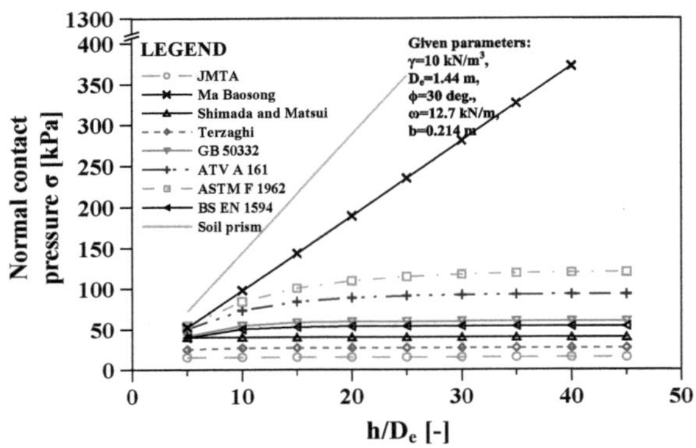

Fig. 8. Variation of normal contact pressure σ' against h/D_e

was not related to the h/D_e value, but related to the values of P_u and b. It can also be seen from Fig. 8 that as $h/D_e < 7$, ASTM F 1962 was the maximum, while JMTA was the minimum. These results matched with those shown in Fig. 7. As ASTM F 1962 made a more conservative assumption that reduces the mobilised shearing resistance along shearing bands by taking only half the friction angle into account, it gave the highest estimation. On the other hand, JMTA calculated the σ' value by introducing the reduced weight of soil prism that takes the overburden height equal to half the outer pipe diameter into account and considered fully developed shearing bands. The aforesaid two assumptions are considered as the two main reasons to lead to the lowest estimation.

3.3 Back-Analysed Frictional Coefficient

Dividing the pipe friction from the difference in the jacking force at both ends of a baseline section by the section length leads to the average jacking force (f_{avg}). Dividing the f_{avg} value by the outer periphery of pipe yields the measured frictional stress (τ). Thus, the frictional coefficient (μ_{avg}) can be back-analysed through dividing the τ value by the σ' value. The three values are summarized in Table 5. In the first and final sections at Drive A, the μ_{avg} values in JMTA were back-analysed to be 0.59 and 0.52, respectively (Table 6), largely in excess of the upper limit of 0.3 suggested by Stein et al. (1989). In contrast with JMTA, the others reported the μ_{avg} values below the lower limit of 0.1. In fact, the higher μ_{avg} values of 0.59 and 0.52 were caused by excessive pipe frictions while tunnelling through two gravel sections where local collapse or ground closure happened. Thus, JMTA gave the more reliable μ_{avg} values.

At Drive B, the μ_{avg} values derived from all of the models were far below the lower limit of 0.1, indicating that Drive B was very well-lubricated. In the second section at Drive C, the μ_{avg} value in JMTA was 0.9, largely in excess of the upper limit of 0.3, suggesting that it was not adequately lubricated. The μ_{avg} values from the other models were within the limits of 0.1 and 0.3 recommended by Stein et al. (1989) for lubricated

Table 5. Summary of Calculated normal contact pressure based on (a) JMTA, (b) Ma Baosong, (c) Shimada and Matsui, (d) Terzaghi, (e) GB 50332, (f) ATV A 161, (g) ASTM F 1962, (h) BS EN 1594

Drives	A	B	C	D
f_{avg} (kN/m)	1: 36.8; 2: 1.6; 3: 32.7	1: 0.4; 2: 3.1; 3: 2.1; 4: 5.4	1: 1.6; 2: 56.5; 3: 0.5; 4: 12.3	1: 2.0; 2: 18.1; 3: 49.0; 4: 3.5
τ (kPa)	1: 8.1; 2: 0.4; 3: 7.2	1: 0.09; 2: 0.7; 3: 0.5; 4: 1.2	1: 0.4; 2: 12.5; 3: 0.1; 4: 2.7	1: 0.4; 2: 4.0; 3: 10.8; 4: 0.8
(a) JMTA	1: 13.8; 2: 14.8; 3: 13.8	1: 13.8; 2: 13.8; 3: 14.8; 4: 13.8	1: 13.8; 2: 13.8; 3: 13.8; 4: 14.8	1: 13.8; 2: 14.8; 3: 14.8; 4: 14.8
(b) Ma Baosong	1: 89.8; 2: 115.8; 3: 89.8	1: 89.8; 2: 89.8; 3: 115.8; 4: 89.8	1: 92.8; 2: 92.8; 3: 92.8; 4: 119.4	1: 92.8; 2: 119.4; 3: 119.4; 4: 119.4
(c) Shimada and Matsui	1: 75.4; 2: 50.9; 3: 75.4	1: 75.4; 2: 75.4; 3: 50.9; 4: 75.4	1: 75.4; 2: 75.4; 3: 75.4; 4: 50.9	1: 75.4; 2: 50.9; 3: 50.9; 4: 50.9
(d) Terzaghi	1: 58.9; 2: 65.3; 3: 58.9	1: 58.9; 2: 58.9; 3: 65.3; 4: 58.9	1: 59.4; 2: 59.4; 3: 59.4; 4: 66.1	1: 59.4; 2: 66.1; 3: 66.1; 4: 66.1
(e) GB 50332	1: 97.7; 2: 86.0; 3: 97.7	1: 97.7; 2: 97.7; 3: 86.0; 4: 97.7	1: 99.8; 2: 99.8; 3: 99.8; 4: 87.7	1: 99.8; 2: 87.7; 3: 87.7; 4: 87.7
(f) ATV A 161	1: 108.4; 2: 99.5; 3: 108.4	1: 108.4; 2: 108.4; 3: 99.5; 4: 108.4	1: 111.1; 2: 111.1; 3: 111.1; 4: 102.0	1: 111.1; 2: 102.0; 3: 102.0; 4: 102.0
(g) ASTM F 1962	1: 118.9; 2: 101.7; 3: 118.9	1: 118.9; 2: 118.9; 3: 101.7; 4: 118.9	1: 122.2; 2: 122.2; 3: 122.2; 4: 104.3	1: 122.2; 2: 104.3; 3: 104.3; 4: 104.3
(h) BS EN 1594	1: 92.4; 2: 84.6; 3: 92.4	1: 92.4; 2: 92.4; 3: 84.6; 4: 92.4	1: 94.2; 2: 94.2; 3: 94.2; 4: 86.3	1: 94.2; 2: 86.3; 3: 86.3; 4: 86.3

Note 1, 2, 3, and 4 represent the numbering of baseline section

Table 6. Summary of the back-analysed μ_{avg} values for the four pipe-jacking drives

Drives	A	B	C	D
μ recommended by Stein et al. (1989)	1: 0.3–0.4; 2: 0.2–0.3; 3: 0.3–0.4	1: 0.3–0.4; 2: 0.3–0.4; 3: 0.2–0.3; 4: 0.3–0.4	1: 0.3–0.4; 2: 0.3–0.4; 3: 0.3–0.4; 4: 0.2–0.3	1: 0.3–0.4; 2: 0.2–0.3; 3: 0.2–0.3; 4: 0.2–0.3
Back-analysed μ_{avg} by JMTA	1: 0.59; 2: 0.03; 3: 0.52	1: 0.007; 2: 0.05; 3: 0.03; 4: 0.09	1: 0.03; 2: 0.9; 3: 0.007; 4: 0.18	1: 0.03; 2: 0.30; 3: 0.73; 4: 0.05
Back-analysed μ_{avg} by Ma Baosong	1: 0.09; 2: 0.003; 3: 0.08	1: 0.001; 2: 0.008; 3: 0.004; 4: 0.01	1: 0.004; 2: 0.14; 3: 0.001; 4: 0.02	1: 0.004; 2: 0.03; 3: 0.09; 4: 0.007
Back-analysed μ_{avg} by Shimada and Matsui	1: 0.11; 2: 0.008; 3: 0.10	1: 0.001; 2: 0.009; 3: 0.01; 4: 0.02	1: 0.005; 2: 0.17; 3: 0.001; 4: 0.05	1: 0.005; 2: 0.08; 3: 0.21; 4: 0.02
Back-analysed μ_{avg} by Terzaghi	1: 0.14; 2: 0.006; 3: 0.12	1: 0.002; 2: 0.01; 3: 0.008; 4: 0.02	1: 0.007; 2: 0.21; 3: 0.002; 4: 0.04	1: 0.007; 2: 0.06; 3: 0.16; 4: 0.01
Back-analysed μ_{avg} by GB 50332	1: 0.08; 2: 0.005; 3: 0.07;	1: 0.001; 2: 0.007; 3: 0.006; 4: 0.01	1: 0.004; 2: 0.13; 3: 0.001; 4: 0.03	1: 0.004; 2: 0.05; 3: 0.12; 4: 0.009
Back-analysed μ_{avg} by ATV A 161	1: 0.08; 2: 0.004; 3: 0.07	1: 0.001; 2: 0.006; 3: 0.005; 4: 0.01	1: 0.004; 2: 0.11; 3: 0.001; 4: 0.03	1: 0.004; 2: 0.04; 3: 0.11; 4: 0.008
Back-analysed μ_{avg} by ASTM F 1962	1: 0.07; 2: 0.004; 3: 0.06	1: 0.001; 2: 0.006; 3: 0.005; 4: 0.01	1: 0.003; 2: 0.10; 3: 0.001; 4: 0.03	1: 0.003; 2: 0.04; 3: 0.10; 4: 0.008
Back-analysed μ_{avg} by BS EN 1594	1: 0.09; 2: 0.005; 3: 0.08	1: 0.001; 2: 0.008; 3: 0.006; 4: 0.01	1: 0.004; 2: 0.13; 3: 0.001; 4: 0.03	1: 0.004; 2: 0.05; 3: 0.13; 4: 0.009
Tunnel bore condition deduced by JMTA	1: Ground closure; 2: Stable tunnel bore; 3: Ground closure	1: Stable tunnel bore; 2: Stable tunnel bore; 3: Stable tunnel bore; 4: Stable tunnel bore	1: Stable tunnel bore; 2: Possible ground closure; 3: Stable tunnel bore; 4: Ground closure	1: Stable tunnel bore; 2: Possible ground closure; 3: Not applicable; 4: Stable tunnel bore

Note 1, 2, 3, and 4 represent the numbering of baseline section

drives. As jacking through the second section at Drive C, the μ_{avg} value of 0.9, the highest μ_{avg} value observed from this study, was most likely ascribed to increasing pipe friction resulting from the excessive pipe deviation greater than the threshold of 60 mm. While the other models resulted in a misleading interference for estimation of the μ_{avg} value.

In the second section of silty gravel or silty sand at Drive D, the μ_{avg} value of 0.3 in JMTA matched with the upper limit of 0.3, indicating that it was moderately lubricated. Whereas the other models reported the μ_{avg} values below 0.1. According to the measurements, the μ_{avg} value of 0.3 was attributed to increasing pipe friction resulting from the pipe deviation being greater than 60 mm. Similar to the other drives, JMTA reported the μ_{avg} values more reliable than the other models. Additionally, the third section at Drive D characterised by JMTA indicated inadequate lubrication, with the μ_{avg} value of 0.73 (the second highest μ_{avg} value observed in this study). The possible reason to lead to this higher μ_{avg} value was not caused by unfavorable pipe friction resulting from the excessive pipe deviation, but by the gravel formation at 32 m not being long enough to establish lower face resistance or total jacking load.

4 Conclusions

(1) JMTA provided the lowest estimation of the normal contact pressure (σ') by introducing the reduced overburden height and assuming fully developed shearing bands. As $h/D_e < 7$, ASTM F 1962 gave the highest estimation as it made a more cautious assumption that reduces the mobilised shearing resistance along the sliding planes by taking only half the friction angle into account.

(2) In JMTA, the back-analysed μ_{avg} value, resulting from jacking through the first and final sections of gravel at Drive A, was 0.59 and 0.52, respectively, while the other models reported the μ_{avg} values below 0.1. The higher μ_{avg} values were most likely due to increasing pipe friction incurred while jacking through the gravel sections where local collapse or ground closure occurred. The μ_{avg} values of 0.9 and 0.3, respectively, induced by tunnelling through the second section of gravel at Drive C and the second section of fine soil governed ground at Drive D were attributed to increasing friction resistance resulting from the excessive pipe deviation greater than the threshold value of 60 mm. While the other models reported the μ_{avg} values within the limits of 0.1 and 0.3 and below 0.1, respectively.

(3) JMTA characterised the pipe-jacking through the third section of silty gravel or clayey gravel at Drive D as inadequately lubricated, with the μ_{avg} value of 0.73. The higher μ_{avg} value was not caused by unfavourable pipe friction resulting from the excessive pipe deviation, but by the gravel formation at 32 m distance not being long enough to establish lower face resistance or total jacking load.

(4) JMTA yielded the μ_{avg} value more realistic than the other models and has shown a potential use in assessing the interface performance during pipe-jacking works.

References

ASTM: Standard Guide for Use of Maxi-Horizontal Directional Drilling for Placement of Polyethylene Pipe or Conduit under Obstacles Including River Crossings. F 1962-11, West Conshohocken, PA (2011)

British Standards: BS EN:1594-09 Gas supply system-pipelines for maximum operating pressure over 16 bar-functional requirements. Brussels, pp. 76–78 (2009)

Cheng, W.C., Ni, J.C., Shen, S.L.: Experimental and analytical modeling of shield segment under cyclic loading. Int. J. Geomech. (2016). https://doi.org/10.1061/(ASCE)GM.1943-5622.0000810

Cheng, W.C., Ni, J.C., Shen, S.L., Huang, H.W.: Investigation into factors affecting jacking force: a case study. In: Proceedings of the Institution of Civil Engineers—Geotechnical Engineering (2017). https://doi.org/10.1680/jgeen.16.00117

German ATV Rules and Standards: ATV-A 161 E-90. Structural calculation of driven pipes. Hennef, pp. 18–20 (1990)

Japan Microtunnelling Association (JMTA): Pipe-jacking Application, Japan Microtunnelling Association (JMTA), Tokyo (2000)

Ma, B.: The Science of Trenchless Engineering. China Communications Press, Beijing (2008)

Pellet-Beacour, A.L., Kastner, R.: Experimental and analytical study of friction forces during microtunnelling operations. Tunn. Undergr. Space Technol. **17**(1), 83–97 (2002). https://doi.org/10.1016/S0886-7798(01)00044-X

Stein, D., Möllers, K., Bielecki, R.: Microtunneling: Installation and Renewal of Nonman-Size Supply and Sewage Lines by the Trenchless Construction Method. Ernst, Berlin, Germany (1989)

Shimada, H., Matsui, K.: A new method for the construction of small diameter tunnels using pipe jacking. In: Proceedings of Regional Symposium on Sedimentary Rock Engineering, pp. 234–239 (1998)

Terzaghi, K.: The shearing resistance of saturated soils and the angle between the planes of shear. In: Proceedings of the 1st International Conference on Soil Mechanics and Foundation Engineering. Harvard University Press, Cambridge, MA, pp. 54–56 (1936)

Terzaghi, K.: Theoretical Soil Mechanics. Wiley, New York (1951)

The Ministry of Construction of China: GB 50332-02, Structural Design Code for Pipeline of Water Supply and Waste Water Engineering. Beijing, pp. 11–12 (2002)

Deformation Law of Surrounding Rock of Expansive Soil Tunnel Based on Dry–Wet Cycle Model Test

Ke Wu[1(✉)], Yajun Wang[2], Yalin Yu[1], Shuaishuai Cui[1], and Qianjin Zhang[1]

[1] School of Civil Engineering, Shandong University, Jinan 250061, China
wk4223@163.com
[2] Geotechnical and Structural Engineering Research Institute, Shandong University, Jinan 250061, China

Abstract. Expansive soil is a kind of geological body with special hazards to the construction engineering. It has the engineering properties of shrinkage, fracture and overconsolidation. Its engineering hazard is rich in water content in soil. The expansive geomaterials can produce wetting and shrinkage deformation, which adversely affect the subgrade, slope and tunnel support structure. In order to solve this problem, the dry and wet cycle model test device was established by simulating the "rain" and "sunshine" cycle, and the typical dry wet circulation, shrinkage and other factors, focusing on the expansion and contraction of soil tunnel rock failure instability development model, the scope and its conditions and the impact of factors such as simulated rainfall under the conditions of the tunnel excavation and support, the stress, strain and deformation of the surrounding rock of the tunnel are analyzed, and the possible damage modes are analyzed, and the mechanism of the surrounding rock disaster in the shrinkage fracture zone is revealed.

Keywords: Dry and wet cycle · Model test · Expansive soil · Tunnel Deformation law

1 Introduction

Water swelling of expansive soil or loss of water shrinkage results in unstable clay, and its distribution is extremely wide. Expansive soil is a special soil and causes harm to project activities on the soil owing to its water swelling and loss of water characteristics. This type of soil usually causes mountain collapses, landslides, debris flows, and other geological disasters. The reasons for the special properties of expansive soil are complicated. Current scholars mainly study from the three aspects of mineral composition, namely, related physical and chemical properties and physical mechanics. The expansive soil after expansion and the structure of interaction are complex, and the scene of influencing factors is difficult to study. Physical similarity simulation is an important, effective, and scientific method for solving complex geotechnical engineering problems [1–3]. The engineering geomechanical model testing technology pioneered by Fumagalli [4] is gradually being recognized by scholars and applied on a

W.-C. Cheng et al. (eds.), *Tunneling in Soft Ground, Ground Conditioning and Modification Techniques*, Sustainable Civil Infrastructures,
https://doi.org/10.1007/978-3-319-95783-8_5

large scale. Ashby [5] used the model test technique to study the theory of slope dumping intensively. Baumgargarter et al. [6–8] also used the model test to conduct relevant research. Patton [9] used gypsum to simulate the climbing angle effect of zigzag structure. Einstein et al. [10] used gypsum to simulate the climbing phenomenon and shear mechanism of zigzag structure. Jacoby and Schmeling [11] used glycerol, fused paraffin, and other simulated mantle convection and plate-related role. Kincaid and Olson [12] also simulated the lithosphere and asthenosphere using similar materials.

For similar materials, domestic scholars have produced numerous research results. Zuo et al. [13] conducted a rock mass material simulation test using quartz sand, gypsum, and cement. Wu et al. [14] performed a weak rock simulation test using barite powder, gypsum, fine sand, laundry liquid, and water. Li et al. [15] used sand, barite powder, talc, cement, petrolatum, silicone oil, and mixed water for the flow-solid coupling of a relevant simulation test. Ye et al. [16] used cement, gypsum, and glycerol to configure similar materials and studied the effect of different mortar ratios and cement contents on the physical and mechanical properties of materials. Liu et al. [17] conducted a related study on the ratio of similar materials in rock mass. Zhang et al. [18] used barite powder, sand, industrial salt, gypsum, and bentonite in a collapsible loess simulation test. However, few studies on the ratio of expansive similar materials exist in China. Lai et al. [19] developed similar materials for expansive rocks using sand, bentonite, lime, admixture, and mixed water. Guo [20] compared the mechanical properties of the similar material with different ratio of illite, kaolinite, chlorite, and bentonite.

In this study, large-scale model tests are carried out using similar materials prepared using bentonite and gypsum (particle size is less than 0.5 mm). The stress–strain and settlement deformations of the tunnel's surrounding rock in an expansive soil area are studied.

2 Method

The test model includes a rain simulation, a rain container, a working table, and a model frame with a length of 2 m and a width of 0.5 m, as shown in Fig. 1. Fine sand is used to fill the right, left, and bottom of test model, with a thickness of 25 cm, to ensure successful permeation of water.

Through a series of tests, expansive soil is simulated using the sands in Huang River, bentonite, and gypsum, with a mass ratio of 8:3:1. The free expansive ratio of the similar material is 68%, and its permeability coefficient is 4.2×10^{-7} cm/s when the degree of compaction is 95% and is 2.53×10^{-8} cm/s when the degree of compaction is 90%.

2.1 Monitoring Element

Three types of monitoring elements are present in the test model, namely, strain brick, earth pressure cell, and multipoint displacement meter. The strain brick is made with similar material, as shown in Fig. 2, which is a cube with 2 cm sides. It is used to

Fig. 1. Test model apparatus

monitor the strain of the specified part. The process of making the strain brick is shown in Fig. 3. The strain bricks are placed in the test model, as shown in Fig. 4, to monitor the strain of the surrounding rock during testing. In Fig. 4, 1 represents the monitoring point at the bottom of the tunnel; 2 and 3 represent the monitoring points at the haunch of the tunnel, with a distance of 11 cm; 4 and 5 represent the monitoring points at the spandrel of the tunnel, with a distance of is 11 cm; and 6 and 7 represent the monitoring points at the top of the tunnel, with a distance of 11 cm. Considering the special case of the test model and its purpose, the location of earth pressure cell is the same as that of the strain bricks, as shown in Fig. 4. The multipoint displacement meters are placed in the test model as shown in Fig. 5. In Fig. 5, 1 represents the monitoring point at the bottom of the tunnel; 2, 3, 4, 5, and 6 represent the monitoring points at the haunch of the tunnel, with a distance of 11 cm; and 7, 8, 9, 10 and 11 represent the monitoring points at the top of the tunnel, with a distance of 11 cm.

2.2 Experimental Process

Firstly, the similar material must be mixed at a ratio of 8(sand from Huang River):3 (bentonite):1(gypsum), which is designed on the basis of a series of tests. The sand from Huang River, bentonite, and gypsum are mixed using an agitator, which can ensure the sufficient mixing of the similar material to satisfy the test demand, as shown in Fig. 6. Secondly, the seepy material must be placed at the bottom, right, and left of the model as the seepy boundary of similar material. In this test, silver sand is used as a type of seepy material, with a thickness of 25 cm. Thirdly, the similar material is poured at a thickness of 30 cm at once and compacted by a hammer. Then, the monitoring element should be placed at the proper place, as shown in Fig. 7.

When all preparatory works are done, the upper bench of the first part of the tunnel should be excavated and protected by gypsum, as shown in Fig. 8. Then, the lower bench of the first part should be excavated 25 min later when the surrounding rock has stabilized and protected by gypsum, as shown in Fig. 8. Subsequently, rain should be

Fig. 2. Strain brick

Fig. 3. Strain brick

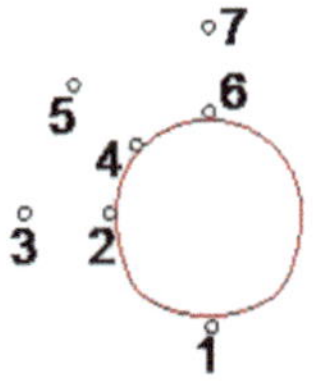

Fig. 4. Location of monitoring points of strain bricks and earth pressure cells

simulated every day until the test model is permeated by water. Then, the second part of the tunnel should be excavated in the same manner as the first part, as shown in Fig. 9. Then, sunshine should be simulated, as shown in Fig. 10. The third part of the tunnel should be excavated after the sunshine is simulated for 10 days, as shown in Fig. 11.

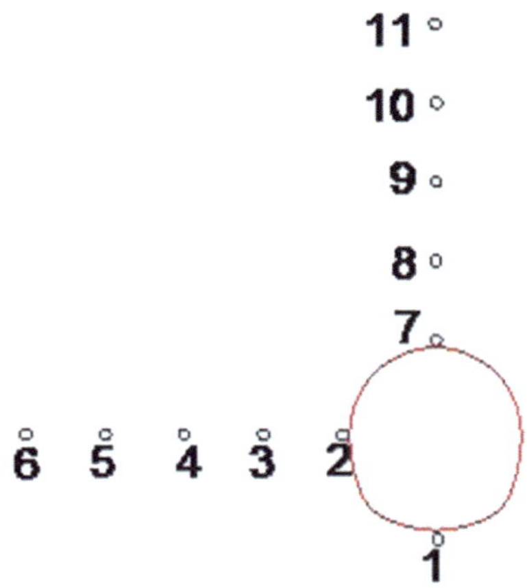

Fig. 5. Location of multipoint displacement meters

Fig. 6. Mixing of similar material

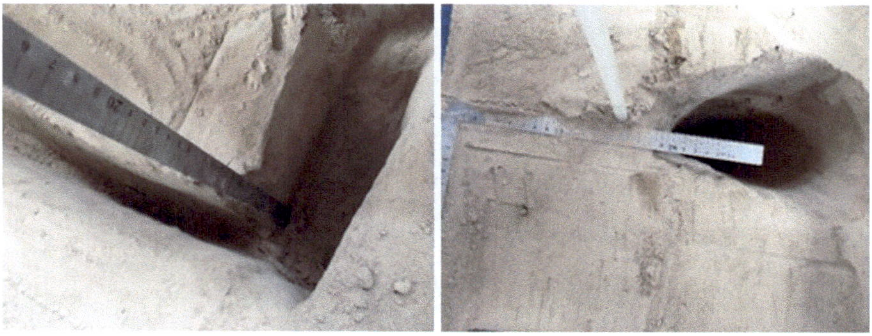

Fig. 7. Location of monitoring elements

Fig. 8. Excavation of the first part of the tunnel

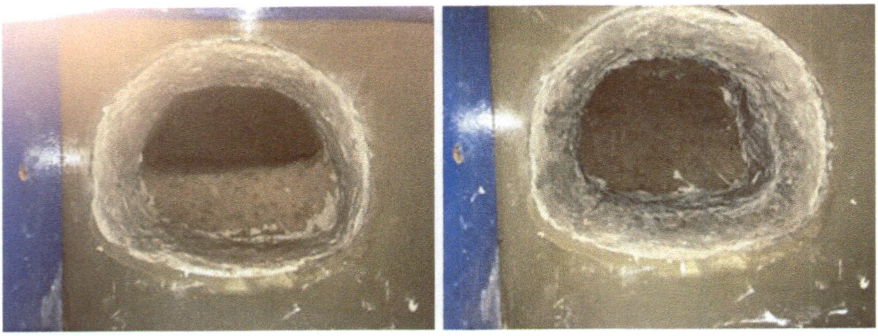

Fig. 9. Excavation of the second part of the tunnel

Fig. 10. Simulation of sunshine

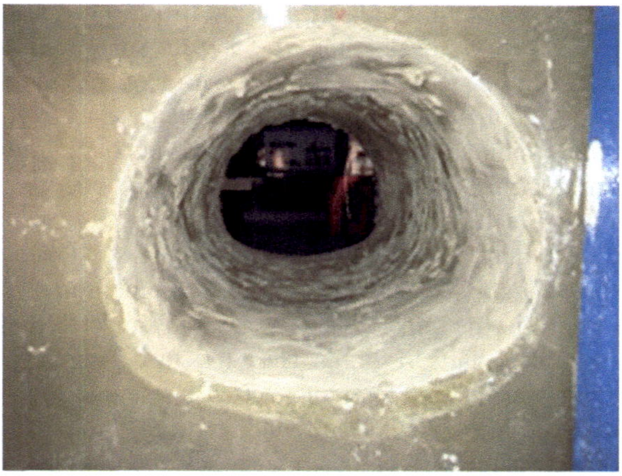

Fig. 11. Excavation of the third part of the tunnel

3 Results

3.1 Strain

For the strain brick, monitoring point 1 at the bottom of the tunnel represents the changing strain of the bottom of the tunnel. With passing time, the strain of monitoring point 1 changes, as shown in Fig. 12. The Fig. 12 shows that the strain of monitoring point 1 increases rapidly and can reach −550 με after the excavation of the first part of the tunnel with the release of the stress of the surrounding rock near the first part of the tunnel. When the strain of monitoring point 1 reaches the peak value, it decreases rapidly and can be close to 0. The reason is that the stress of the surrounding rock near the first part of the tunnel is released, the surrounding rock deforms, and then the model reaches a new balance, therefore the strain of the surrounding rock reaches its minimum value. When it rains, load can be generated by the rainwater on the entire model. Accordingly,

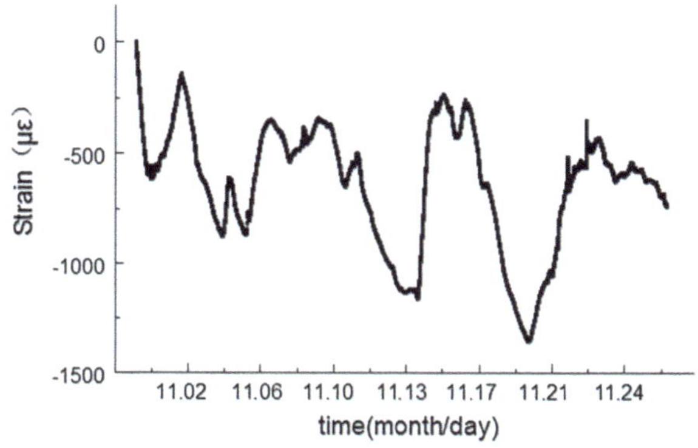

Fig. 12. Strain of monitoring point 1

the strain of monitoring point 1 increases with the permeating water. The strain of monitoring point 1 reaches its peak value of −900 με when rain water reaches the bottom of the model. Subsequently, the strain value of monitoring point 1 waves in approximately 200 με and then decreases to the minimum value, which may be caused by the expansion of the similar material with water immersion. Furthermore, the entire model reaches a new balance. When rainwater reaches the bottom of the model, the simulation of rain should be stopped and the second part of the tunnel should be excavated. Then, the strain of monitoring point 1 decreases rapidly until −300 με. Thereafter, sunshine should be simulated. With sunshine simulated, the strain of monitoring point 1 remains the same. After five days of simulated sunshine, the similar material loses its water and starts to shrink. The stress in the surrounding rock changes rapidly, and the strain value of monitoring point 1 changes rapidly from −300 to −1450 με. At this time, crevices develop in the surrounding rock. With passing time, the stress of the surrounding rock is released and the model gets a new balance. After the excavation of the third part of the tunnel, the strain of monitoring point 1 decreases rapidly to −500 με.

The strain values of monitoring points 2 and 3, as shown in Figs. 13 and 14, respectively, represent the strain of the haunch of the tunnel. The corresponding graphs show that the trend of the strain data of monitoring points 2 and 3 is similar to the monitoring point 1 at the bottom of arch. The change patterns of monitoring point 2 are similar with those of monitoring point 1. The change rules of monitoring points 2 and 1 are the same and the only difference is the strain value. Monitoring point 2 achieves a minimum value of −500 με and a maximum value of 800 με for strain. Given that monitoring point 3 is 11 cm from the tunnel arch position, its corresponding strain curve shape differs from those of monitoring points 1 and 2, mainly because that the strain of monitoring point 3 changes less when similar material being permeated. Monitoring point 3 does not show any fluctuation in the process of water infiltration because it is 11 cm from the outside of the tunnel arch on the vertical section and 15 cm on the horizontal section from the exposed surface of the first ring. The distance

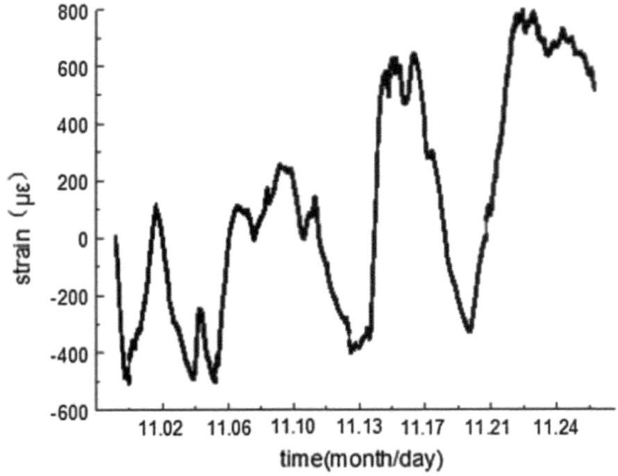

Fig. 13. Strain of monitoring point 2

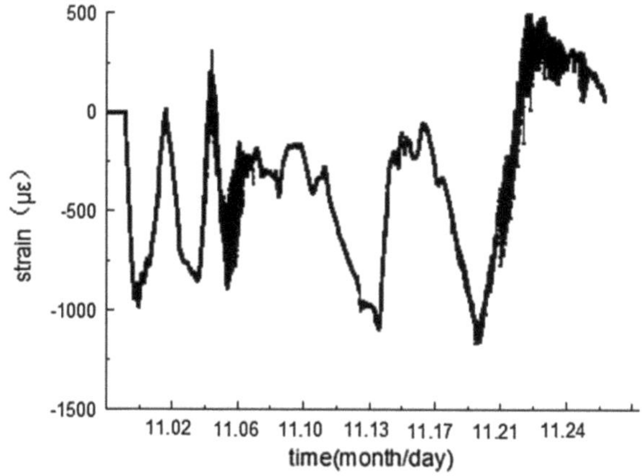

Fig. 14. Strain of monitoring point 3

is still relatively far. And the effect of excavation of first ring tunnel coupled with the internal stress redistribution caused by the infiltration of water on monitoring point 3 is relatively less than the monitoring point 2.

The strain values of monitoring points 4 and 5, as shown in Figs. 15 and 16, respectively, represent the strain of the haunch of the tunnel. Monitoring points 4 and 5 are located at the 45° direction of the haunch and the top of the tunnel, and the curves of monitoring points 4 and 5 show that monitoring point 4 is the same as monitoring point 5 and also as the monitoring points 1 and 2. And the total change range of monitoring points 4 and 5 with a maximum value of 200 με and a minimum value of −1000 με is relatively smaller than that of monitoring point 2.

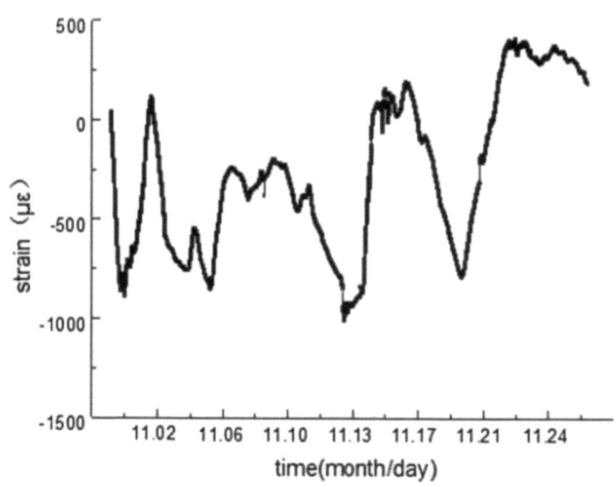

Fig. 15. Strain of monitoring point 4

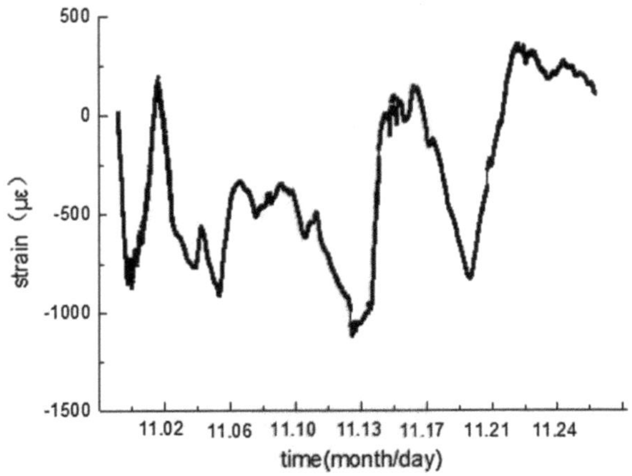

Fig. 16. Strain of monitoring point 5

The strain values of monitoring points 6 and 7, as shown as Figs. 17 and 18, respectively, represent the strain of the haunch of the tunnel. Monitoring point 6 is the same as monitoring point 7 and also same as monitoring points 1, 2, 4, and 5, with the only difference of the specific strain values. Monitoring point 6 possesses a minimum strain value of −1600 με and a maximum value of −150 με, while monitoring point 7 possesses a minimum strain value of −2800 με and a maximum value of 800 με. Monitoring point 6 is located at the tunnel dome, and the change in the strain change value is not too much from the change value of the other position. Monitoring point 7 is located 11 cm above monitoring point 6 and is largely affected by the model surface relative to the other monitoring points. This effect is reflected in the simulated rainfall and sunshine. Monitoring point 7 significantly changes in range and fluctuations.

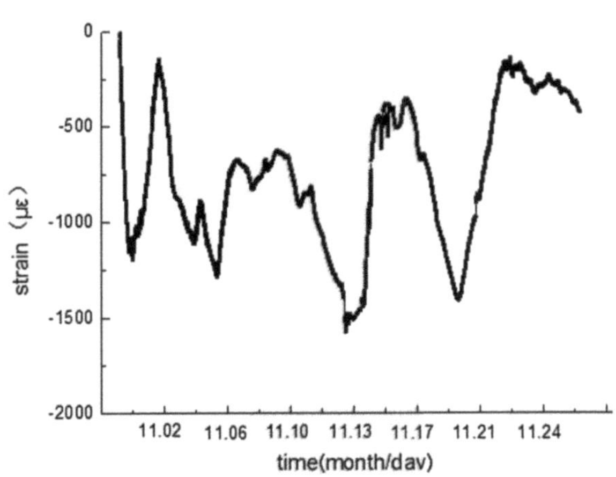

Fig. 17. Strain of monitoring point 6

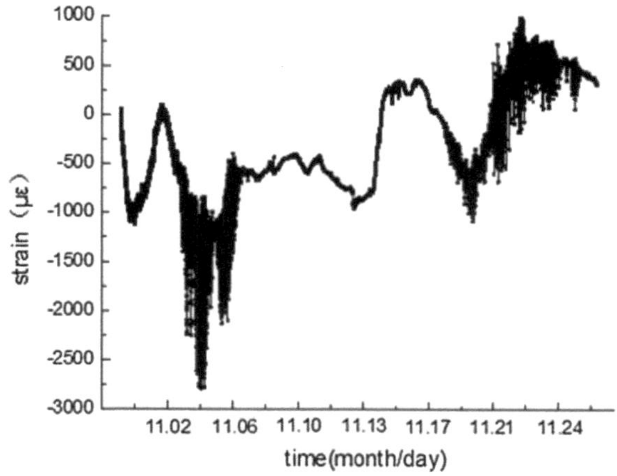

Fig. 18. Strain of monitoring point 7

3.2 Stress

For earth pressure cells, monitoring points 2 and 3 at the haunch of the tunnel represent the stress change of the haunch of the tunnel. With time passing by, the stresses of monitoring points 2 and 3 change, respectively, as shown in Figs. 19 and 20. The data of the earth pressure sensor 2 is not so good due to several conditions in the experiment, but the curve shape is very similar to that of earth pressure sensor 3. For the arch pressure analysis of the location of the arch, the earth pressure sensor 3 is used. The curve of earth pressure sensor 3 shows that the absolute value of the pressure at the earth pressure monitoring point 3 increases after the tunnel is excavated, and the maximum value is −120 Pa under the simulated rainfall. After the simulated rainfall, the absolute value of the earth pressure at monitoring point 3 gradually decreases to approximately 30 Pa because of the rainfall. Thus, the water load on the top of the expansive soil's similar material gradually decreases. With the penetration of rainwater in the similar material, the stress distribution in the expansive soil is completed. Furthermore, the absolute value of the earth pressure increases first to approximately 250 Pa and then decreases at the full penetration of rain, and the absolute value of the earth pressure decreases to approximately 0 Pa. After the simulated rainfall on November 6, the second ring is excavated and then sunshine is simulated. With the water swelling of expansive soil, the absolute value of earth pressure gradually increases to approximately 70 Pa. However, the expansion of the expansive soil is completed. With the further development of the simulated sunshine, the cracks inside the similar materials of the expansive soil gradually develop and the stress is further redistributed. Finally, the stress value is maintained at −25 Pa. When the tunnel is excavated, the absolute value of the earth pressure fluctuates by approximately 120 Pa. This finding is due to that the excavation of the last ring further breaks the stress balance within the similar material, the pressure of the expansive soil is released, and the moisture can be removed from the tunnel through evaporation. The gradual

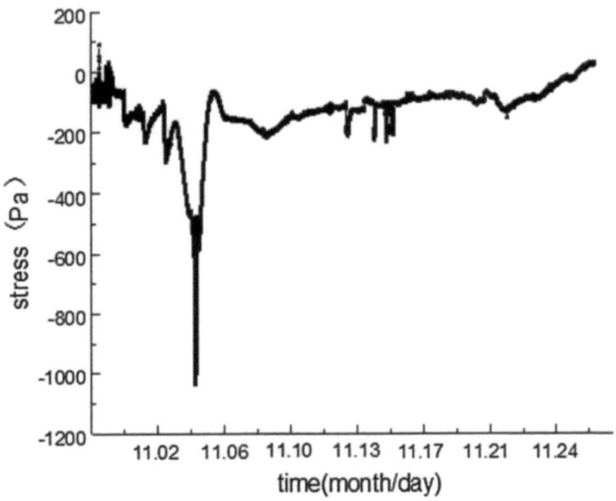

Fig. 19. Stress of monitoring point 2

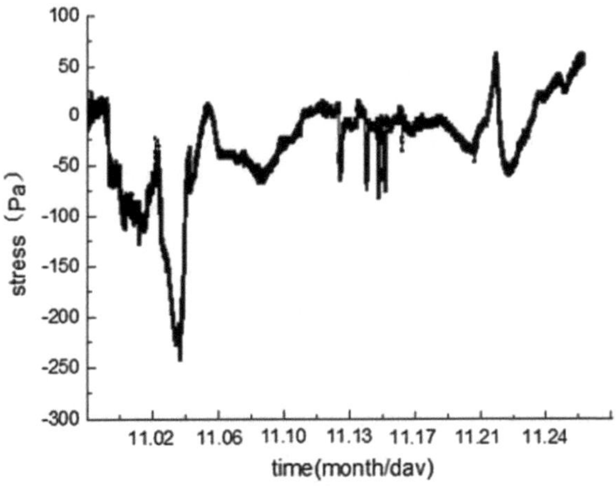

Fig. 20. Stress of monitoring point 3

dehydration contraction in the tunnel around the material results in fluctuations in the curve of the situation. For similar materials, cracks develop further and eventually penetrate.

Figure 21 shows that the fluctuation of monitoring point 5 is smaller than that of monitoring points 3 and 2, but it can be seen that the change shape is similar. The difference is that impact of the tunnel excavation on the monitoring point 5 is not as obvious as that of monitoring points 2 and 3 because its distance from the tunnel position is further. At the beginning of the test, the change trend of monitoring point 5 is the same as that of monitoring point 3. However, when the simulated sunshine is carried out, the earth pressure value of monitoring point 5 gradually increases and

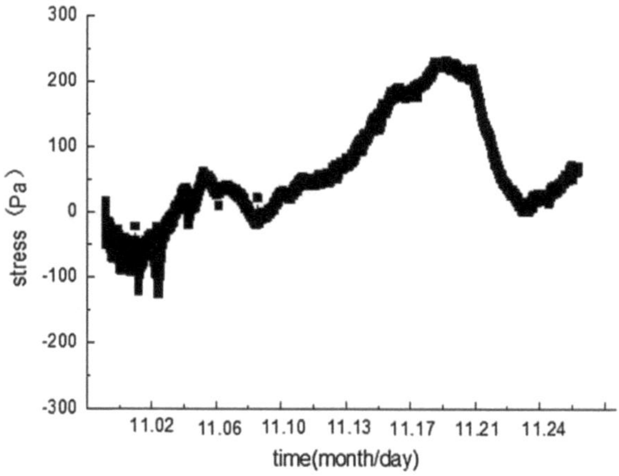

Fig. 21. Stress of monitoring point 5

finally reaches approximately 200 Pa. And then the earth pressure value of monitoring point 5 gradually decreases to 0 Pa. With the development of the apparent fractures, the earth pressure value increases again.

The graphs of earth pressure monitoring points 6 and 7 (Figs. 22 and 23) show that the curve shape is substantially the same as that of monitoring point 5, except that the specific values of the earth pressure are different. As the position of the monitoring point gradually is closing to the top of the expansive soil, the stress value gradually decreases. Earth pressure monitoring point 6 can reach a maximum value of approximately 150 Pa and a minimum value of approximately −100 Pa. The pressure value of the earth pressure monitoring point 7 is smaller than that of monitoring point 6, and its maximum value is 100 Pa and minimum value is approximately 50 Pa.

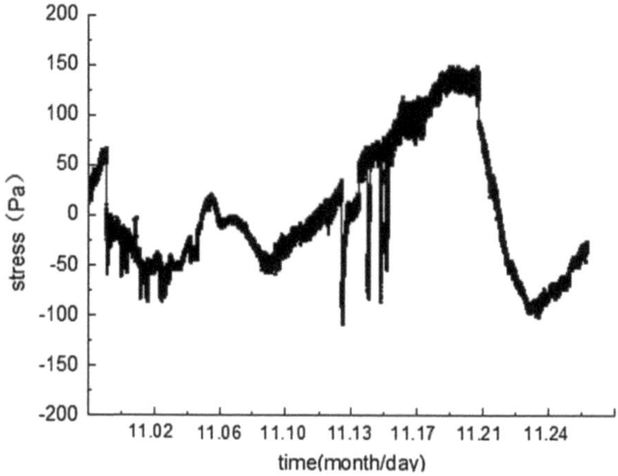

Fig. 22. Stress of monitoring point 6

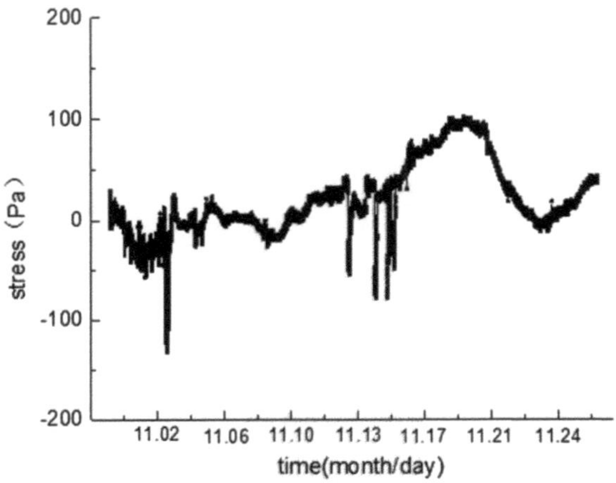

Fig. 23. Stress of monitoring point 7

3.3 Multipoint Displacement Meter

Figure 24 shows that, after excavation of the first ring on November 1, the bottom of the arch exhibits an uplift phenomenon. After the simulated rainfall on November 6, the expansion in the model of expansive soil gradually develops. When the material gradually penetrated completely, the displacement of the bottom of the arch reaches a maximum value of 0.3 mm. In the early stage, the bottom of the arch presents an uplift phenomenon mainly due to the excavation of the first ring. Then, with the gradual infiltration of rainfall in the expansive soil similar material, water swelling gradually increases the amount of uplift. As a result, the arch uplift gradually increases to 0.3 mm in the simulation of sunshine. On November 14, displacement increases due to the construction crane operation. The sunshine exhibits slight effect on the displacement of

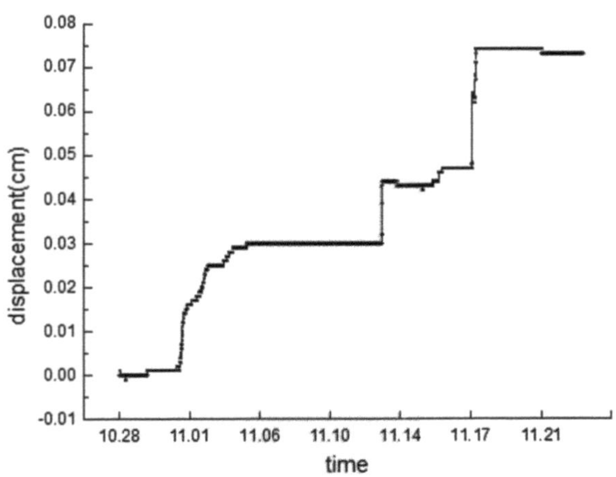

Fig. 24. Displacement of monitoring point 1

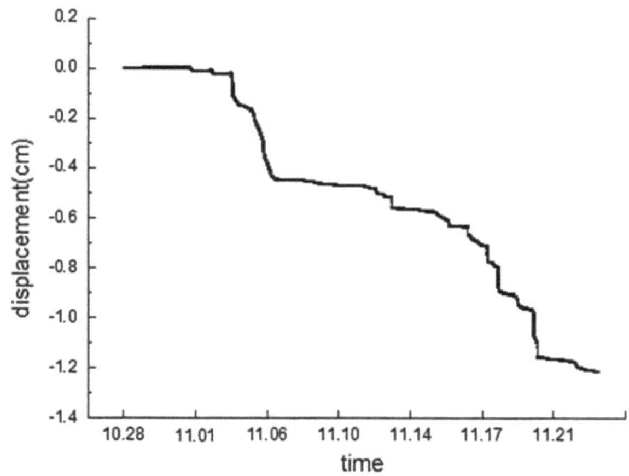

Fig. 25. Displacement of monitoring point 2

the arch. In the final ring, the uplift of the arch increases from 0.45 to 0.7 mm. The effect of sunshine on the displacement of the arch is very small, and the effect of rainfall on the displacement of the arch is relatively larger.

Figures 25, 26, 27 and 28 show that, the influence of excavation of the first ring on the displacement of the monitoring points 2 to 5 is very small because the position of the monitoring section has not been excavated. However, after the rainfall, the displacement on the monitoring point 2 and 3 represent upheaval, while the monitoring point 4 and 5 represent settlement. After the second ring is excavated, the absolute displacement value of monitoring point 2 increases to approximately 0.45, the absolute displacement value of monitoring point 3 increases to approximately 0.2, and the displacement absolute values of monitoring points 4 and 5 are negligible. After the excavation of the second ring, the displacement of the monitoring point 2, 3, 4 and 5

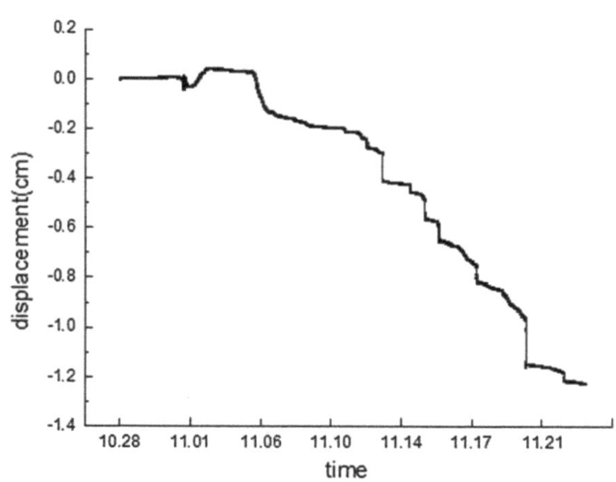

Fig. 26. Displacement of monitoring point 3

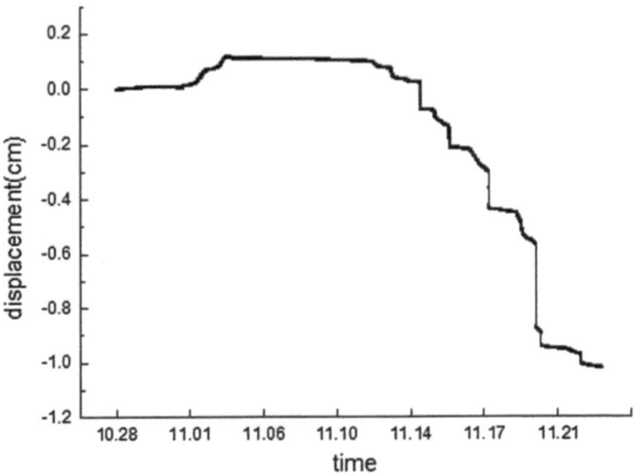

Fig. 27. Displacement of monitoring point 4

gradually develops under simulated sunshine. Monitoring points 2 and 3 reach approximately 1 cm before the third ring excavation, While the absolute displacement values of monitoring points 4 and 5 reach 0.6 and 0.7 cm, respectively. After the excavation of the third ring, the displacement value of each monitoring point of the arch gradually increases. Moreover, the displacement values of monitoring point 2 and 3 reach 1.3 cm, while those of monitoring points 4 and 5 reach 1.1 and 0.9 cm, respectively.

From the displacement curve of the monitoring points of the vaults, shown as Figs. 29, 30 and 31, the displacement of monitoring points does not increase immediately after the excavation of the first ring of the tunnel. However, with the gradual infiltration of the rainfall in the similar material, the settlement values of monitoring points 7 and 8 reach 0.8 cm and the settlement of monitoring point 9 reaches 0.5 cm.

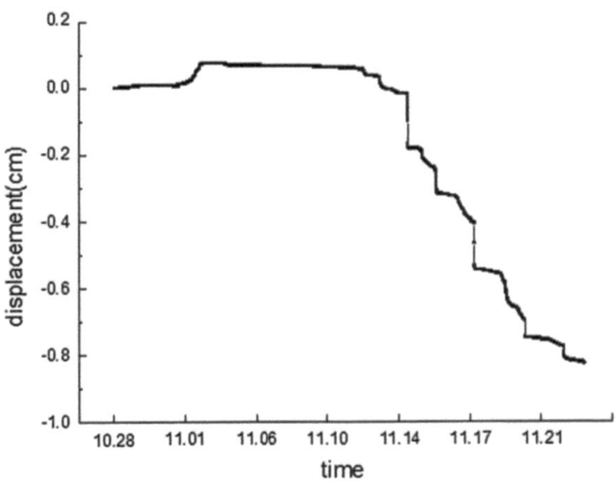

Fig. 28. Displacement of monitoring point 5

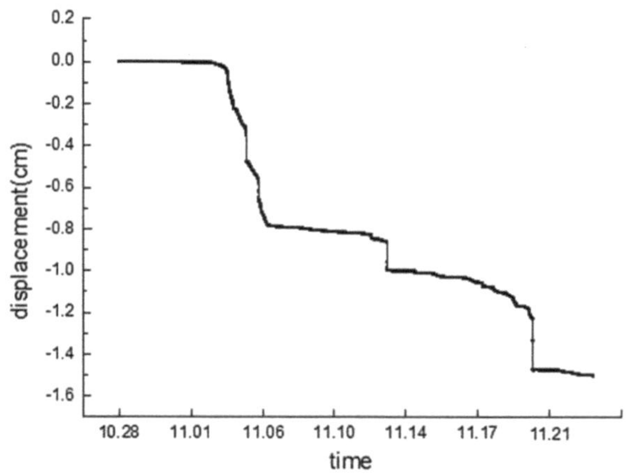

Fig. 29. Displacement of monitoring point 7

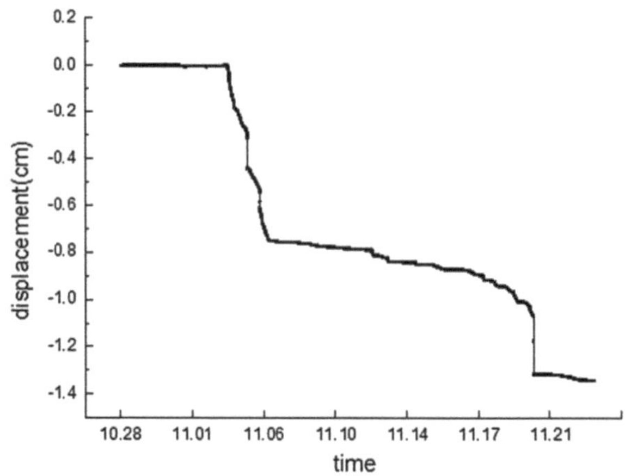

Fig. 30. Displacement of monitoring point 8

Given that the second ring of tunnel is far away from the monitoring part, its influence on the displacement of monitoring points of the vault is relatively small. Before the excavation of the third ring, the settlement values of monitoring point 7, 8, and 9 reach 1.2, 1.0, and 0.7 cm, respectively. After the excavation of the third ring, the settlement of each monitoring point obviously increases. This result is due to the relatively small boundary of the model and the large settlement deformation of the arch after the tunnel has been passed through. Finally, the settlement values of monitoring points 7, 8, and 9 reach 1.5, 1.3, and nearly 1.0 cm, respectively.

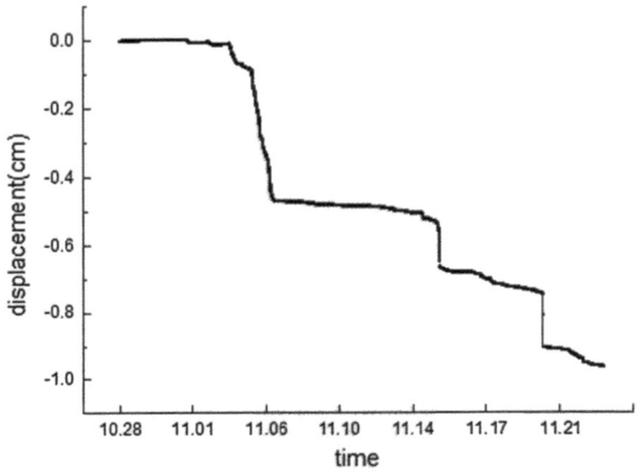

Fig. 31. Displacement of monitoring point 9

4 Conclusion

From the simulation of excavation, rainfall, and sunshine of expansive soil tunnel, the following conclusions are obtained:

(1) After tunnel excavation, the stress release level of the test model is relatively high;
(2) The influence of rainwater infiltration on the stress redistribution of expansive soils is relatively large;
(3) The influence of sunshine on the stress distribution within similar materials is obvious;
(4) The impact of rainwater infiltration on the displacement of the vaults of the expansive soil tunnel is obvious and extremely unfavorable.

References

1. Wang, H., Li, S., Zhang, Q., et al.: Development of a new geomechanical similar material. Chin. J. Rock Mech. Eng. **25**(9), 1842–1847 (2006)
2. Chen, F., Kusaka, H., Bornstein, R., et al.: The integrated WRF/urban modelling system: development, evaluation, and applications to urban environmental problems. Int. J. Climatol. **31**(2), 273–288 (2011)
3. Ren, W., Guo, C., Peng, Z., et al. Model experimental research on deformation and subsidence characteristics of ground and wall rock due to mining under thick overlying terrane. Int. J. Rock Mech. Min. Sci. **47**(4), 614-624 (2010)
4. Fumagalli, E.: Model Simulation of Rock Mechanics Problem. Rock Mechanics in Engineering Practice. London (1968)
5. Ashby, J.: Sliding and toppling modes of failure in model and jointed rock slope. M. S. thesis, Imperial College Royal School of Mines, London (1971)

6. Baumgargarter, P., Stimpson, B.: Development of a tiltable base friction frame for kinematics studies of caving at various depths. Int. J. Rock Mech. Min. Sci. Geomech. Abstr. **16**(4), 265–267 (1979)
7. Bray, J.W., Goodman, R.E.: The theory of base friction models. Int. J. Rock Mech. Min. Sci. Geomech. Abstr. **18**(6), 453–468 (1981)
8. Stewart, D.P., Adhikary, D.P., Jewell, R.J.: Study on the stability of model rock slope. In: Proceedings of the International Conference Centrifuge, vol. 94, pp. 629–634. Singapore (1994)
9. Patton, F.D.: Multiple modes of shear failure in rock. In: Proceedings of the 1st Congress of International Society of Rock Mechanics, pp. 509–513. Lisbon (1966)
10. Einstein, H.H., Veneziano, D., Baecher, G.B., et al.: The effect of discontinuity persistence on rock slope stability. Int. J. Rock Mech. Min. Sci. Geomech. Abstr. **20**(5), 227–236 (1983)
11. Jacoby, W.R., Schmeling, H.: Convection experiments and driving mechanism. Geologische Rundsch. **70**(1), 207–230 (1981)
12. Kincaid C., Olson, P.: An experimental study of subducting slab migration. J. Geophys. Res. **92**(3), 13832–13840 (1987)
13. Zuo, B., Chen, C., Liu, C., et al.: Research on similar material of slope simulation experiment. Rock Soil Mech. **25**(11), 1805–1808 (2004)
14. Wu, B.-T., Zhu, H., Xu, Q., et al.: Experimental study of similar material for weak surrounding rock mass of class IV. Rock Soil Mech. **34**(1), 109–116 (2004)
15. Li, S., Zhou, Y., Li, L., et al.: Development and application of a new similar material for underground engineering fluid-solid coupling model test. Chin. J. Rock Mech. Eng. **31**(6), 1128–1137 (2012)
16. Ye, Y., Shi, Y., Wang, Q., et al.: Test model research on low strength similar material of Shanghengshan multilayer shale deposit. Rock Soil Mech. **35**(2), 114–120 (2014)
17. Liu, X., Sheng, Q., Chen, J., et al.: Seismic shaking table test for large-scale underground cavern group (I): Proportioning test on similar materials of surrounding rock. Rock Soil Mech. **36**(1), 83–88 (2015)
18. Zhang, Y., Wang, X., Liang, Q., et al.: Development of model test similar material of collapsible loess. Chin. J. Rock Mech. Eng. **32**(2), 4019–4024 (2013)
19. Lai, J., Diao, X.: Experimental investigation of similar material of swelling rock on the basis of orthogonal design. J. Lanzhou Univ. Technol. **40**(4), 131–135 (2014)
20. Guo, Z.: Artificial Preparation and Experimental Study on Physical and Mechanics Properties of Hanzhong Expansive Soil. Xi'an Technological University, Xi-an (2015)

Improved Approach for Determining Pile Length of Group Pile Using Complex Continuous Wavelet Transform

Sheng-Huoo Ni[1(✉)], Yu-Zhang Yang[1], Pei-Hsun Tsai[2], and Wei-Hsiang Chou[1]

[1] Department of Civil Engineering, National Cheng Kung University, Tainan, Taiwan
tonyni@mail.ncku.edu.tw
[2] Department of Construction Engineering, Chaoyang University of Technology, Taichung, Taiwan

Abstract. This paper investigates the integrity test ability of the sonic echo (SE) test by using a field constructed group pile foundation. Due to the energy reflected stress wave is generally faded with wave travel path for the testing of a pile with pile cap, a signal process techniques were generally used to enhance the signal interested. The purpose of this paper is to study an improved signal processing approach to detect the pile length of a pile with pile cap. This paper utilized the amplitude and phase message of complex continuous wavelet transform to determine pile length of pile foundations by analyzing the time-frequency-phase angle diagram in the different frequency band. Two- pile group and four-pile group pile were tested. Six piles with different types of defects were installed and tested to verify the proposed approach in this study. The results show that complex continuous wavelet transform is able not only to provide high-resolution results in different frequency bands but also to simplify the identification of the reflection of defects using 3D phase spectrogram. The location of reflected stress wave of pile toe can then be determined using phase diagram.

1 Introduction

There are many non-destructive techniques which had been used to evaluate the pile length of pile foundation. These techniques include sonic echo (SE) method, impulse response (IR) method, parallel seismic method, ultra-seismic method, and cross-hole sonic method, etc. (Lin et al. 1991; Davis 2003; Finno and Gassman 1998; Hertlein and Davis 2006; Hola and Schabowicz 2010). Among these methods, the sonic echo method is the most popular one due to its simplicity and low cost. Due to the energy reflected stress wave is generally faded with wave travel path for the testing of a pile with pile cap, a signal process techniques were generally used to enhance the signal interested. These signal process techniques include applying exponential magnification, a continuous wavelet transforms, the Hilbert-Huang transform, the Wigner-Ville Distribution, etc. (Moubarik et al. 1993; Bouden et al. 2012; Ni et al. 2007). Ni et al. (2008, 2012) performed a series of tests using continuous wavelet transform

W.-C. Cheng et al. (eds.), *Tunneling in Soft Ground, Ground Conditioning and Modification Techniques*, Sustainable Civil Infrastructures,
https://doi.org/10.1007/978-3-319-95783-8_6

(CWT) with piles that were not yet installed in the soil, and the results were prominent. However, the method failed to detect the reflection signal from defects and pile toe after the piles were installed into the soil since the reflection signal becomes to be faded, especially for a long pile. Park and Kim (2001, 2006) used harmonic wavelet analysis of wave (HWAW) with SE method to find reflection signals of different mode numerically. Li et al. (2007) used one-dimensional complex continuous wavelet transform (CCWT) to determine the pile length and defects. However, they lack practical cases to verify the applicability of their method.

The purpose of this paper is to provide the case studies of a signal processing approach that determines the pile length in the group pile from the SE method with using the CCWT. The CCWT is a type of wavelet signal processing technique (Li et al. 2007; Jiang et al. 2012; Ni et al. 2017). In this approach, it utilizes the feature that phase angle is able to detect effectively the continuity of the pile even it is only with the small magnitude of reflected energy. The CCWT was used to decompose a single time domain signal into several time domain signals of different frequencies. These decomposed time domain signals also contained phase angle information that allows one to determine the frequency of reflected wave amplitude to perform the phase angle viewing.

2 Methodology

2.1 Sonic Echo Method

The traditional SE test method provides a way to determine the pile integrity from reflection waves in piles. It determines the location of impedance change by finding the time difference between direct wave and reflection wave from impact pulse using its time history curve. The schematic drawing of this test is shown in Fig. 1. In current practice, the low-strain integrity testing of foundation piles involves the interpretation of results from the SE method. The technique is an example of impact hammer testing, where the response of a pile head to an instrumented hammer blow is measured while the movement of the pile head is recorded with an accelerometer (or geophone). The reflected wave/signal is the wave traveling from the pile head to the pile toe, which then reflects back to the pile head in the pile. In the case of known velocity, based on the one-dimensional wave propagation theory, the pile length can then be calculated by the travel time required for one round trip of a wave.

2.2 Propagation of a Stress Wave in the Pile

A stress wave is propagated by changes in the stress and strain state in an elastic medium. It is a complex behavior when a wave propagates in a bounded elastomer such as piles, slabs, etc. (Miklowitz 1978). In the SE method, only an echo wave from the pile toe (or interface of impedance change) is useful. Others are too complex to be used as a means by which to recognize the echo signals.

A wave generated by an impact force on the surface of a pile head is mainly a compression wave (P-wave). The amplitude of a reflected wave depends on its

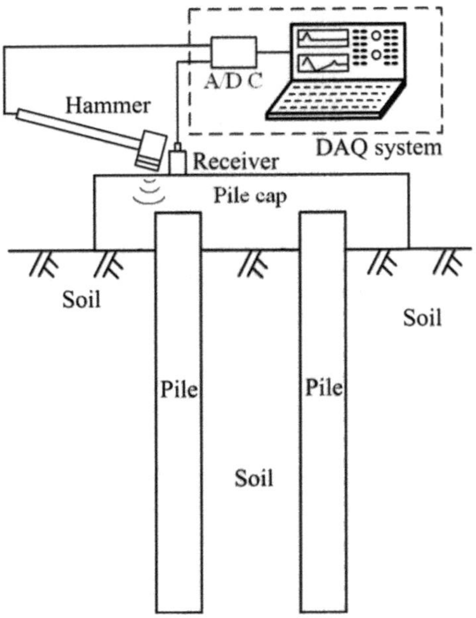

Fig. 1. Schematic drawing of sonic echo test of group pile

damping and dispersive characteristic, and the phase angle is mainly affected by interface boundary condition of the pile toe. The pile toe conditions can be divided into two types:

1. Free end: The pile is in the soil or in voids below the pile end. The impedance ratio between the materials is less than 1, so the reflected wave from the free end is in phase with the incident wave. The difference in the phase angle between the incident and reflected wave is 0°, i.e. they are in phase.
2. Fixed end: Proper contact is made between bedrock and pile end. The impedance ratio of the materials is greater than 1. Incident and reflected waves are anti-phase and the difference in the phase angle is 180° out of phase.

2.3 Wavelet Transform Method

A wavelet transforms analysis is a time-frequency analysis of a signal. It is widely used in different engineering fields. One major advantage provided by wavelet transform (WT) is the ability to perform a local analysis, i.e., to analyze only a section with a long duration signal. The WT method has been found particularly useful when analyzing periodic, noisy, intermittent, and transient signals.

Consider a real or complex-value continuous-time function $\psi(t)$ with two properties (Daubechies 1992; Grossmann and Morlet 1984). The function integrates to zero, and its square integral, or, equivalent, has finite energy. The function $\psi(t)$ is called a mother wavelet, or wavelet if it satisfies these two properties. Let $f(t)$ be any square integral function. The continuous wavelet transform (CWT) with respect to a wavelet $\psi(t)$ is defined as

$$W(a,b) = \frac{1}{\sqrt{a}} \int_{-\infty}^{+\infty} f(t)\psi^{*}(\frac{t-b}{a})dt \tag{1}$$

where a and b are real and * denotes the complex conjugate. The $\psi_{a,b}(t)$ is set to

$$\psi_{a,b}(t) = \frac{1}{\sqrt{|a|}}\psi^{*}\left(\frac{t-b}{a}\right) \tag{2}$$

Then, combining Eqs. (1) and (2), yields

$$W(a,b) = \int_{-\infty}^{+\infty} f(t)\psi_{a,b}(t)dt \tag{3}$$

The variable b represents the time shift or translation, and the variable a determines the amount of time scaling or dilation. Since the CWT is generated using dilates and translates of the signal function, $\psi(t)$, the wavelet for the transformation is referred to the mother wavelet.

In order to successfully apply the CWT, it is important to select the most appropriate (or optimal) mother wavelet function to perform the analysis. Complex-valued wavelets provide phase information and are therefore very important in the time-frequency analysis of non-stationary signals. They can be adapted to analyze the characteristics of the signal. In general, there is a trade-off between spatial and spectral resolution inherent in the choice of the wavelet, and this function is popular in vision analysis. Based on theory and a performance comparison, it is clear that the WT based on complex Gaussian function is suitable for signal analyses since it can achieve excellent time and frequency concentration, and is able to track the frequency trend at the local time better than other functions. The truncated complex Gaussian function $\Psi_G(t)$, as shown in Fig. 2, is given by the Matlab Wavelet-Toolbox (Misiti et al. 2015):

$$\Psi_G(t) = C_p \cdot e^{-t^2} \cdot e^{-jt} \tag{4}$$

By taking the p-th derivative of $\Psi_G(t)$, the integer p is the parameter of this family, and from the previous formula, C_p must be a constant to maintain $\left|\Psi_G(t)^{(p)}\right|^2 = 1$, where $\Psi_G(t)^{(p)}$ is the p-th derivative of $\Psi_G(t)$. The Eq. (3) can be written as:

$$W(a,b) = \int_{-\infty}^{+\infty} f(t)\Psi_{G,ab}(t)dt. \tag{5}$$

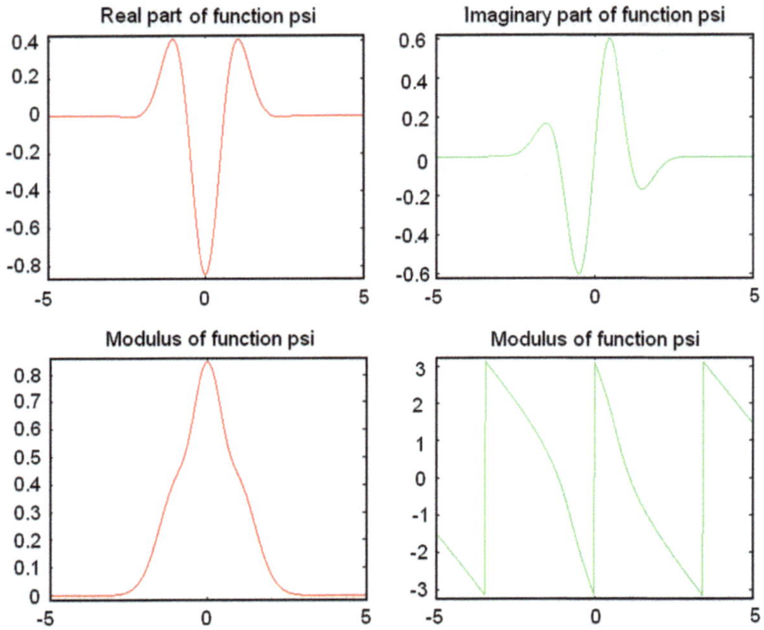

Fig. 2. Display of 2nd derivative of complex Gaussian function

Then, the instantaneous phase angle $\phi(a, b)$ of $W(a, b)$ is calculated as follows:

$$\varphi(a,b) = \arctan\left(\frac{W_I(a,b)}{W_R(a,b)}\right) \tag{6}$$

where $W_I(a, b)$ and $W_R(a, b)$ are the imaginary part and the real part of $W(a, b)$, respectively.

From above equations, the CCWT uses two real wavelets in the real part and imaginary part to transform signals at the same time and then obtain the wavelet coefficients. The real part and imaginary part of a complex wavelet are orthogonal. In contrast to a real wavelet transform, which only obtains the signal amplitude in one space, the CCWT provides additional phase angle message in two orthogonal spaces. Then, the four basic figures, the real part, the imaginary part, the amplitude, and the phase angle are obtained with the CCWT, which is helpful to more accurately detect the signal.

The amplitude from wavelet transform is a magnitude of signal intensity which is proportional to the square root of reflected energy. When the structure has an obvious interface, the significant amplitude change of interface position is observed in the amplitude spectrum. The phase angle is a continuous measure of the phase angle spectrum. Its characteristic is regardless of the intensity of the reflected energy, the phase change can be displayed. It is a continuous phase angle if the wave propagation in an isotropic and homogeneous medium. In the case of abnormal position, there is a significant change in phase angle spectrum and its position can be determined.

3 Test Material and Equipment

3.1 Test Layout of the Experiment

The testing site is located at National Cheng Kung University, Taiwan, next to the civil engineering building. A pile caps were installed on top of the single piles to form a group piles. Two pile caps, one large and one small, were installed for this study. One is two-pile group while the other is a four-pile group. The large pile cap is 1.8 by 1.8 m in dimension, and the small one 1.8 by 0.9 m in dimension. Both pile caps are 0.4 m in thickness. The drawings of the two pile caps were shown in Fig. 3. The two group piles foundation with six hollow precast concrete (PC) piles were constructed to investigate the ability of the stress wave-based methods with regard to evaluating the integrity of the pile foundation. Both group piles (piles with pile cap) tests were performed to detect the pile length under pile cap in this study.

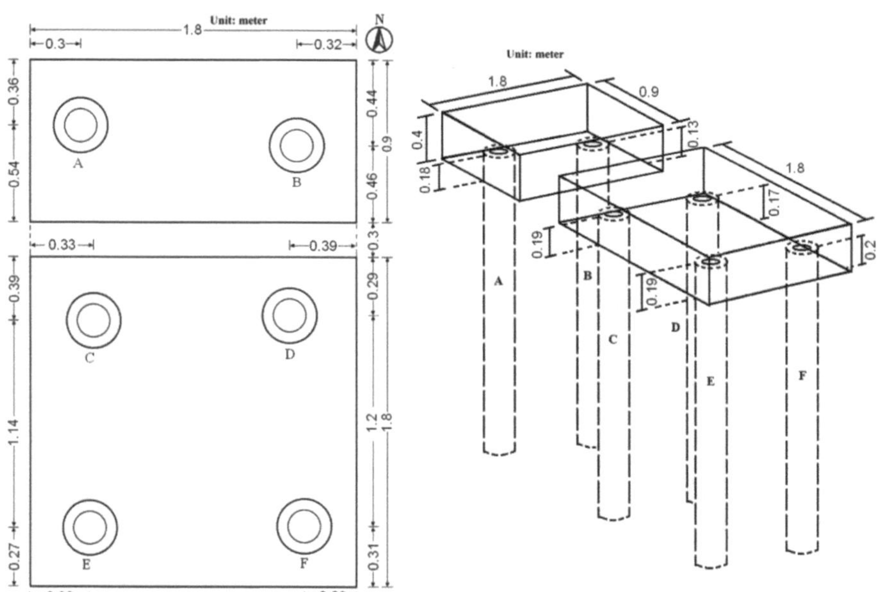

Fig. 3. Layout of the experiment test site

All six piles are six meters long, with outside diameters of 0.3 m and inside diameters of 0.18 m. The piles were labeled A, B, C, D, E, and F. Except for pile A, which is an intact pile, all other piles have differently designed defects. The pile A was designed as an intact pile, while others contained defects at various depths (Huang and Ni 2012). The wave velocity of each pile measured before installation is shown in Table 1.

Table 1. Pile wave velocity evaluated from group piles

Pile no.	Pile wave velocity before installation (m/s)	Pile wave velocity determined by CCWT (m/s)	Error of pile length evaluated (%)	Wave velocity of pile cap (m/s)
A	4365	4291	−1.7	4034
B	4153	4258	2.5	4040
C	4282	4109	−6.1	3883
D	4320	4320	0	4032
E	4226	4261	0.8	4032
F	4264	4254	0.7	4369

3.2 Testing Equipment

In order to obtain high-quality signals for an in situ sonic echo test, the optimal configuration of the hammer force source, sensor and signal capture facility are needed. A typical set of equipment is shown in Fig. 4. The equipment consists of an instrumented impact hammer, geophone sensors, and a computer-controlled signal capturer/analyzer. In this test program, a geophone with a natural frequency of 4.5 Hz was used. A dynamic signal analyzer was used to capture the signal. The impulse hammer with a weight of 53.4 N (12 lbs) was used to create the pulse source. The transient force was applied to the surface of the pile cap closing to the pile testing.

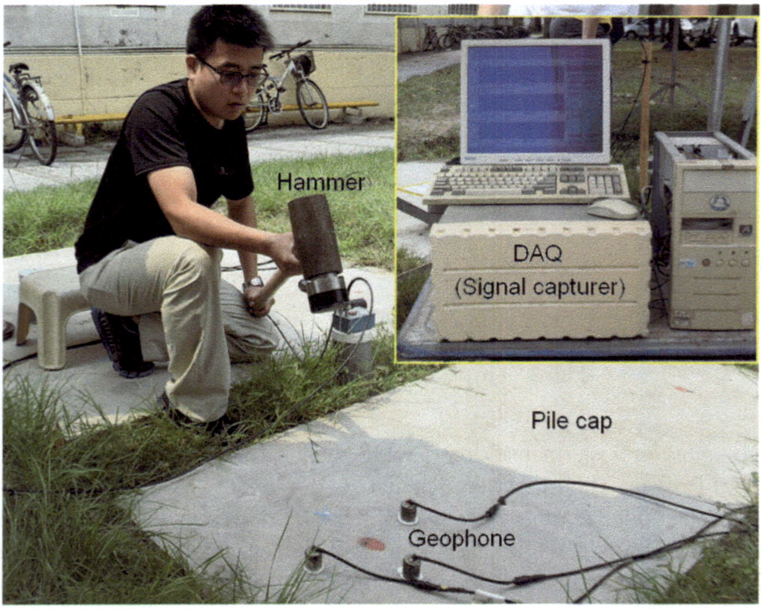

Fig. 4. Setup of equipment for the SE testing of pile

4 Analysis Method

As described in the above sections, the pile length L can be calculated from the time domain using the doubled travel time from the top to the bottom after the SE method is performed. However, if the CCWT in the time-frequency domain and the phase angle spectrum at the frequency corresponding to the maximum amplitude are used, the length of the pile (L) can be combined and written as follows:

$$L = \frac{1}{2} \times V \times \Delta t \tag{7}$$

where V is the wave velocity, Δt is the double travel time from top to bottom, calculated from phase spectrum.

To present the analytical procedure of the CCWT approach, the result of SE test performed on the small pile cap with a hollow PC pile A is illustrated. The typical time history signal obtained for the intact pile A is shown in Fig. 6a. The resulting three-dimensional time-frequency-amplitude diagram (simply called 3D amplitude spectrogram) is shown in Fig. 5. The 3D amplitude spectrogram can be more clearly re-plotted as a 2D manner shown in Fig. 6b. Then, the corresponding 3D phase angle spectrogram (three-dimensional time-phase angle-frequency relationship) can be plotted in Fig. 6c. The magnitude of the phase angle is presented in the manner of the gray level image. The white color is 180° (π) of phase angle and black color is −180° ($-\pi$) of phase angle in the gray level image of 3D phase angle spectrogram. Noting that the energy concentration point is located at the shifting line of phase angle from π to $-\pi$ in the 3D phase angle spectrum (see Fig. 6c).

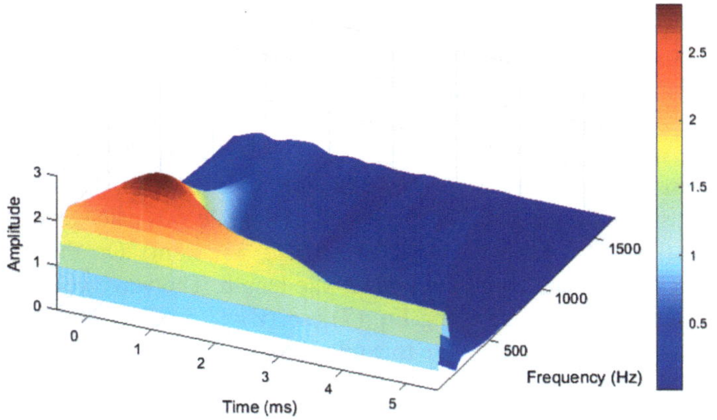

Fig. 5. 3D amplitude spectrogram of the group pile A

To determine the pile length (or depth) is illustrated. The line marked as A (simply called line A) shown in Fig. 6b are the lines passing the maximum energy (amplitude) concentration (MA) point. The energy concentration point for Line B is not obvious.

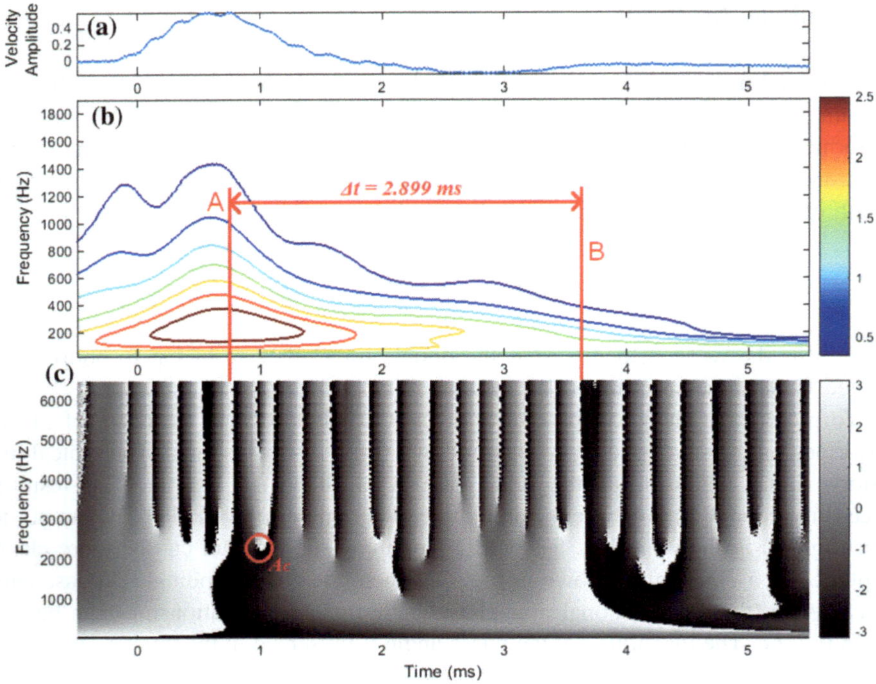

Fig. 6. Test result of pile A. **a** Time history obtained from SE test. **b** 3D amplitude spectrogram. **c** 3D phase angle spectrogram

However, Line B can be found by all frequency waves and will be reflected as the waves arrive at the pile toe due to the changes in the impedance, i.e. the in-phase line of π to $-\pi$ will be continued for all frequencies, as shown in Fig. 6c. The time between the two lines is the doubled travel time for the wave propagating from the pile top to the pile toe. Therefore, the double reflected time (Δt) of the wave can be calculated from Fig. 6c, i.e., $\Delta t = t_{B\text{-line}} - t_{A\text{-line}} = 2.899$ ms. The length of the pile could then be calculated with the known stress wave propagation speed, which was measured before the pile was installed in the ground. The length of the pile was calculated to be 6.11 m using Eq. (7) (i.e. 1/2 * 4365 * 2.899/1000 − 0.22 = 6.11 m). Or, the wave velocity of the pile can be calculated with the known pile length. The wave velocity of the pile was calculated to be 4291 m/s using Eq. (7) (i.e. $V = 2 * 6.22/2.899 * 1000 = 4291$ m/s). This result was very close to the true pile velocity of 4365 m/s. The error is 1.7%. The depth of pile cap can generally be calculated from the time domain data. This is because the reflected energy will not be faint shown as the small fluctuation in the time history data in Fig. 6a. However, to calculate the depth of pile cap using the phase, one needs to find the phase reverse (PR) point (i.e. phase angle stop extending with time to the low-frequency part and inverting back to the high-frequency part, e.g. the point A_c as shown in Fig. 6c. The circled point A_c in Fig. 6c is the wave reflection from the bottom of pile cap. The double travel time is 0.1983 ms. The wave velocity of pile cap can be calculated as 2 * 0.4/0.1983 * 1000 = 4034 m/s.

The step procedure is summarized as follows.

1. Find the area of the concentrated energy in the 3D amplitude spectrogram shown in Fig. 6b. Then, find the phase angle changes from π to −π (or where the phase angle equals to 0), which are corresponding to the concentrated energy area. The time points of the lines (A-line and B-line in Fig. 6) in 3D phase angle spectrogram are corresponding to signal for the incident wave and the reflected wave from the pile bottom, respectively.
2. Calculate the average travel time of wave $\Delta t = t_{B\text{-}line} - t_{A\text{-}line}$ in the Fig. 6c. The reflected wave of the lower frequency is no pile toe reflection signal (its wavelength is too long, the frequency should be greater than *V*/2L) and consider the mixing phenomenon of the adjacent frequency. If the frequency is less than c/L, the travel time will not be included in the evaluation. Moreover, the reflected wave of higher frequency signal carry less energy, in order to avoid the influence of noise, the travel time is not included in the evaluation.
3. The length/wave velocity of the pile can then be calculated using Eq. (7) since the traveling time was obtained.

5 Test Results

In time domain the echo signal is so weak it is difficult to use the traditional time-domain analysis method to determine the arrival time or maximum amplitude of reflected P-wave. In contrast, through the proposed method, the clear arrival time position of the reflected wave is obtained by applying the 3D amplitude spectrogram and the 3D phase angle spectrogram. Then, the length of the pile is calculated using the aforementioned steps. The typical results of the two piles, pile B, and D, are presented in this paper. The other piles with similar results will be omitted. From the 3D phase angle spectrogram of these piles, the phase angle in the location of the pile head and the pile toe are in phase (i.e. no phase difference). The feature of phase characteristic display the pile end condition is a free end, which is in accordance with the characteristic of friction pile.

5.1 Pile B

The result of SE test performed on the small pile cap with a hollow PC pile B is shown in Fig. 7. The time history signal obtained for the pile B is shown in Fig. 7a. The 3D amplitude spectrogram is shown in Fig. 7b. Then, the corresponding 3D phase angle spectrogram can be plotted in Fig. 7c. Lines A and B shown in Fig. 7b are the lines passing the two energy concentration points. The travel time of wave ($\Delta t = t_{B\text{-line}} - t_{A\text{-line}} = 2.945$ ms) can be calculated from Fig. 7c. Time $t_{A\text{-line}}$ and time $t_{B\text{-line}}$ correspond to the locations of the time points along lines A and B in the phase angle diagram. The time between the two lines is the doubled travel time for the wave propagating from the pile top to the pile toe. The wave velocity of the pile

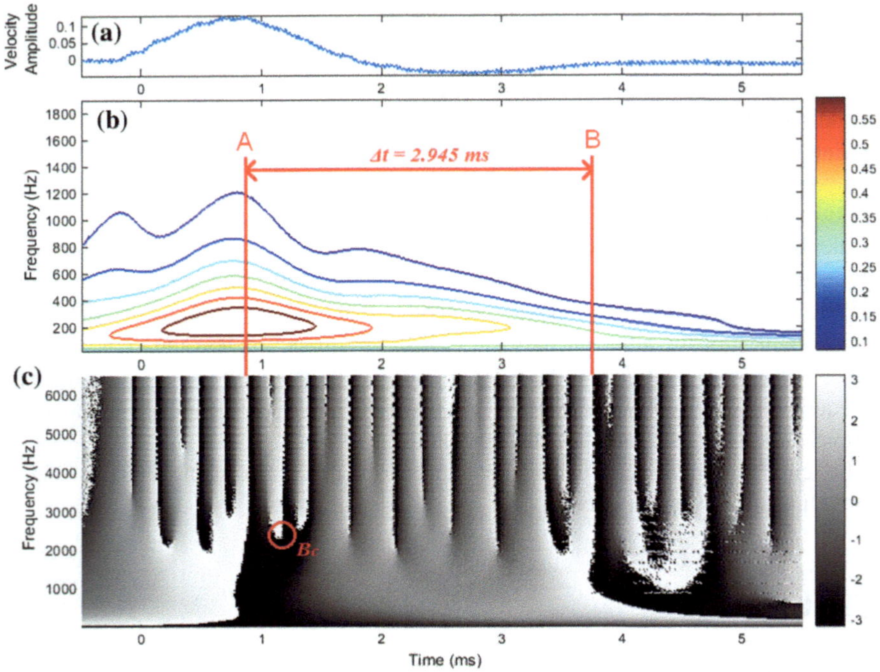

Fig. 7. Results of CCWT analysis of pile B. **a** Time history obtained from SE test. **b** 3D amplitude spectrogram of test. **c** 3D phase angle spectrogram of test

can be calculated with the known pile length. The wave velocity of the pile was calculated to be 4258 m/s using Eq. (7) (i.e. $V = 2 * 6.27/2.945 * 1000 = 4258$ m/s). This result was very close to the true pile velocity of 4153 m/s. The error is 2.5%. The circled point B_c in Fig. 7c is the wave reflection from the bottom of pile cap. The double travel time is 0.198 ms. The wave velocity of pile cap can be calculated as $2 * 0.4/0.198 * 1000 = 4040$ m/s.

5.2 Pile D

The result of SE test performed on the large pile cap with a hollow PC pile D is shown in Fig. 8. The time history signal obtained for the pile E is shown in Fig. 8a. The 3D amplitude spectrogram is shown in Fig. 8b. Then, the corresponding 3D phase angle spectrogram can be plotted in Fig. 8c. Lines A and B shown in Fig. 8b are the lines passing the two energy concentration points. The travel time of wave ($\Delta t = t_{B\text{-line}} - t_{A\text{-line}} = 2.884$ ms) can be calculated from Fig. 8c. Time $t_{A\text{-line}}$ and time $t_{B\text{-line}}$ correspond to the locations of the time points along lines A and B in the phase angle diagram. The time between the two lines is the doubled travel time for the wave propagating from the pile top to the pile toe. The wave velocity of the pile can be calculated with the known pile length. The wave velocity of the pile was calculated to be 4320 m/s using Eq. (7) (i.e. $V = 2 * 6.23/2.884 * 1000 = 4320$ m/s).

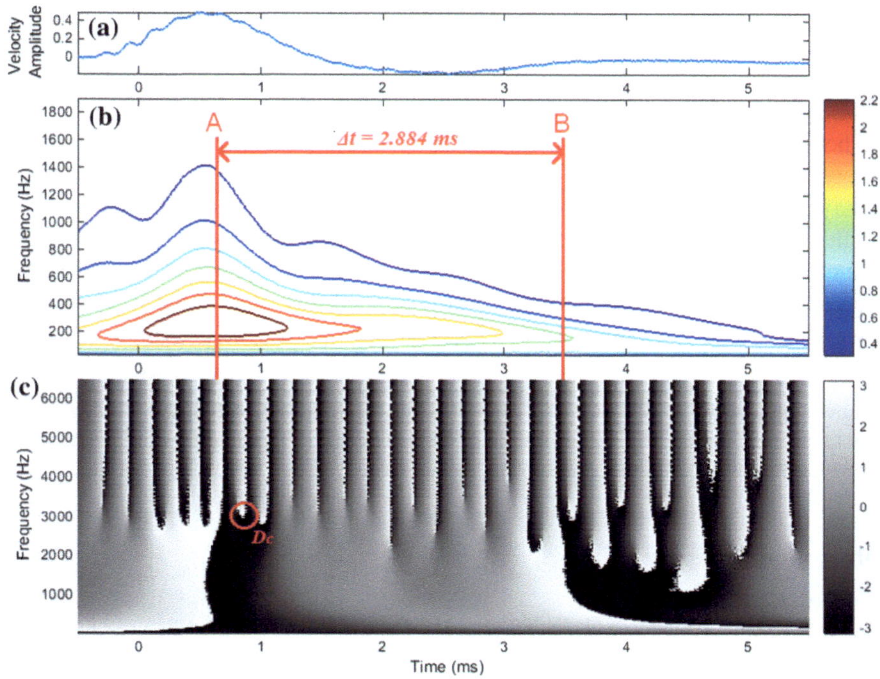

Fig. 8. Results of CCWT analysis of pile D. **a** Time history obtained from SE test. **b** 3D amplitude spectrogram. **c** 3D phase angle spectrogram

This result was very close to the true pile velocity of 4320 m/s. The error is 0%. The circled point D_c in Fig. 8c is the wave reflection from the bottom of pile cap. The double travel time is 0.1984 ms. The wave velocity of pile cap can be calculated as $2 * 0.4/0.1984 * 1000 = 4032$ m/s.

The testing results of the two group piles are obtained from the test site. The statistical results of each pile are organized in Table 1. As shown in the table, the errors of pile length evaluated are within 6.1%. The resulting accuracy is good. This result represents that the accuracy and consistency of pile length are not affected by the pile with some defects. The results obtained are good for the general purpose of engineering. From the above experimental tests, even only the weak signal is captured, the suggested approach is still able to identify the pile length, and effectively determine the exact reflected wave position of the pile toe. These results enhance the usefulness of the suggested method in this study.

6 Conclusions

An improved approach for integrity test of group pile using sonic echo method with complex continuous wavelet transform is proposed. The following conclusions can be drawn from this study.

1. The pile wave velocities can be determined from phase spectrum with complex continuous wavelet transform in the case of group piles with the error less than 8%.
2. Test results show that the accuracy and consistency of pile length are not affected by the pile with some defects.
3. The depth or wave velocity of a pile cap can be determined.
4. The proposed procedure can reduce the chance of man-made error, and it improves the accuracy of determination of the pile length of a pile under a pile cap.
5. The non-destructive testing of piles with pile cap is more difficult than piles without pile cap because the energy would mainly dissipate on pile cap. However, if the complex continuous wavelet transform is applied, the time of the weak reflection wave from both the bottom of pile cap and pile toe may be determined.

Acknowledgements. This study was supported by the Ministry of Science and Technology under Grant No. MOST 105-2221-E-006-044, Taiwan, R.O.C. The authors would like to express their special thanks to all other participants in this project.

References

Bouden, T., Djefi, F., Dib, S.D., Nibouche, M.: Hilbert Huang transform for enhancing the impact-echo method of nondestructive testing. J. Auto Syst. Eng. **6–4**, 172–184 (2012)

Daubechies, I.: Ten Lectures on Wavelets. SIAM Bookstore (1992)

Davis, A.G.: The nondestructive impulse response test in North America: 1985–2001. NDT E Int. **36**, 129–136 (2003)

Davis, A.G., Dunn, C.S.: From theory to field experience with the nondestructive vibration testing of piles. Proceed. Inst. Civ. Eng. Part 2 **57**, 571–593 (1974)

Finno, R.J., Gassman, S.L.: Impulse response evaluation of drilled shafts. J. Geotech. Geoenviron. Eng. **124**(10), 965–975 (1998)

Grossmann, A., Morlet, J.: Decomposition of hardy functions into square integrable wavelets of constant shape. SIAM J. Math. Anal. **15**, 723–736 (1984)

Hertlein, B., Davis, A.G.: Nondestructive Testing of Deep Foundations. Wiley, Chichester, England (2006)

Huang, Y.H., Ni, S.H.: Experimental study for the evaluation of stress wave approaches on a group pile foundation. NDT E Int. **47**, 134–143 (2012)

Jiang, X., Ma, Z.J., Ren, W.X.: Crack detection from the slope of the mode shape using complex continuous wavelet transform. Comput. Aided Civ. Infrastruct. Eng. **27**, 187–201 (2012)

Li, Q.F., Miao, X.X., Xu, J.H.: Data processing based on complex continuous wavelet transform in project test. J. China Univ. Min. Technol. **36**(1), 23–26 (2007) (in Chinese)

Lin, Y., Sanalone, M., Carino, N.J.: Impact echo response of concrete shafts. Geotech. Test. J. **14** (2), 121–137 (1991)

Miklowitz, J.: The Theory of Elastic Waves and Waveguides. North Holland Publishing Company (1978)

Misiti, M., Misiti, Y., Oppenheim, G., Poggi, J.M.: Wavelet Toolbox: For Use with MATLAB (2015)

Moubarik, S.E., De Vadder, D., Benoist, P.: Wavelets and nondestructive evaluation. In: Review of Progress in Quantitative Nondestructive Evaluation, pp. 727–734 (1993)

Ni, S.H., Charng, J.J., Lo, K.F.: Nondestructive evaluation of in-isolation pile shaft integrity by Wigner-Ville distribution. J. Mech. **23**(1), 15–21 (2007)

Ni, S.H., Lo, K.F., Lehmann, L., Huang, Y.H.: Time-frequency analyses of pile integrity testing using wavelet transform. Comput. Geotech. **35**, 600–607 (2008)

Ni, S.H., Isenhower, W.M., Huang, Y.H.: Continuous wavelet transform technique for low-strain integrity testing of deep drilled shafts. J. Geoengin. **7**, 97–105 (2012)

Ni, S.H., Yang, Y.Z., Tsai, P.H., Chou, W.H.: Evaluation of pile defects using CCWT analysis. NDT E Int. **87**, 50–59 (2017)

Olson, L.D.: Nondestructive testing of deep foundations with sonic methods. In: Foundation Engineering, pp. 1173–1183 (1989)

Park, H.C., Kim, D.S.: Non-destructive pile integrity test using HWAW method. In: Key Engineering Materia, pp. 363–366. Trans Tech Publication (2006)

Park, H.C., Kim, D.S.: Evaluation of the dispersive phase and group velocities using harmonic wavelet transform. NDT E Int. **34**, 457–467 (2001)

Application of Displacement Direction Angle Theory on Excavation Methods Conversion Opportunity of Zi-Zhi Tunnel

Wei Wang, Mingjun Hu, and Hengwen Zhang(✉)

School of Civil Engineering, Central South University, Hunan 410075, China
wangweicsu@csu.edu.cn, 1214731683@qq.com,
345461777@qq.com

Abstract. Stratum in shallow buried excavation section of Zi-Zhi Tunnel is generally upper-soft and lower-hard. To reduce the rock disturbance and ensure the construction progress, excavation method conversion from Cross Diaphragm with Step method to bench method should be conducted ahead of schedule. To determine the excavation methods conversion opportunity, considering different positions between tunnel face and the interface of stratum, based on displacement direction angle theory, the change laws for the displacement direction angle and three-dimensional deformation of surrounding rock under different conditions were analyzed by $FLAC^{3D}$. Then, an excavation methods conversion opportunity is derived. Meanwhile, the rationality of the opportunity is validated by the field monitoring data. The results indicate that: when the distance of working face cutting through the interface is 1.5B (B = tunnel width), displacement direction angle of surrounding rock reaches a stable state and the longitudinal stress concentration is weakened. It is conducive to utilize the self-bearing capacity of the rock mass. Hence, it is suggested to convert the excavation methods when the distance is greater than 1.5B; the stable value and the convergence rate of vault subsidence in site were both below warning value, which can meet the safety and economic benefits of engineering.

Keywords: Shallow-buried tunnel · Displacement direction angle
Upper-soft and lower-hard · Excavation methods conversion

1 Introduction

Tunnel construction is mostly conducted in a single stratum. However, when tunnel passes through upper-soft and lower-hard formation, single excavation method could not meet the requirements of the construction. In this case, excavation method should be adjusted dynamically according on the geological conditions in front of the tunnel face [1–3]. Using Reasonable excavation methods conversion opportunity could reduce the disturbance of surrounding rock. Therefore, it is significantly important to study on the excavation methods conversion opportunity in upper-soft and lower-hard formation.

Currently, the study on tunnel excavation method in upper-soft and lower-hard formation focused more on the comparisons [4, 5] and safety of excavation methods [6], while less on the impact of conversion opportunity on surrounding rock stability.

W.-C. Cheng et al. (eds.), *Tunneling in Soft Ground, Ground Conditioning and Modification Techniques*, Sustainable Civil Infrastructures,
https://doi.org/10.1007/978-3-319-95783-8_7

Moreover, in construction practices, the conversion opportunity is mostly based on engineering experiences, though calculation method of relaxation pressure of shallow large span tunnel in up-soft and low-hard rock stratum has been obtained [7].

In order to determine the excavation methods conversion opportunity from Cross Diaphragm with Step method to bench method of Zi-Zhi Tunnel, according to the geographical characteristic that the stratum is generally upper-soft and lower-hard, based on the displacement direction angle theory, this paper reveals the change laws of the displacement direction angle theory in different intersection positions between tunnel face and the interface of stratum. Engineering practice has verified the rationality of the opportunity. The results can provide references for future similar projects.

2 Engineering Survey

Zi-Zhi Tunnel is the longest urban underground highway tunnel of Asia for its length, 14.4 km. The project is divided into three sections: U-shape section, open cut section and shallow buried excavation section. In the shallow buried excavation section which is considered in this paper, the length, height and width of the tunnel are 877.1, 9.7 and 12.8 m, respectively. Generally, stratum in this section is upper-soft and lower-hard, and longitudinal profile is shown in Fig. 1. The design excavation methods of this shallow buried excavation section are Cross Diaphragm with Step method and bench method, corresponding design mileage is also shown in Fig. 1.

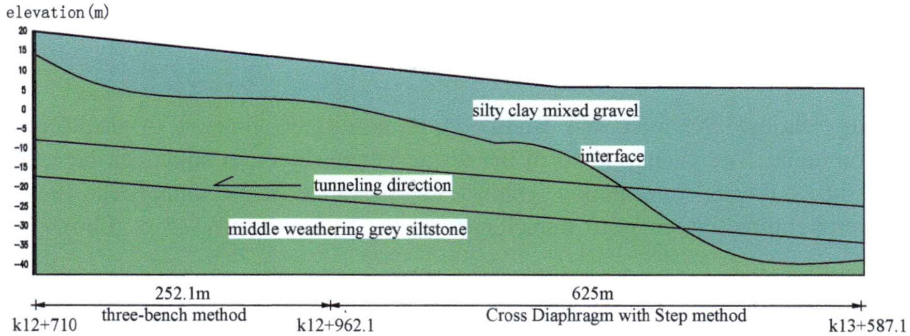

Fig. 1. The longitudinal profile of geology

With tunnel face moving forward, middle weathering grey siltstone (hereinafter referred to as the "bedrock") gets into the tunnel face and the invading bedrock depth of the tunnel bottom is increasing. In Cross Diaphragm with Step method, whose excavation sequences and designed support parameters are shown in Fig. 2 and Table 1, respectively, it indicated that too many excavation steps resulted in multiply disturbing on surrounding rock. On the other hand, tunnel face is divided into several smaller parts, which leads to small operation space and could only fit for artificial or light mechanical excavation and impedes excavation progress. To ensure the construction period, excavation methods conversion from Cross Diaphragm with Step method to bench method should be conducted ahead of schedule.

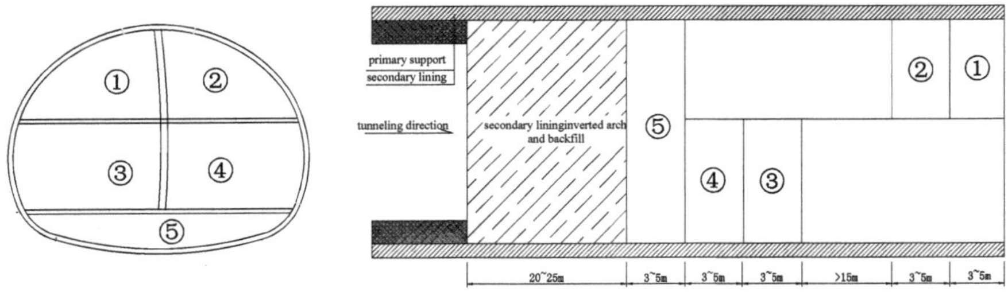

Fig. 2. Excavation sequences of cross diaphragm with step method

Table 1. Support types and parameters of cross diaphragm with step method

Support types	Support parameters
Primary support	C25 thick: 30 cm I22 steel arch space: 50 cm
Secondary lining	C35 thick: 60 cm

3 Approach of Construction Method Conversion Opportunity

3.1 Displacement Direction Angle Theory

Austrian scholars Schubert and Budil [8] proposed the concept of displacement direction angle. The change trend of the displacement direction angle can be used to reflect the geological information within a certain range of the tunnel face [9]. The displacement direction angle of tunnel surrounding rock is expressed as follows:

$$\tan\alpha = \frac{\Delta L}{\Delta S}$$

In the formula, α is the displacement direction angle of the tunnel surrounding rock, ΔL is the increment of longitudinal displacement of tunnel vault, and ΔS is the increment of vertical displacement of the tunnel vault.

The physical meaning of the displacement vector azimuth trend line can be illustrated by Fig. 5. When excavating to the section position 1, the displacement vector at the vault is V_1, and the corresponding initial vertical and longitudinal deformation components are S_1 and L_1 respectively. When excavating to the section position 2 (assuming that the displacement between section 1 and section 2 is 1.0B, B = tunnel width), the displacement vector of the vault in section 1 is V_2, and the corresponding secondary vertical and longitudinal deformation components are S_2 and L_2 respectively. It means that when tunnel passes through 1.0B, the displacement vector of the measuring point of the vault is changed by $\Delta V = V_2 - V_1$, and the corresponding vertical and longitudinal deformation components are $\Delta S = S_2 - S_1$ and $\Delta L = L_2 - L_1$ respectively. The process brings the displacement direction angle is α. The angle α is based on the measuring point as the reference point and the vertical line as the baseline. The clockwise rotation is negative. Or, α is positive when it rotates in an anticlockwise direction. Each monitoring section has a set of displacement direction angle α corresponding to the measuring point at the 1.0B in the rear of excavation face, and the curve formed by the connection is the displacement direction angle variation trend line (Fig. 3).

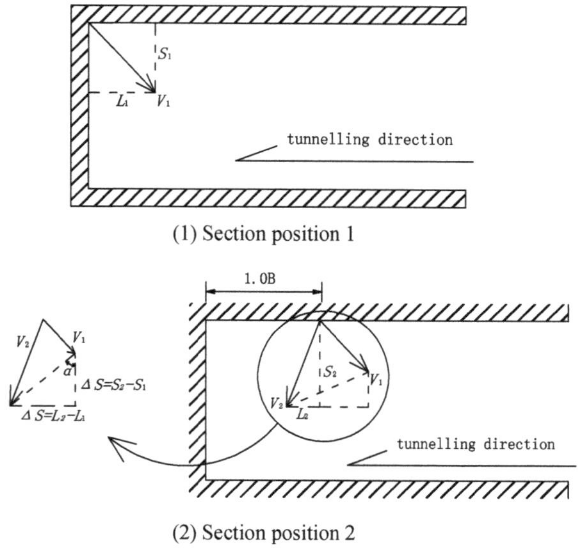

Fig. 3. Schematic diagram of displacement direction angle

When the tunnel passes through the single stratum condition, the trend line of the displacement direction angle is a horizontal line. At this time, it can be called "normal state". As shown in Fig. 4 (1), when the tunnel is excavated from hard rock to soft rock, the change law of the displacement direction angle firstly increases to the peak and then decreases, and the direction of the displacement direction angle is opposite to the excavation direction of the tunnel. As shown in Fig. 4 (2), when the tunnel is excavated from soft rock to hard rock, the change law of the displacement direction angle is reduced to the peak and then increases. The trend line of the tunnel displacement direction angle can be used as the basis for predicting the geological condition in front of excavation face, and provide reference for the advance conversion of excavation method.

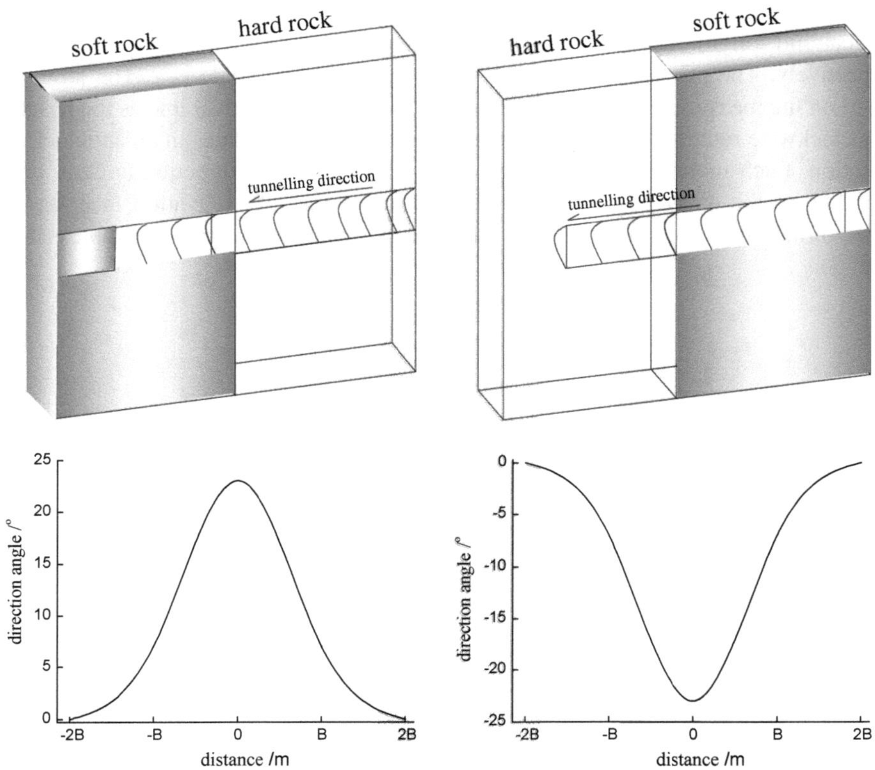

(1) from hard rock to soft rock

(2) from soft rock to hard rock

Fig. 4. The variation rule of displacement direction angle

3.2 Calculation Model and Condition Design

According to the geological characteristics shown in Fig. 1, the inclination of the interface of stratum is 45°, and half of the tunnel is modeled. FLAC^{3D} is used to establish the three-dimensional calculation model and its transverse length is 4B = 51.2 m, vertical length is 8B = 102.4 m, longitudinal length is 12B = 153.6 m. The model is shown in Fig. 5. The surrounding rock of tunnel is simulated by solid element and the rock-soil mass is subject to the Mohr-Coulomb yield criterion. The model is composed of 9680 nodes and 103,764 elements. The bottom of the model adopts the fixed constraint, and the four sides of the model are all restrained by normal constraint. According to the results of geological survey, soil parameters are shown in Table 2. Tunnel construction use the original design of the Cross Diaphragm with Step method and the excavation footage cycle is 0.5 m.

In order to avoid the boundary effect, a series of numerical calculation monitoring points are set up within the range of [−3B, 3B], and the layout of the monitoring points are shown in Fig. 6. The initial vertical displacement value and the initial longitudinal displacement value of the tunnel vault are obtained when the monitoring point is

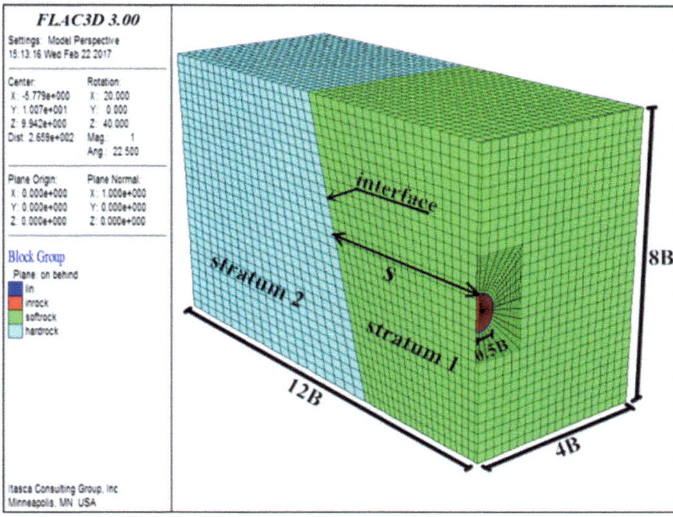

Fig. 5. Numerical calculation model

Table 2. Calculation parameters in different stratum

Stratum	Density (kg/m^3)	Elastic modulus (GPa)	Poisson ratio	Friction angle (°)	Cohesion (KPa)
1	1980	0.2	0.29	17	45
2	2320	0.2	0.23	36	200

Note Stratum 1-Silty clay mixed gravel, stratum 2-Middle weathering grey siltstone

located on the tunnel face. When the monitoring point is at 0.5B (B = tunnel width) in the rear of the tunnel face, the secondary vertical displacement value and the secondary longitudinal displacement value of the tunnel vault deformation are obtained. With the tunnel continues to dig forward, the vault monitoring points in the rear of tunnel face is continuously established and the displacement value of the tunnel surrounding rock is continuously obtained.

In order to estimate the change law of the surrounding rock displacement direction angle during the different positional relationship between the tunnel face and the interface of stratum, assume that *s* refers to the distance between the interface of strata and tunnel arch. *s* is negative when tunnel arch did not reach the interface of stratum, otherwise, *s* is positive. Calculation conditions are shown in Table 3.

3.3 The Variation Law of Displacement Direction Angle

When tunnel face is in interface of stratum, namely No. 7 calculation condition, the vertical displacement nephogram at the −0.5B is shown in Fig. 7. The vertical displacement contour map of the part of the dashed frame is obtained using telcopt10.0. It can be seen that the vertical displacement at the vault is −15.6 mm. By that analogy,

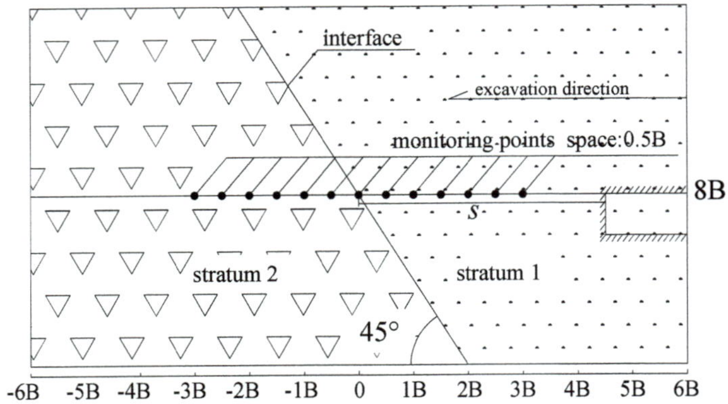

Fig. 6. Numerical monitoring points

Table 3. Calculation conditions

Calculation conditions	1	2	3	4	5	6	7	8	9	10	11	12	13
s/B	3	2.5	2	1.5	1	0.5	0	−0.5	−1	−1.5	−2	−2.5	−3

Note B is tunnel width

Fig. 7. Vertical displacement of No. 7 condition

the vertical displacement and the longitudinal displacement of the condition No. 1 to the condition No. 7 are shown in Fig. 8.

It can be seen from Fig. 8 that the vertical displacement and longitudinal displacement of the surrounding rock of the tunnel vault at the tunnel face and the rear of the tunnel face are obviously increased when the tunnel is excavated from the soft rock stratum 1 to the hard rock stratum 2. Compare to longitudinal displacement, the vertical deformation changes rapidly in the range of [0B, 3B]. Therefore, when the tunnel is excavated from the soft rock stratum 1 to the hard rock stratum 2, special attention

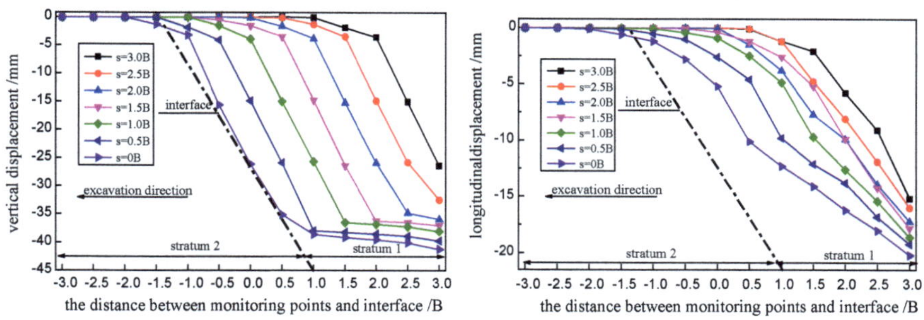

(a) vertical displacement curve (b) longitudinal displacement curve

Fig. 8. Displacement curve of conditions

should be paid to controlling the vertical displacement of the tunnel in the soft rock stratum.

According to the calculation method of the displacement direction angle, the displacement direction angle under all working conditions are calculated, and the calculation result is shown in Fig. 9.

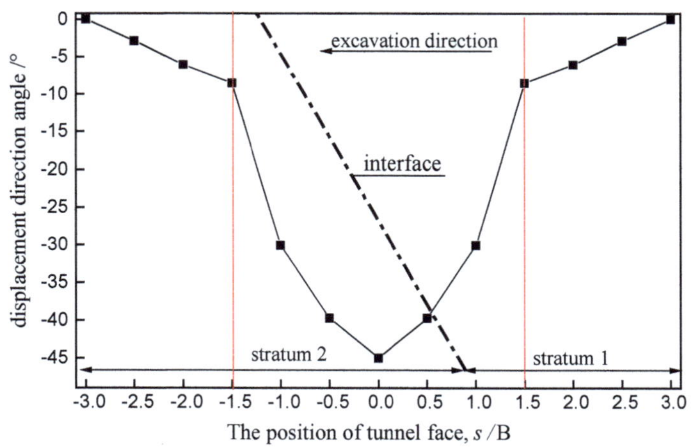

Fig. 9. Curve of displacement direction angle under different conditions

As shown in Fig. 9, with the continuous forward tunneling of the tunnel, the change law of the displacement direction angle trend line can be roughly divided into three stages in the process of tunneling from the silty clay stratum to the middle weathering grey siltstone. The three stages are the linear slow descent stage, the concave rapid change stage and the linear slow rise stage.

Linear slow descent stage: When the distance between the tunnel face and the stratum interface is in the range of [1.5B, 3B], the displacement direction angle decreases slowly and the trend line of displacement direction angle changes linearly. It shows that the strength of the surrounding rock changed little in this range and the surrounding rock strength is low, and the longitudinal stress concentration of the surrounding rock is large, which restrains the longitudinal displacement of the surrounding rock.

Concave rapid change stage: When the distance between the tunnel face and the stratum interface is in the range of [−1.5B, 1.5B], the trend line of displacement direction angle changes concavely and rapidly. The displacement direction angle begins to increase greatly when it is about 1.5B apart from the stratum interface, and reaches the maximum at the center line of the stratum interface, then begins to decrease and gradually tends to be stable when the distance from the stratum interface is −1.5B. It can be interpreted as: when tunnel face approaches stratum interface, tunnel longitudinal stress concentration phenomenon is weakened, making the direction of displacement angle change toward the excavation direction and displacement vector orientation angle increases. When the tunnel face passes through the strata interface, the formation strength increases, the longitudinal stress concentration phenomenon increases, and the displacement direction decreases rapidly.

Linear slow rise stage: When the distance between the tunnel face and the stratum interface in the range of [−1.5B, −3B], the displacement direction angle increases slowly and the trend line of displacement direction angle changes linearly. It shows that the strength of the surrounding rock changed little in this range and the surrounding rock strength is high, and the longitudinal stress concentration of the surrounding rock is large, which restrains the longitudinal displacement of the surrounding rock.

According to above analysis, with tunneling from the silty clay stratum to the middle weathering grey siltstone, when the face pass 1.5B through the stratum interface, the direction of displacement of surrounding rock angle tended to be stable, longitudinal stress concentration decreased, and surrounding rock strength in front of the tunnel face is high. It is conducive to utilize the self-bearing capacity of the rock mass. Hence, it is suggested to convert the excavation methods when the tunnel face passed 1.5B through the stratum interface.

4 Validation of Construction Method Conversion Opportunity

According to the construction method conversion opportunity above mentioned, the project department converts the Cross Diaphragm with Step method to three-bench excavation method. Actual conversion mileage is K13 + 198.6, and site actual tunnel excavation methods corresponding mileages are shown in Table 4.

Table 4. Tunnel excavation methods corresponding mileage in site

Excavation methods	Cross diaphragm with step method	Bench method
Start mileage	K13 + 587.1	K13 + 198.6
Terminal mileage	K13 + 198.6	K12 + 710
Length (m)	388.5	488.6

In order to figure out the conversion effect, crown settlement and its settlement rate in K13 + 198.6 are measured. Deformation curve and deformation rate curve of crown settlement are shown in Figs. 9 and 10, respectably.

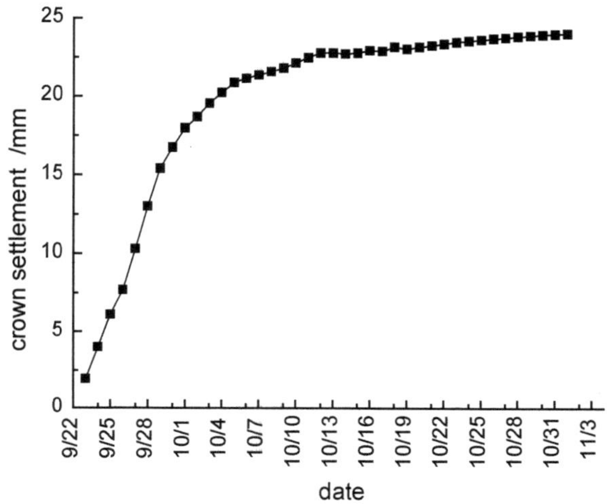

Fig. 10. Deformation curve of crown settlement

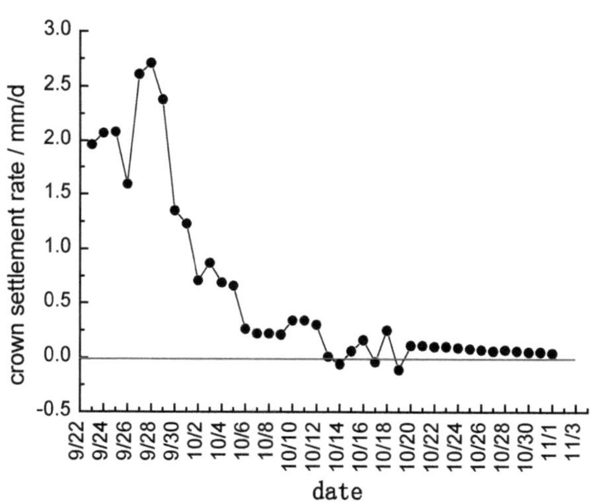

Fig. 11. Deformation rate curve of crown settlement

Figures 10 and 11 indicate that cumulative deformation rapidly increases and the crown settlement rate is large from an early time; after 15 days, the crown settlement rate slows down and cumulative deformation tends to be stable. The stable value of vault subsidence is 23 mm, which is lower than warning value, 30 mm. The rate of vault subsidence stable value is 0 mm/d, which is also lower than warning value 1 mm/d. The results show that the opportunity of construction method conversion can meet the safety benefit. The shallow buried excavation section of Zi-Zhi Tunnel is completed on time because the conversion from Cross Diaphragm with Step method to three-bench method is conducted ahead of time.

5 Conclusion

(1) In the shallow buried excavation section of Zi-Zhi Tunnel, stratum is generally upper-soft and lower-hard. To reduce the disturbance of surrounding rock and ensure the construction progress, excavation methods conversion from Cross Diaphragm with Step method to bench method should be conducted ahead of schedule.

(2) Based on displacement direction angle theory, the change laws for the displacement direction angle of surrounding rock under different conditions indicate that when the face pass 1.5B through the stratum interface, the direction of displacement of surrounding rock angle tended to be stable, longitudinal stress concentration decreased, and surrounding rock strength in front of the tunnel face is high. It is conducive to utilize the self-bearing capacity of the rock mass. Hence, it is suggested to convert the excavation methods when the tunnel face passed 1.5B through the stratum interface.

(3) According to the construction method conversion opportunity above mentioned, the vault subsidence stable value and its rate are both below warning value in cross-section of K13 + 198.6, which can meet the safety benefit of engineering. Construction period of the shallow buried excavation section is shortened because the conversion from Cross Diaphragm with Step method to three-bench method is conducted ahead of time.

Acknowledgements. The authors appreciate the support of Science and Technology Plan Projects of Hunan Province, China (2010GK3173) and co-sponsored by the China Railway Tunnel Group Science and Technology Innovation Project (2013–30).

References

1. Lunardi, P.: The Design and Construction of Tunnels Using the Approach Based on the Analysis of Controlled Deformation in Rocks and Soils. T&T International ADECO-RS Approach (May 2000)
2. Zhao, C.: Research on key technology of shallow-buried tunnel's construction method conversion in upper-soft lower-hard strata. J. Railw. Sci. Eng. **6**, 1444–1451 (2015)
3. Xiao, J.Z., Dai, F.C., Wei, Y.Q., et al.: Comparative analysis of excavation schemes for a tunnel constructed through loose deposits. J. Perform. Constr. Facil. **30**(4) (2015)
4. Diao, Z., Li, C.: Study on construction method of large-section tunnels in upper soft and lower hard layers. Tunn. Constr. **S2**, 433–436 (2007)

5. Zhang, J., Guo, F., Lei, G.: Study on construction methods of shallow-buried tunnel under condition of upper-soft lower-hard ground by numerical simulation. Chin. J. Undergr. Space Eng. **S1**(7), 1416–1421 (2011)
6. Liu, T., Zhao, S., Sun, F.: Security analysis of dismantling temporary supports of large-span shallow tunnel in upper-soft and lower-hard stratum. Rock Soil Mech. **S1**, 306–310 (2014)
7. Wang, Z., Qiao, C., Song, C.: Calculation method of relaxation pressure of shallow large span tunnel in up-soft/low-hard rock stratum. Rock Mech. **35**(8), 2342–2352 (2014)
8. Schubert, W., Budil, A.: The importance of longitudinal deformation in tunnel excavation (1995)
9. Zhang, Y., Wu, X., He, C.: Study on application of displacement vector orientation trend line to wushaoling tunnel. Chin. J. Rock Mech. Eng. **26**(7), 1448–1456 (2007)

The Effects of Local Cavities on the Cracking Performance of an Existing Tunnel Lining

Jinyang Fu[1,2(✉)], Jiawei Xie[1], Junsheng Yang[1], Shuying Wang[1], and Feng Yang[1]

[1] School of Civil Engineering, Central South University, 410075 Changsha, China
jy.fu@csu.edu.cn

[2] National Engineering Laboratory for High Speed Railway Construction, Central South University, 410075 Changsha, China

Abstract. Local cavities around the tunnel lining due to imperfect construction or ground water erosion will lead to local contact loss and discontinuity in the ground-lining interaction. This paper evaluated the effect of local cavities on the mechanical and cracking performance of an existing tunnel lining. The geometry and position of the local cavity were defined according to the results of field investigation. Numerical analyses were performed to investigate how the local cavities affect the stress state of the concrete lining and to check the stability of the tunnel. Extended Finite Element Method (XFEM) was used to illustrate the development of cracking directions and patterns on the existing tunnel lining in the vicinity of the cavity. The above results was then verified by comparing with field observations. The presented study suggested an effective way for modelling tunnel lining cracking, and shown a systematic method to preliminary evaluate the cracking performance of an existing tunneling lining with local cavities behind.

1 Introduction

Tunnels have the advantages of shortening the distance and improving the safety when transportation line crossing the mountain area. In order to maintain the safety of the mountain tunnels, it is essential to keep the effective contact between lining and surrounding rock. However, local cavities around the tunnel lining due to imperfect construction or ground water erosion will lead to local contact loss and discontinuity in the ground-lining interaction. Such issues have been discussed in several studies. Lee [1] proposed a new system identification approach based on tunnel deformation data to find the damage in the tunnel liner. Zhang [2] found the local cavities due to imperfect construction can influence the stress state of a tunnel structure and plays an important role in the stability of the operating tunnel. Zhang [3] indicated that the finite discrete element method technique could be benefit to quantitatively evaluate the stability and damage level of tunnel lining system. Meguid [4] found that depending on the flexibility ratio between the lining and the surrounding soil, the cavity size can have a significant impact on the circumferential stresses in the tunnel lining. However, most of them are

W.-C. Cheng et al. (eds.), *Tunneling in Soft Ground, Ground Conditioning and Modification Techniques*, Sustainable Civil Infrastructures,
https://doi.org/10.1007/978-3-319-95783-8_8

purely theoretical studies and lack of physical testing. Leung [5] set up an experimental to investigate the effect of local contact loss between a tunnel lining and the surrounding ground on the earth pressure distribution acting on the tunnel liner. The study suggests that the presence of a small lining area that is not in direct contact with the surrounding ground can have a significant impact on the performance of the tunnel lining. Lee [6] investigated the characteristics of some typical cracks, characterises their spatial distribution digitally and developed an approach for diagnosing causes of lining cracks in mountain tunnels, but the development of cracking directions and patterns on the existing tunnel lining is rarely reported.

In this paper, a simulation methodology was developed to evaluate the effect of local cavities on the mechanical and cracking performance of an existing tunnel lining. An operational tunnel with cracks in the lining was studied by field investigation and numerical analysis, in which a 5 m long cavity was field-investigated behind the tunnel vault. The cracking performance resulting from the numerical analysis was found to be in good agreement with the behaviors observed in the field investigation.

2 Project Overview

An operational single-rail track tunnel which passes through the mountain area with a stratum of mudstone intercalated sandstone in southwest China is discussed in this paper. Field investigation shows that the surrounding ground is nearly horizontally layered with developed joints but there are not much groundwater. The tunnel was constructed by the conventional drilling and blasting method and supported by an integral cast-in-place concrete lining system.

The tunnel depth is about 30–50 m (except for the entrance and exit section) from the surface. The studied section of the surrounding rock is 284 m long and is specified as grade IV which is classified by the degree of rock hardness and the integrity of the rock according to the Chinese code for designing railway tunnels. This kind of surrounding rock has relatively complete rock mass, and the elastic velocity of longitudinal wave of the surrounding rock is between 1.5–3.3 km/s. The tunnel was first supported with a 10 cm-thick shotcrete in the arch and the walls and then it was supported by cast-in-place symmetrically with the help of the steel formwork. The spatial bolted shotcrete supporting structure is used only when necessary. The minimum thickness of plain concrete at the arch and the invert are 55 and 45 cm, respectively. Finally, about 1.77 m thick C20 concrete is backfilled in the invert area. The longitudinal steel bar and ring bar were placed in the vault area with spcing 25 cm. The parameters for the reinforcement of the supporting system are shown in Table 1. As shown in Fig. 1, the tunnel width is 8.87 m and the height is 10.11 m which result in a span ratio (height/width) about 1.35.

Table 1. Parameters for ground and lining reinforcement for the study section

Rock grade	Lattice girder			Forepoling			Bolt		
	Position	Parameters		Position	Form	Spacing (m)	Parameters		
		Spacing (m)	Position				Position	Spacing (m)	Length (m)
IV	Arch and wall when necessary	1.0	Arch in soft rock	Arch in soft rock	Bolt or small pipe L = 3.5	Ring 0.4	Arch and wall	1.2	3.0

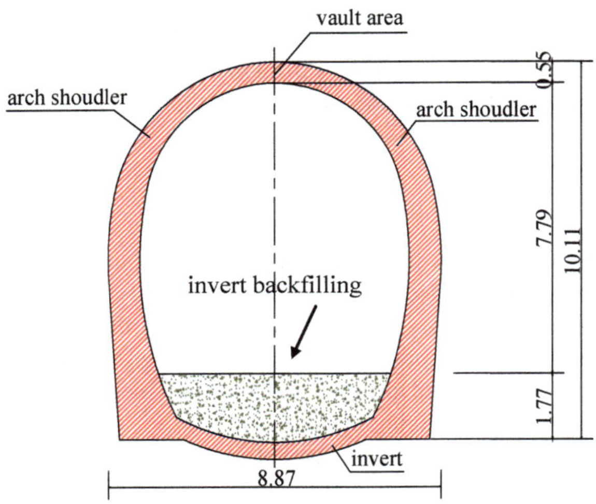

Fig. 1. Tunnel transverse cross section (unit in meter)

3 Field Investigation on the Tunnel Damages

In the field investigation, slight cracks were found in the tunnel lining (width less than 1 mm) after about nine years of operation. However, a large scale of severe cracking was found in some sections during the tunnel lining inspection in one year. In typical sections, the tunnel lining is characterized by symmetrical tension cracks along the arch shoulder, as shown in Fig. 2a and crush at the vault area due to the compressive stress as shown in Fig. 2b. The corresponding location of the observed cracking area is shown in Fig. 1. According to the observation and statistics results, the cracks are majorly developing along the tunnel longitudinal direction. The observed maximum length, width and the detected depth of cracks are 30 m, 10 mm and 15 cm, respectively. Moreover, nondestructive examination reveals the length of intrados of serious cavity behind the lining in the typical sections to be more than 5 m.

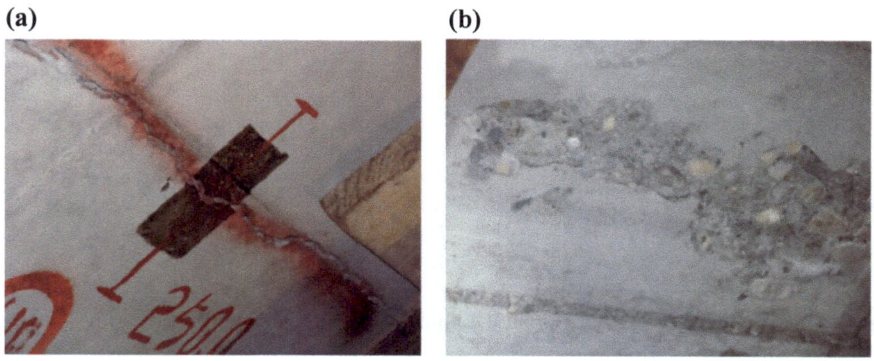

Fig. 2. Longitudinal crack at intrados of tunnel lining **a** observed longitudinal crack in tunnel arch shoulder (mileage DKI68 + 340); **b** observed longitudinal crush at the vault area (mileage DKI68 + 349)

4 Numerical Investigation on the Cracking Performance of Tunnel Lining

4.1 Numerical Modelling with XFEM

In order to evaluate the effect of local cavities on the mechanical and cracking performance of the existing tunnel lining, numerical analyses were performed to investigate how the local cavities affect the stress state of the concrete lining and to check the stability of the tunnel. Three numerical models (see Table 2) are modelled respectively using ABAQUS software. A two-dimensional plane strain model was considered to investigate the continuous damage occurring in cross section of the concrete lining.

Table 2. Numerical models of the studied cases

	Case 1	Case 2	Case 3
Situation	Without a cavity	With a cavity behind the vault area	With a cavity behind the arch shoulder

As shown in Fig. 3a, the cavities behind the lining are generally with irregular geometry. The characterization of cavities' geometric features can be simplified for the two-dimensional analysis. Therefore, a model of the circumferential angle of $\theta = 60°$ (using the same center of circle with the vault arc) and radial depth of d = 0.9 m was used to simulate the cavity shape based on the results of field investigation, as shown in Fig. 3b.

In order to avoid the boundary effect of the model, the distance between the boundaries and center of the tunnel should be 3–5 times of the tunnel span [7]. The calculation range in longitudinal direction, the distance between the center of the tunnel and the boundary is 4.5 times of the tunnel span. And in the vertical direction, the

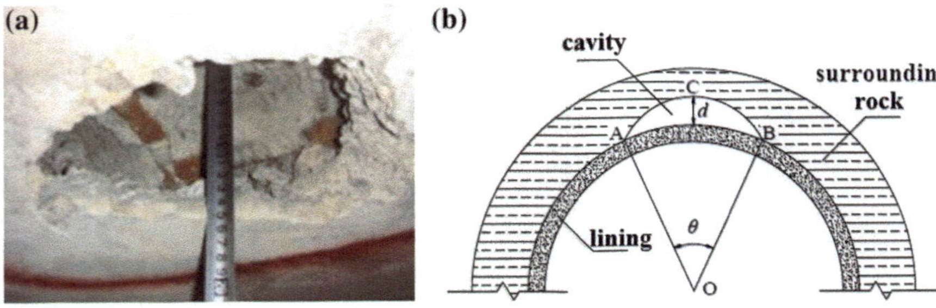

Fig. 3. Cavity condition **a** cavity behind the lining of the site; **b** simplified model of the cavity

distance between the center of the tunnel and the boundary is 7 times of the tunnel span. The two lateral boundaries are constrained in the horizontal direction while the bottom boundary is fixed both in the longitudinal and the vertical. The numerical model for case 2 and case 3 are shown schematically in Fig. 4.

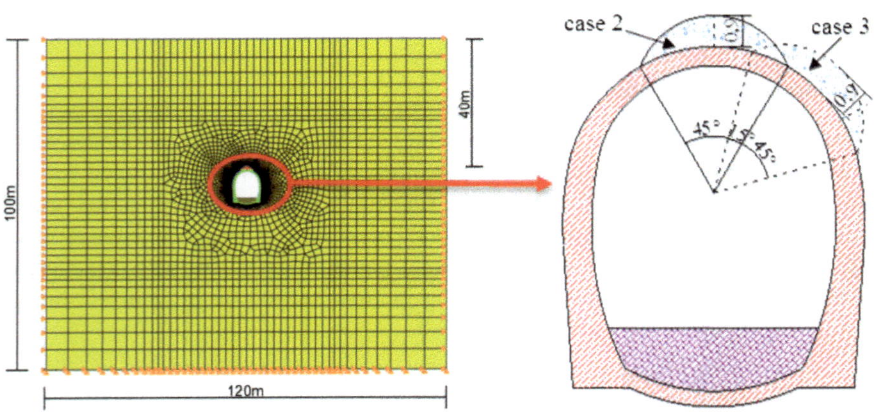

Fig. 4. Numerical analysis model in case 2 and case 3

As the tunnel was constructed by conventional tunneling method and the surrounding rock is nearly horizontally layered with developed joints, there is a great potential of over-break at the vault area. Such over-break was usually not enough back infilled. Moreover, due to the low workmanship of the nozzle man, the placement of the shotcrete at the vault area is not very perfect. Furthermore, the cast-in-place concrete lining was placed by first pouring the concrete at the vault area and then let it move downward to fill the wall area. There would be void forming behind the arch area due to a possible early leave of the concrete pump, insufficient infill of the concrete at the vault area due to gravity effects and the shrink of cast-in-place concrete. All above imperfect construction factors would lead to cavities behind the lining vault area.

As field investigation shows that the surrounding rock mass is relatively complete and there is not much groundwater exist. The specific parameters of the surrounding rock are derived from the results of geological investigation. The physical and mechanical parameters of the surrounding rock and the lining are listed in Table 3. The surrounding rocks was simulated following the Mohr-Coulomb yield criterion. The cavity was modelled by elastic fictitious material with very low modulus and regardless of its gravity. The interaction between the lining and surrounding rocks was modelled by setting a series of elastic springs [8].

Table 3. Parameters of the surrounding rock and the lining for the studied section

Material	Density ($\gamma/kN\,\mathrm{m}^{-3}$)	Elastic modulus (E/GPa)	Poisson's ratio	Cohesive force (c/kPa)	Friction angle
Surrounding rock	2200	0.5	0.35	200	30
Lining	24	28	0.2	–	–

Moreover, in order to illustrate the development of cracking directions and patterns on the existing tunnel lining, Extended Finite Element Method (XFEM) [9, 10] was used in modeling the lining area. Such technique has been widely used in the study of cracks and can successfully reproduce many experimental results [11], and it is acknowledged as an elegant tool to model cracks within a finite element without the requirement of remeshing. Comparing to the finite element method, this method provides a mathematical model for describing the interaction mechanism of crack surfaces and does not require to add the interface elements. The method can be easily used since it has been integrated in general finite element software like Abaqus. The crack damage initiates when the concrete's maximum principal stress reaches the critical value. The damage initiation criterion is satisfied when $1.0 \leq \frac{\langle\sigma_{\max}\rangle}{\langle\sigma^0_{\max}\rangle} \leq 1.0 + f_{tol}$, where $\sigma^0_{\max}$ is the critical value of the maximum principal stress (C20 concrete value as $\sigma^0_{\max} = 1.5\,\mathrm{MPa}$) and f_{tol} is a user-specified tolerable value which is default to be 0.05. The crack plane is perpendicular to the direction of the maximum principal stress which is used to illustrate the development of cracking directions and patterns [12].

4.2 Comparison with Field Observations

The crack distributions of the concrete lining are visualized by using the PHILSM and STATUSXFEM output variables in ABAQUS. Crack plane is perpendicular to the direction of the maximum principal stress which is used to illustrate the development of cracking directions and patterns on the existing tunnel lining. The crack on the exterior and interior side of the tunnel lining can be illustrated in the numerical model (see Fig. 5b). The acquired crack direction and location was found to be in good agreement with the field observations which clearly demonstrate that cracks are majorly longitudinal at the intrados of tunnel arch shoulder symmetrically (see Fig. 5a). Moreover, the cracking potential at the extrados of tunnel crown was highlighted by the numerical analysis and should be pay great attention as it cannot easy to be detected in the field.

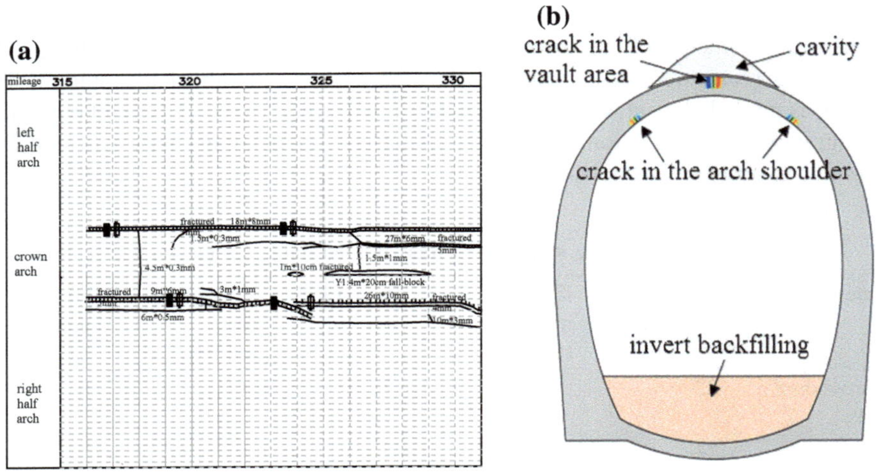

Fig. 5. **a** The distribution of observed tunnel lining cracking performance at mileage DIK 68 + 781 − 796; **b** modelled crack directions and patterns in case 2

The numerical results shows that the crack initiation due to existing of cavity behind the tunnel lining vault starts at the exterior side of the tunnel lining. Such phenomenon suggests that the actual cracking performance of the tunnel lining should be more severe than that of observed at the interior side tunnel lining surface.

4.3 Comparison for Cases with and Without Cavities

The maximum principal stress of the tunnel lining without cavities is shown in Fig. 6a. It is clear that the maximum principal stress in each part of the lining is below the critical value of 1.5 Mpa and the maximum value appears in the corner between the wall and invert. However, the principal stress in the tunnel lining is redistributed with the emergence of cavity especially increasing rapidly in the vault area. The presence of cavity change the deformation direction of the vault area from converging towards tunnel centre to squeeze outwards to the surrounding rock, and meanwhile the deformation direction of the arch shoulder behaves in an opposite way. The contour of tunnel deformed and undeformed shape is shown in Fig. 6b with the amplification factor of 100. With the increase of the maximum principal stress, the crack appears initially interior of the vault and propagates towards the surface of the arch shoulder. The stress fields of the lining with the cavity is shown in Fig. 7a, b. The stress concentration near the crack tip is very obvious and the apex value of the maximum principal stress grows by 28 times. The results indicate that the presence of the cavity behind the lining has great influence on the stress state of the tunnel lining and would lead to development of crack on the tension area of tunnel lining.

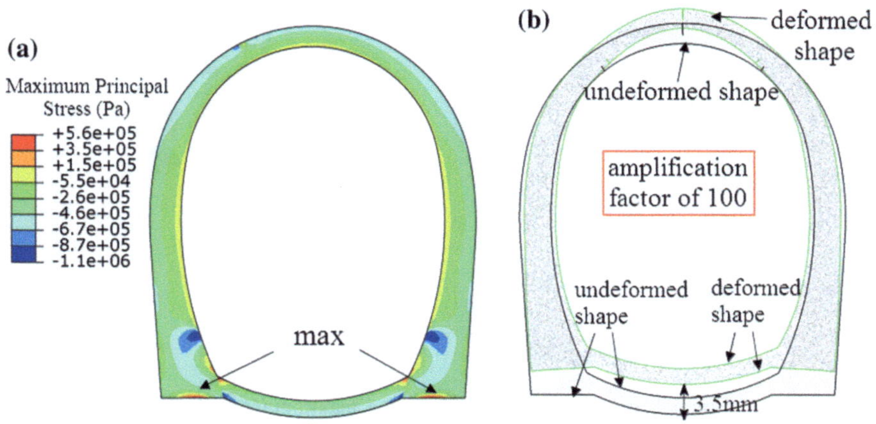

Fig. 6. **a** Distribution of maximum principal stress on the tunnel lining without cavity; **b** the contour of tunnel deformed and undeformed shape

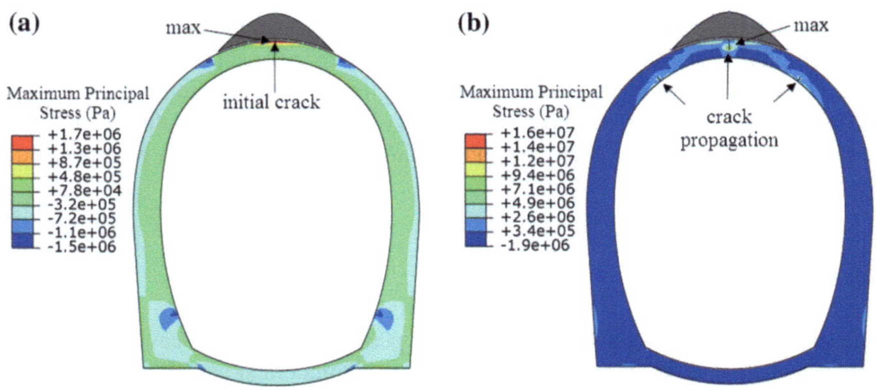

Fig. 7. Distribution of maximum principal stress on the tunnel lining with a cavity behind the vault area: **a** crack initiation; **b** crack propagation

4.4 Influence of the Cavity Location on the Cracking Performance of Tunnel Lining

An important role of the lining is to constrain the deformation of surrounding rock. When the cavity behind the lining appears in different positions, the stress state of the lining structure result in different distributions. A cavity behind the tunnel arch shoulder will cause bias effect and local stress concentration of the lining structure. The lining structure will shift from full section compression into local tension near the hole and lead to lining cracking behavior. The maximum principal stress of the tunnel lining with a cavity behind the tunnel arch shoulder is shown in Fig. 8. It is interest that the cracking behavior occurs at the exterior side lining close to the cavity center.

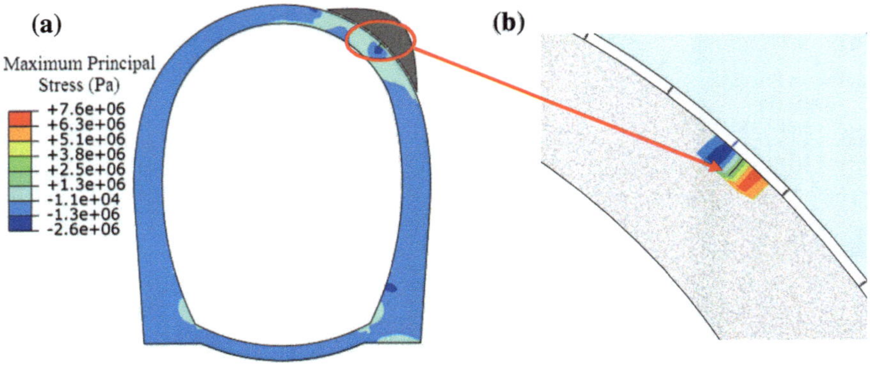

Fig. 8. Numerical model: **a** Numerical model and the maximum principal stress distribution; **b** the crack patterns in case 3

Obviously, the cavity behind the arch shoulder led to different lining stress distributions and the apex of the maximum principal stress is a little lower than that with a cavity behind the vault area. The result of the numerical simulation shows that the cavity behind the arch shoulder cause the crack in the exterior side of the arch shoulder. As the cracking initiate in the exterior side of the lining is difficult to be detected in the field, it is suggested that engineers should pay great attention to the cracks at the exterior side the lining once cavity behind the lining arch shoulder area was revealed by inspection.

5 Conclusions

In this paper, the cracking performance of an operating tunnel with cavities behind the lining was evaluated based on field investigation and numerical analysis. The following conclusions were obtained.

(1) The presence of cavities behind the lining is one of essential factor for lining cracking. The stress distributions of the tunnel lining with a cavity behind are different from the lining which is full in contact with the surrounding rock. Due to loss of reaction force at the cavity area, the tunnel lining deforms squeezing outwards to the cavity. Such behavior lead to increase of bending tension stress of the lining at the cavity area so that stress concentration near exterior of tunnel lining was very obvious and crack is initiated when maximum principal stress higher than the tolerable tensile strength of concrete.
(2) Cracks are majorly developed at the intrados of tunnel arch shoulder symmetrically which is found to be in good agreement with the field observations. There is a potential that the crack initiation starts at the extrados of tunnel crown when the cavities are majorly behind the vault area. The cracking initiate at the exterior of the tunnel lining was highlighted and should be pay great attention as such cracking performance is difficult to be detected in the field.

(3) When the cavities behind the lining form in different positions, the stress state of the lining structure will result in different cracking performance. The apex value of maximum principal stress in the lining with a cavity behind the vault area was a little higher than that with a cavity behind the tunnel arch shoulder.

Acknowledgements. This paper was supported by the National Key R&D Program of China (No. 2016YFC0802504) and National Natural Science Foundation of China (No. 51608539). The authors also appreciate the support of the China Postdoctoral Science Foundation (No. 2016M592451, No.2017T100610) and the Fundamental Research Funds for the Central Universities of Central South University.

References

1. Lee, J.S., Choi, I.Y., Lee, H.U., et al.: Damage identification of a tunnel liner based on deformation data. Tunn. Undergr. Space Technol. **20**(1), 73–80 (2005)
2. Zhang, Y., Shi, Y., Zhao, Y., et al.: Damage in concrete lining of an operational tunnel. J. Perform. Constr. Facil. **31**(4), 06017002 (2017)
3. Feng, J.: Modelling of cracking in concrete tunnel lining by continuous-discontinuous approach tunnel lining by continuous-discontinuous approach. Isrm Congress 2015, The International Congress of Rock Mechanics (2015)
4. Meguid, M.A., Dang, H.K.: The effect of erosion voids on existing tunnel linings. Tunn. Undergr. Space Technol. **24**(3), 278–286 (2009)
5. Leung, C., Meguid, M.A.: An experimental study of the effect of local contact loss on the earth pressure distribution on existing tunnel linings. Tunn. Undergr. Space Technol. **26**(1), 139–145 (2011)
6. Lee, C.H., Wang, T.T.: Diagnosing causes of lining cracks in mountain tunnels. Proc. Inst. Civ. Eng. Forensic Engineering **167**(4), 151–164 (2014)
7. Möller, S.C.: Tunnel induced settlements and structural forces in linings. University of Stuttgart, Institute of Geotechnical Engineering, Stuttgart, Germany (2006)
8. Zhang, Y., Yang, J., Yang, F.: Field investigation and numerical analysis of landslide induced by tunneling. Eng. Fail. Anal. **47**, 25–33 (2015)
9. Belytschko, T., Black, T.: Elastic crack growth in finite elements with minimal remeshing. Int. J. Numer. Meth. Eng. **45**(5), 601–620 (1999)
10. Dolbow, J., Belytschko, T.: A finite element method for crack growth without remeshing. Int. J. Numer. Meth. Eng. **46**(1), 131–150 (1999)
11. Ren, Q.W., Dong, Y.W., Yu, T.T.: Numerical modeling of concrete hydraulic fracturing with extended finite element method. Sci. China Ser. E: Technol. Sci. **52**(3), 559–565 (2009)
12. HKS Inc.: ABAQUS analysis user's manual. HKS Inc., Providence, RI, USA (2004)

Integration of Fault Tree and Bayesian Network for Falling Risk of the Bridge Project—Precasting Prestressing Segmental Construction Method

Ying-Chun Hung[1], Tung-Tsan Chen[2(✉)], and Ting-Yu Yue[2]

[1] Department of Urban Planning and Landscape, National Quemoy University, No. 1. University Rd., Jinning, Kinmen 892, Taiwan (R.O.C.)
hij@nqu.edu.tw

[2] Department of Civil Engineering and Engineering Management, National Quemoy University, No. 1. University Rd., Jinning, Kinmen 892, Taiwan (R.O.C.)
tungtsan@nqu.edu.tw, a109710518@gmail.com

Abstract. The current practice of safety management in bridge construction depends on the voluntary effort of the contractors and relevant government agencies. Due to varying degree of experience and knowledge of the inspectors, results of bridge inspection could not be analyzed in a systematic and consistent way. This study focuses on the use of Fault tree and Bayesian-network to analyze and generate a risk analysis model for falling risk in Precasting Prestressing Segmental Bridges Construction Method. After comparing the risk analysis model with the traditional inspection method, it is found that the risk analysis model is consistent with the traditional inspection method in their ability to predict falling hazards.

Keywords: Bayesian network · Fault tree · Falling risks · Bridge construction Precasting prestressing segmental construction method

1 Introduction

Construction of transportation systems is an important index for a nation's growth. Under the fastidious effort of the government, in recent years many bridge projects has been completed. This is especially important to urban areas, where road systems had already been fully developed.

Since the topography of Taiwan is marred with rivers and mountains, which hindered the accessibility, bridges in Taiwan are important infrastructures that shorten the distances between two urban areas. In Kinmen, one of Taiwan's offshore islands, a cross-strait bridge construction is also underway, linking Kinmen to its smaller neighboring island, Lieyue. It can be seen that bridges are very vital to economic growth and transportation in Taiwan.

On the other hand, bridges now are designed with increasing complexity. As a result, the construction has become more difficult and complex as well. A design less

W.-C. Cheng et al. (eds.), *Tunneling in Soft Ground, Ground Conditioning and Modification Techniques*, Sustainable Civil Infrastructures,
https://doi.org/10.1007/978-3-319-95783-8_9

than optimum may result in lower quality, higher cost, and even accidents. In recent years a surge of construction related accidents in bridge construction has been witnessed. The causes are thought to be unsafe working environment, unsafe conduct of workers, inadequate safety equipment, tight budget, or tight schedule.

These accidents not only cause deaths and injuries, but also delay the construction progress, which in turn increases cost and lowers quality of construction. Therefore, construction firms now face an important problem of how to best lower the frequency of construction related accidents.

According to recent government study on construction site safety, falling accounts for the most frequent among all construction related accidents (MOL 2013). Therefore, the risk of falling is treated as the focus of this study. To more closely assess the problem of falling, the construction cases discussed in this study are carried out by Precasting Prestressing Segmental Construction Method (P.P.S. Construction Method).

2 Literature Review

2.1 Fault Tree Analysis

Fault Tree Analysis (FTA) is seen by the industry as an effective way to assess the probability and causal relationship of an (usually undesirable) event. However, the accuracy and reliability of this analysis method is decreased if fuzziness is introduced to the system (such as human error). Rodak and Sillima applied a combination of Fault Tree (FT) and probabilistic risk analysis (PRA) to the management of underground water resource. Factors such as chemical property of pollutants, land use, method of reclamation are among some of the uncertainties in quality protection of underground water. Rodak and Sillima concluded that the PRA-FT is an effective tool for the protection of underground water quality (Rodak and Silliman 2012).

FTA is also used widely in the aerospace and national defense industry. In the 1970s, FTA is introduced to the nuclear power plants, chemical processing plants, and construction management to assess risks (Griffin 1973; Hadipriono and Wang 1986; Millet 1999; Morcos and Singh 1995; Pan et al. 2004; Powers and Tompkins 1974).

In the past decade in Taiwan, accidents related to retainment wall construction has occurred quite frequently. Due to a lack of historical data and human negligence, computation of the probability of these accidents occurring proves to be difficult. Chen had proposed a method which combines FTA, FS, and AHP that assesses the risk during retainment wall construction. He checked his method against the cases of retainment wall failure that took place in Kaohsiung, and concluded that this combined method could effectively help the safety managers analyze such failures, thus reducing the chance of accidents taking place (Chen 2005).

FTA is a widely accepted analysis method. It could assess accidents in a quantitative way, providing information about the significance of and relationship among every factor. In addition, FTA is not only used for the prediction of accidents, but could be used in post-accident analysis as well.

2.2 Bayesian Network

Since the 1990s, BN has been applied to troubleshooting systems concerning a wide range of fields, including computer processor production, wastewater treatment, and health monitoring (Graves et al. 2007; Jitwasinkul and Hadikusumo 2011; Chong and Walley 1996). Microsoft has applied Bayesian Network (BN) to the trouble shooting of personal computers (Heckerman et al. 1995). Also, the use of BN is becoming more common in the field of medicine (Ogunyemi 1999).

Since BN is highly accurate and highly compatible with other analytical methods, it is used in many fields such as construction (Huang 2009; Chang 2007), medicine, ecology, education, and business. It is also very flexible as it could be used in accordance with experts' opinions or other programs to overcome the problem of lack of information.

3 Research Method

The research method used in this study can be broken into three phases- Fault Tree Analysis, Bayesian Network, and transformation of FTA into BN. First, FTA is used to build an analytic framework for the relationship among construction processes related to falling. The second step involves the computation for the probability of falling using BN. Since the setting up of BN is quite complicated, the last step involves the transformation of FTA into BN, combining with experts' opinions, and transforming nodes of BN and CPT.

3.1 Fault Tree Analysis

FTA is a type of Logical Cause and Effect Relationship Graph; it shows the status of the system (top event) based on the bottom condition (basic event). FTA is built from top to bottom, according to the relationship between events (Griffin 1973; Franke et al. 2009). It employs a graphical model method, allowing for the prediction of a failure event, as shown in Table 1.

In application to safety management, the undesirable event such as an accident is placed as the top event, and its causes are looked into afterwards. The top event must be described in terms of time, location, and environment. Regarding possible safety hazards, every possible cause should be looked into. However, regarding events which has already taken place, causes could usually be deduced from on-site situation and construction schedule. This study focuses on the use of FTA on the risk of falling in construction projects using the P.P.S. Construction Method.

3.2 Bayesian Network

BN is an expert system that combines probability and graphical analysis. It is especially useful for analyzing highly complex but interrelated events. It can also be the way to find the uncertainty and accuracy of the inferred uncertainty, especially the complex system with a high degree of relevance. BN uses the graphics pattern to show its

Table 1. Common model path of FTA

Symbol shape	Gate type	Description
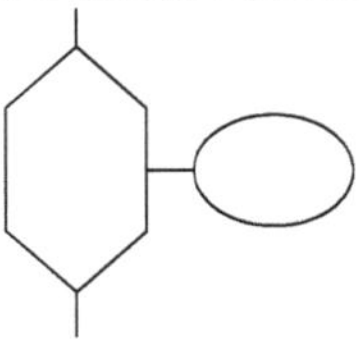	(Inhibit gate) 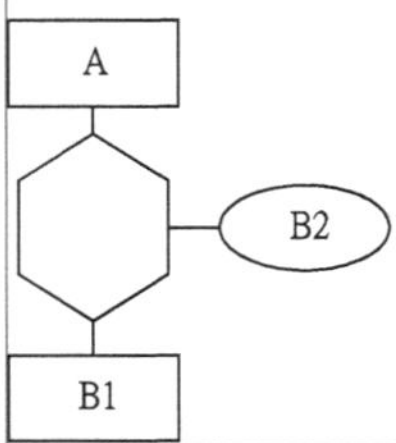	P(A = 1 | B1 = 1, B2 = 1) = 1 P(A = 1 | else) = 0 It can lead to the occurrence of the output event (A) above, then the input event (B1) must meet the limit condition (B2)
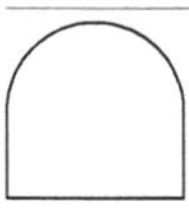	(AND gate) 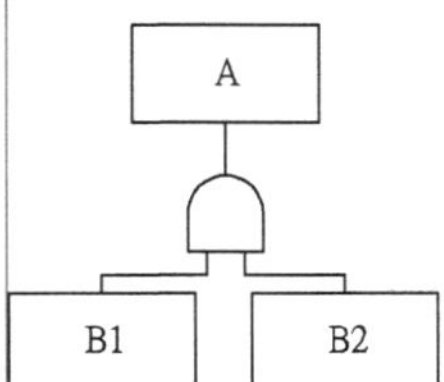	P(A = 1 | B1 = 1, B2 = 1) = 1 P(A = 1 | else) = 0 B1, B2 must exist at the same time then A will be established. If B1, B2 one does not exist, then A is not established, so A = B1 ∩ B2 (intersection)
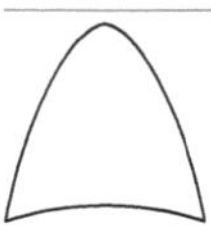	(OR gate) 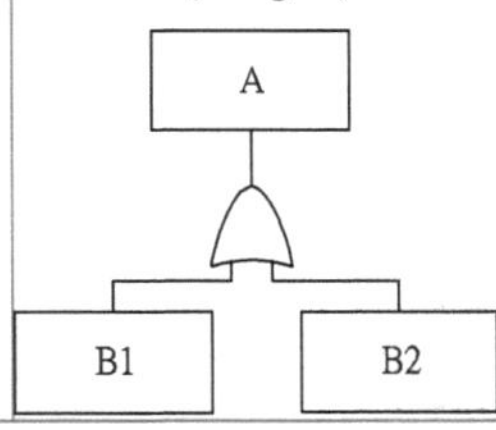	P(A = 1 | B1 = 0, B2 = 0) = 0 P(A = 1 | else) = 1 B1, B2 as long as one of them exists, A will be established, so A = B1 U B2 (Union).

relevance. It is a graphs model of random variables and conditional dependency probability, which also explains the degree of their interaction with the probability of the relationship to infer its importance.

In recent years, the BN method has been widely used in various fields as an effective analytical method. However, in terms of construction projects, which are faced with many security risks in the uncertainties, and in the case of lack of available information, how to use effective and high accuracy methods to obtain the probability of occurrence of each factor has become an increasingly important engineering practice topic (Xiao et al. 2008; Doguc and Ramirez-Marquez 2009; Chen and Wang 2015).

The advantage of BN compared to FTA is that it has the ability to deal with uncertainty problems, especially the ability to handle incomplete data. When the information collected by the user is insufficient for the problem, the characteristic of the BN method allows the system constructor to assume the value of the parameters in the network and is then amended by the process of learning.

3.3 Fault Tree Transformation Bayesian Network

In summary, FT logic relationships are converted to Bayesian network processes with specifications as follows (Li et al. 2002, 2003; Chen 2010; Khakzad et al. 2011; Chen and Leu 2014):

(1) All the basic events in the FT are expressed as the root nodes in the BN. If the basic events in the FT are repeated, they can only be expressed as single root node in the BN.
(2) The probability of all the basic events in the FT is assigned directly to the corresponding root node in BN as a prior probability.
(3) Each logical gate in the FT corresponds to one node in the BN. The node conditional probability value is consistent with the output event of the logic gate in the FT.
(4) Connect the nodes in BN according to the relationship between the logic gate and the basic event expressed in FT. The direction of the directional edge of the

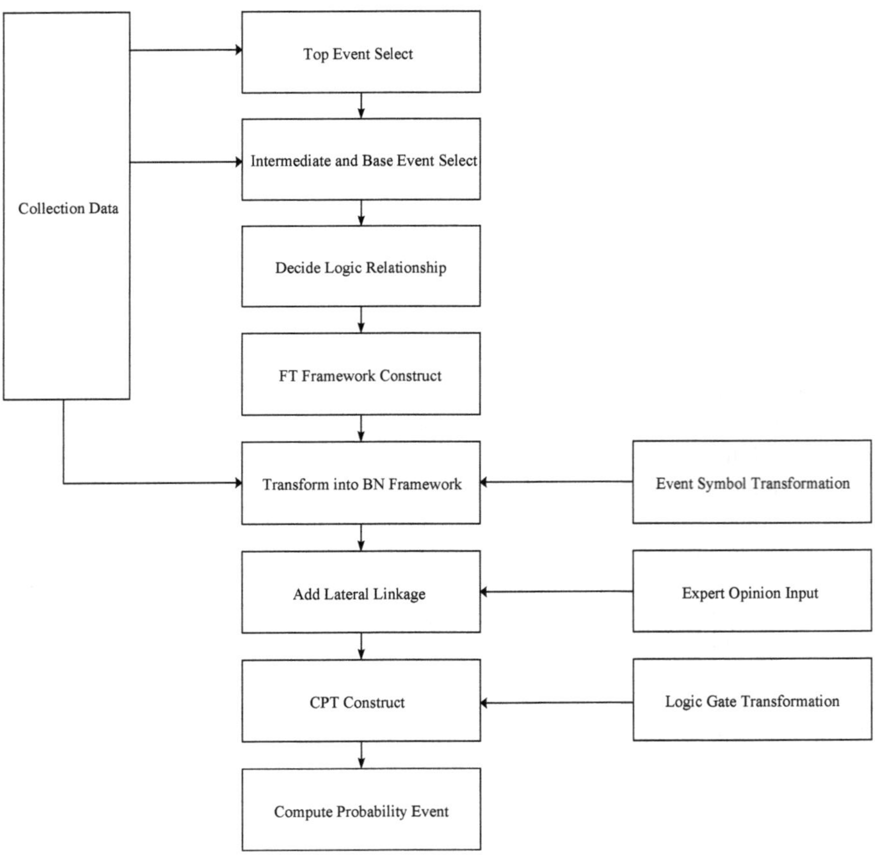

Fig. 1. Transformation flow chart from FT to BN

connected node corresponds to the input and output relationship of the logic gate in FT.

(5) The logical relationship of the logic gates in the FT is expressed as the conditional probability table (CPT) of the corresponding node in BN.

Figure 1 summarizes the simplified procedure of mapping FTs into BNs.

4 Practical Case Discussion Analysis

In this study, the fall of the occupational disaster type in the construction industry is the type of disaster in this study. The engineering type of the civil engineering is bridge engineering. The construction method is the prefabricated preloading section construction method in the bridge construction.

4.1 Fault Tree Analysis to Explore the Bridge Fall Accident

Of most of the bridge construction projects accidents, "falling" is one of the major types of disasters, and therefore selected as the top event of FT in this study. The direct cause of the fall in the work can be divided into the accident location and status (such as bridge deck attached works, section construction, hoisting vehicle operation, etc.) and their details, indirect causes and their details (such as unsafe movements, unsafe equipment and unsafe environments, etc.), as well as the root causes (such as improper safety planning policies, poor environmental maintenance, inadequate safety education and poor management, etc.). In this paper, the relationship between these causes and the top event is connected by the logic gate. The P.P.S. Construction Method is used as an example. This study is based on the construction method, the related literature, and the practical case, and is carried out with the practical experts and scholars repeated amendments to complete the initial FT structure, as shown in Fig. 2.

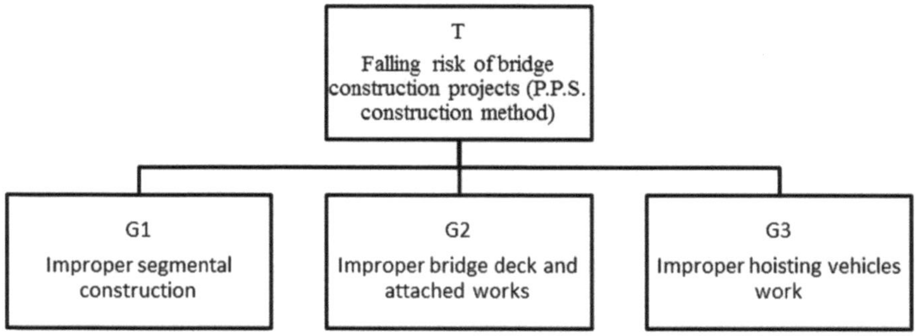

Fig. 2. FT of falling accidents at bridge construction projects (P.P.S. construction method)

The main reason for the falling accidents were improper segmental construction, improper bridge deck and attached works, improper hoisting work vehicles operation. In the case of falling accidents caused by segmental construction, through interviews with experts and literature review, the surrounding operations can play a role in triggering. There are two main operations that may occur in the fall:

(1) Segmental casting
(2) Segmental lifting.

If necessary, the job can be broken down into more detailed sub-operations. In addition, these indirect causes may lead to analysis item-by-item in segmental construction.

In the case of falling accidents with bridge deck works improperly, through expert interviews and literature review, the surrounding operations can play a trigger role, so that the situation of falling accidents can be analyzed in more detail. There are three main operations that may occur in the fall:

(1) Bad steel works
(2) Bad template works
(3) Poor concrete pouring operation.

If necessary, the job can be broken down into more detailed sub-operations. In addition, these indirect causes may lead to analysis item-by-item in improper bridge deck works.

Another example is the accidental fall accident of the improper hoisting work vehicles operation. Through expert interviews and literature review, the surrounding operations can play a trigger role, so that the situation of falling accidents can be analyzed in more detail. There are three main operations that may occur during the fall:

(1) Improper assembly
(2) Improper Push forward
(3) Improper disassembly.

If necessary, the job can be broken down into more detailed sub-operations. In addition, these indirect causes may lead to analysis item-by-item in the improper hoisting work vehicles operation.

According to the concept of occupation (labor) safety and health management and literature. The four basic causes of occupational disaster accidents are: poor safety and health management, improper safety and health plan, poor maintenance of the site environment, and lack of safety education and training. Based on records of accidents, safety and health theory and expert interviews, the overall FT model is built, as shown in Fig. 3.

4.2 Development Questionnaire

(1) Conditional probability table calculation

The constructed BN architecture can use the software AgenaRisk to calculate the conditional probability table. The expression of functional types for the high risk value, moderate risk value and low risk value, etc. was selected for the function parameters based upon the definition of FT logic gates. Furthermore, questionnaires were designed to collect the information about the relative weights of parent nodes on their child

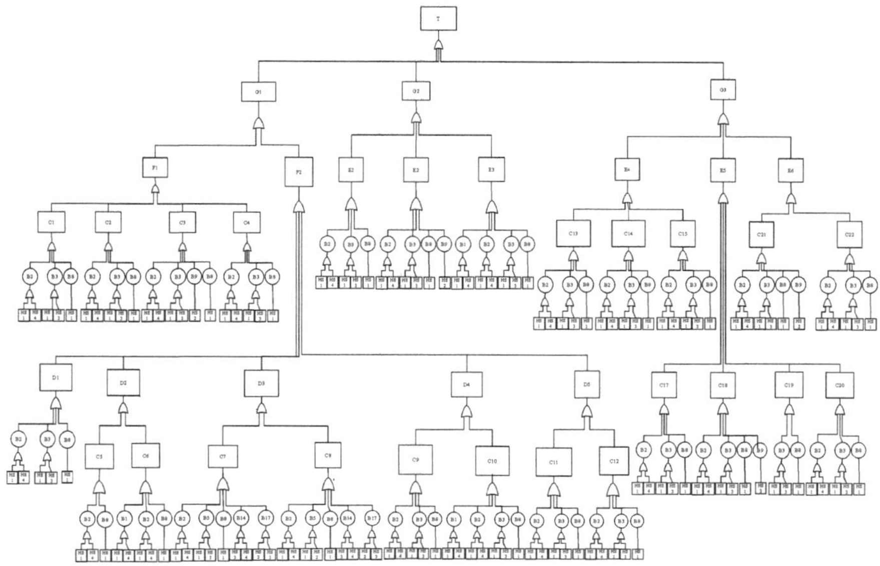

Fig. 3. Overall FT of falling accidents at bridge construction projects (P.P.S. construction method)

nodes. The questionnaire example is depicted in Table 2. For example, there are three influencing factors (G1, G2, and G3) of falling risk at the P.P.S. construction method projects (T). Based upon their impacts to the falling risk, the weight scores in column 3 are defined and filled (5 means the greatest influence, and 0 is minimal influence).

Table 2. Engineering experts' profiles

No.	Department	Title	Numbers	Working years
1	Industrial safety and QC dept.	Manager	2	20, 25
2	Construction office	Executive V.P.	3	25, 28, 33
3	Construction project site office	Director	4	22, 25, 25, 30
4	Construction project site office	Vice director	4	10, 15, 16, 18
5	Construction project site office	Senior engineers	10	15–30
6	Construction planning	Junior manager	1	20
7	H/S section	Section manager	1	18
8	H/S office	Specialist	2	12, 19
9	Environmental safety office	Senior manager	1	25
10	Labor safety and QC office	Supervisor	2	17, 20

This study invited 30 experts (Table 3) to evaluate 134 questions based on their own practical experience and to conduct statistical analysis with their responses. By using the above-mentioned input data, the CPTs for all the arcs in the BN were

Table 3. Questionnaire example of relative weights of BN arcs

AgenaRisk id		Levels of influence						Cause	AgenaRisk code	Q&A no.
		High		Medium		Low				
		5	4	3	2	1	0			
T	Falling risk of bridge construction projects			x				Improper segmental construction	G1	Q1
				x				Improper bridge deck and attached work	G2	Q2
				x				Improper hoisting vehicles work	G3	Q3
G1	Improper segmental construction			x				Segmental casting	F1	Q4
				x				Segmental lifting	F2	Q5
G2	Improper bridge deck and attached works			x				Bad steel operation	E1	Q6
				x				Bad formwork job	E2	Q7
				x				Bad concrete pouring operation	E3	Q8
G3	Improper hoisting vehicles work			x				Improper assembled	E4	Q9
				x				Improper push forward	E5	Q10
				x				Improper disassembly	E6	Q11
F1	Segmental casting			x				Concrete pouring is not correct	C1	Q12
				x				Improper formwork assembly operation	C2	Q13
				x				Improper formwork disassembly operation	C3	Q14
				x				Improper steel operation	C4	Q15
F2	Segmental lifting			x				Improper segmental lifting stacking	D1	Q16
				x				Improper segmental joint operation	D2	Q17
				x				Measurement operation is not correct	D3	Q18
				x				Closed section of the poor operation	D4	Q19
				x				Improper prestressing operation	D5	Q20
E1	Bad steel operation			x				Operation procedure is not correct	B2	Q21
				x				Dangerous procedures or methods	B3	Q22
				x				Operation error	B8	Q23
B5	Not familiar with the environment			x				H/S insufficient education and training	HS1	Q127
				x				H/S environment is not properly maintained	HS2	Q128
B8	Operation error			x				H/S insufficient education and training	HS1	Q129
B9	Lack of safety facilities			x				H/S environment is not properly maintained	HS2	Q130
B14	Lacks personal protective equipment			x				H/S is improper planned	HS3	Q131
				x				H/S is poorly managed	HS4	Q132
B17	Lack of security signs			x				H/S environment is not properly maintained	HS2	Q133
				x				H/S is improper planned	HS3	Q134

Note x is 5–0; 5 represents the greatest influence; 0 is minimal influence successfully calculated. Finally, the subsequent probabilities of the BN nodes (including both leaf nodes and the intermediate nodes) are generated once the probabilities of root causes (i.e. prior probabilities of BN) are assessed and inputted into the complete BN (the prior probability of BN is shown in Appendix 1).

(2) Assessment of prior probabilities

There are four significant primary causes defined in the model and they are H/S insufficient education and training, H/S environment is not properly maintained, H/S is improper planned, H/S is poorly managed. In order to assess the prior probabilities of these causes, a safety performance evaluation table was established in this study (sample as shown in Appendix 2). Significant safety performance checklist items of each primary cause were summarized and listed in the table. The more items marked based upon the site investigation, the higher probability of a poor performance on the root cause evaluation. By inputting prior probabilities into BN, fall risks at bridge construction projects and their significant causes can be identified through this model.

4.3 Fault Tree Conversion Bayesian Network

The top events, intermediate events, and root events within the FT are directly transformed into the leaf nodes, intermediate nodes, and root nodes of the BN. So the basic arrow between the BN nodes follows the definition of the event relationship within the FT. Convert to BN from FT, as shown in Fig. 4.

After collecting relevant information and interviews with experts and scholars, we first use the FT construction framework to analyze the correlation between the related works of the fall accident in the construction process of the bridge, and the analysis was carried out with BN to calculate the probability of related activities of the falling accident occurred in the construction of the bridge. That is, the first to establish the FT analysis system architecture, and then convert to BN from FT architecture, combine with the expertise of the project engineers, in the BN root node input prior probability of management factors (root event), and the intermediate node input relative weight data to calculate the subsequent probability of each node.

4.4 Bayesian Network Analysis Fall Occupational Hazards

(1) Basic information and results of projects

This study investigates the results of the actual labor safety inspection records and the subsequent probability of the BN leaf node events in the five bridge construction projects. The BN model was used to verify the actual project case. The basic information of the five bridge construction projects is shown in Table 4. The probability of calculating the results with BN, and the actual project on-site labor safety inspection performance evaluation records in the five bridge construction projects is summarized in Table 5.

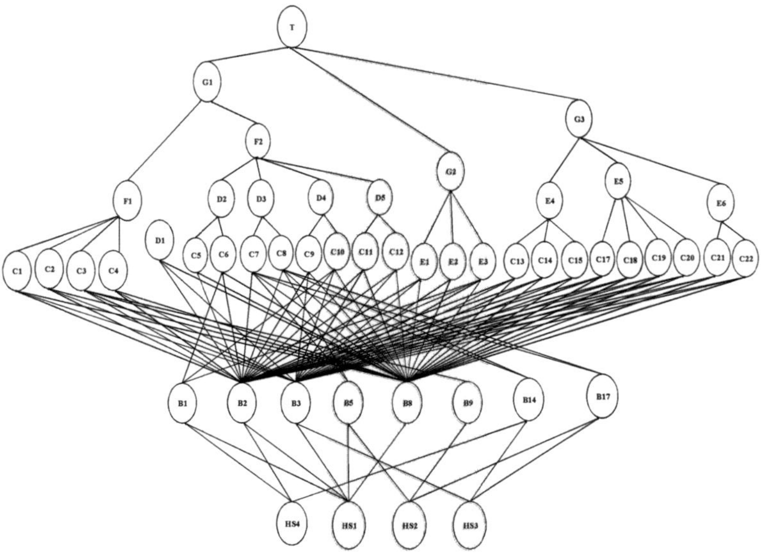

Code	Risk Factor	Code	Risk Factor
T	Falling risk of bridge construction projects	C10	Improper concrete pouring operation
G1	Improper segmental construction	C11	Pre-tube grouting poor
G2	Improper bridge deck and attached works	C12	Improper pull pre-stress
G3	Improper hoisting vehicles work	C13	Ground assembly is bad
F1	Segmental casting	C14	Support tripod assembly bad
F2	Segmental lifting	C15	Improper steel beams and trusses hanging
E1	Steel operation is bad	C17	Jack improper operation
E2	Formwork job is bad	C18	Steel trusses are improperly moved
E3	Concrete pouring operation is bad	C19	Improper adjust the elevation and position
E4	Improper assembled	C20	Fixture is improperly operated or insufficient
E5	Improper push forward	C21	Improper steel beam truss disassembly
E6	Improper disassembly	C22	Improper hanging to the ground operation
D1	Improper segmental lifting stacking	B1	Improper control procedures
D2	Improper segmental joint operation	B2	Operation procedure is not correct
D3	Measurement operation is not correct	B3	Dangerous procedures or methods
D4	Closed section of the poor operation	B5	Not familiar with the environment
D5	Improper prestressing operation	B8	Operation error
C1	Concrete pouring is not correct	B9	Lack of safety facilities
C2	Improper formwork assembly operation	B12	Lack of safety equipment
C3	Improper formwork disassembly operation	B14	Lacks personal protective equipment
C4	Improper steel operation	B17	Lack of security signs
C5	wet joint is not correct	HS1	H / S insufficient education and training
C6	Epoxy resin improper bonding	HS2	H / S environment is not properly maintained
C7	Improper segmental measurement height	HS3	H / S is improper planned
C8	Deck measurement is not correct	HS4	H / S is poorly managed

Fig. 4. Bayesian network of falling accidents at bridge projects (P.P.S. construction method)

Further explore the key factors that influence the occurrence of the top event (fall) in the bridge project. Sensitivity analysis was performed to summarize the key factors of its sorting, and the top ten risk factors for falling risk are shown in Table 6.

(2) Discuses and analysis

In this study, the safety performance of each project was assessed by a safety performance evaluation form, and the prior probability of four root causes is based on

Table 4. Basic information of five bridge projects. (P.P.S. construction method)

Project no.	Scope of works	Size	Period (days)	Cost (NT$)	Construction method
A	11,080M North line STA.N47K +950～N59K +030	Bridge 7785M/Road 3295M	981	4,310,000,000	P.P.S. construction method
B	12,026M North line STA.N59K +031～N71K +057	Bridge 8716M/Road 3310M	974	4,229,309,405	P.P.S. construction method
C	12,030M South line STA.S58K +990～S71K +020	Bridge 8701M/Road 3331M	973	4,428,000,000	P.P.S. construction method
D	12,010M South line STA.S71K +021～S83K +031	Bridge 8685M/Road 3325M	960	4,425,300,000	P.P.S. construction method
E	11,050M South line STA.S83K +032～S94K +082	Bridge 7767M/Road 3283M	948	4,338,650,000	P.P.S. construction method

Table 5. Comparison between BN and real site assessment

Project no.	Fall risk (%) from BN	(Risk rank By BN)	Real site assessment (score)	Safety performance rank
A	24.241	3	3 (81.48)	3
B	16.329	5	1 (88.65)	1
C	38.657	1	5 (72.38)	5
D	21.894	4	2 (85.72)	2
E	32.735	2	4 (77.24)	4

results of the subjective judgments in the occupation (labor) safety and health manager of project. The relative weight of each intermediate node is based on the project engineers on the relative impact degree of activity, and enter the relative weight. Finally, AgenaRisk software can calculate the probability of each node.

This study is based on the five projects of the P.P.S. construction method in Taiwan. The results of daily routine safety inspection performance and BN's

Table 6. Top 10 falling risk sensitive factors in bridge construction project

Rank	Code	Factors	Total risk (%)
1	E5	Push forward improperly	12.265
2	C19	Adjust the elevation and position improperly	11.807
3	G3	Hoisting vehicles work improperly	11.249
4	E6	Disassembly improperly	7.587
5	E4	Assembled improperly	7.33
6	G2	Bridge deck and attached works improperly	6.737
7	C18	Steel trusses are improperly moved	6.563
8	C14	Support tripod assembly bad	6.56
9	C20	Fixture is improperly operated or insufficient	6.501
10	C12	Pull pre-stress improper	6.444

calculation inference are validated. First, develop a safety performance checklist and the subjective to assess the probability of root causes, then through the BN to calculate the subsequent probability of leaf node (probability of falling risk), and the results of the actually daily labor safety record of the bridge construction project. The results are shown in Table 5. From the results, it can be observed that the ranks of probabilities from the BN model are highly consistent with the ranks of safety performances obtained from the real site assessment records.

Five bridge construction projects monthly labor safety inspection records were investigated in detail. The top priority of the project is indeed excellent in site safety management, safety performance without any bad record, such as security fines and attempted events is very low compared to the other four projects. In contrast, the final project has a poor implementation of its labor safety management, and the BN model shows that the value of falling risk probability is 38.657%. Through the actual assessment and review of the five bridge construction projects, it shows that the proposed BN-based falling risk assessment model has the accuracy and applicability that can be used as a fall risk assessment method for bridge construction projects (P.P.S. construction method).

The main causes of the falling accidents at bridge construction projects were the improper operation of adjusting the elevation and location, improper advance, improper demolition, improper assembly, as well as improper bridge deck and ancillary works operation, etc., indicating that labor in the construction process may not be in accordance with the construction process procedures, or that the operation location is not appropriate in the operation of mechanical equipment, and in turn causes accidents. Therefore, according to the evaluation model and sensitivity analysis, the relevant units of the project management and on-site project managers can strengthen the construction procedures and management in the field to reduce the risk of falling accidents at the site of the bridge project.

5 Conclusion and Suggestion

In this study, the bridge construction method is P.P.S. construction method. A bridge falling occupational hazard model is constructed by FT, and constructs a risk assessment model for the fall risk with BN.

In order to facilitate the construction of BN, FT is constructed to analyze the association of the related operations of the fall (roll-off) occupational disaster accident during the construction of the bridge. FT's top event, intermediate event and basic event are directly transferred to the leaf node, intermediate, and root node of BN. The causes of the falling in the work can be divided into accident location and status: such as bridge deck ancillary works, section construction, hoisting vehicle operations, etc., indirect reasons: unsafe action, unsafe environmental or unsafe equipment, the basic reasons: poor safety and health management, improper safety and health plan policies, poor on-site environmental maintenance and inadequate safety education and training. The relationship between the three basic events, intermediate events and top events, along with the logic gate connection are used to build FT, and then converted to BN from FT to make any analyzation, and to explore the probability of calculating the related activities of the fall (roll) occupational disaster accident. The transition to BN from FT shows that it can effectively construct a realistic and accurate bridge fall accident risk assessment model.

From the FT conversion to BN method, one can get a risk assessment method. But after the conversion of BN, one still needs expert knowledge (subjective probability) input basic management reasons (root node of BN) to obtain conditional probability table. According to the data provided by different experts and knowledge, it will affect the BN in the accuracy of calculation and the quality of the evaluation. In the future, research should pay more attention to expert inspiration.

In addition, BN can be learned from the original data, with the objective BN architecture and parameters, one can use this method to establish and explore. In the future, different types of occupational hazards can be explored on bridge projects, such as electricity, objects flying and collapsing, etc. Due to the different risk factors, the structure of the BN model will be different. The BN model can provide safety assessment diagnosis of bridge engineering project, strengthen the site management and safety equipment, and reduce the bridge project fall occupational disaster risk in the field site.

Acknowledgements. This research was supported by Basic Science Research Program through the Ministry of Science and Technology (NSC102-2221-E-507-007), Thanks to the Ministry of Science and Technology funding, the research process can be successfully implemented. We also thank the experts in the field of bridge engineering for providing valuable experience and recommendations for this study.

Appendix 1: Checklist of Prior Probabilities for Root Node of BN

HS1									
Type	8 options		Parameters		Upper limit	Lower limit	State probability		
	Score	Mean	Alpha	Beta	X1	X2	Low	Medium	High
Beta	8	0.164	0.630	3.210	0.413	0.587	0.90	0.07	0.03
Beta	7	0.248	1.379	4.180	0.413	0.587	0.83	0.13	0.04
Beta	6	0.332	2.268	4.564	0.413	0.587	0.70	0.22	0.08
Beta	5	0.416	3.138	4.420	0.413	0.587	0.51	0.32	0.17
Normal	4	0.500			0.413	0.587	0.30	0.40	0.30
Beta	3						0.17	0.32	0.51
Beta	2						0.08	0.22	0.70
Beta	1						0.04	0.13	0.83
Beta	0						0.03	0.07	0.90
HS2, HS3									
Type	10 options		Parameters		Upper limit	Lower limit	State probability		
	Score	Mean	Alpha	Beta	X1	X2	Low	Medium	High
Beta	10	0.164	0.630	3.210	0.413	0.587	0.90	0.07	0.03
Beta	9	0.231	1.217	4.056	0.413	0.587	0.85	0.11	0.04
Beta	8	0.298	1.905	4.482	0.413	0.587	0.76	0.18	0.06
Beta	7	0.366	2.633	4.564	0.413	0.587	0.63	0.26	0.11
Beta	6	0.433	3.332	4.361	0.413	0.587	0.47	0.34	0.19
Normal	5	0.500			0.413	0.587	0.30	0.40	0.30
Beta	4						0.19	0.34	0.47
Beta	3						0.11	0.26	0.63
Beta	2						0.06	0.18	0.76
Beta	1						0.04	0.11	0.85
Beta	0						0.03	0.07	0.90
HS4									
Type	16 options		Parameters		Upper limit	Lower limit	State probability		
	Score	Mean	Alpha	Beta	X1	X2	Low	Medium	High
Beta	16	0.164	0.630	3.210	0.413	0.587	0.90	0.07	0.03
Beta	15	0.206	0.988	3.800	0.413	0.587	0.870	0.096	0.034
Beta	14	0.248	1.379	4.180	0.413	0.587	0.83	0.13	0.04
Beta	13	0.290	1.829	4.460	0.413	0.587	0.77	0.17	0.06
Beta	12	0.332	2.268	4.564	0.413	0.587	0.70	0.22	0.08
Beta	11	0.374	2.730	4.550	0.413	0.587	0.61	0.27	0.12
Beta	10	0.416	3.138	4.420	0.413	0.587	0.51	0.32	0.17
Beta	9	0.458	3.573	4.203	0.413	0.587	0.41	0.35	0.24
Normal	8	0.500			0.413	0.587	0.30	0.40	0.30
Beta	7						0.24	0.35	0.41

(*continued*)

(*continued*)

HS4									
Type	16 options		Parameters		Upper limit	Lower limit	State probability		
	Score	Mean	Alpha	Beta	X1	X2	Low	Medium	High
Beta	6						0.17	0.32	0.51
Beta	5						0.12	0.27	0.61
Beta	4						0.08	0.22	0.70
Beta	3						0.06	0.17	0.77
Beta	2						0.04	0.13	0.83
Beta	1						0.034	0.10	0.87
Beta	0						0.03	0.07	0.90

Appendix 2: Checklist of Construction Safety Performances

Project:	Date of inspection:
Category (HS1) health and safety training	
Inspection items 1. Holding general H/S training workshop 2. Daily education training before workers enter into job sites 3. Holding H/S training for special operation workers 4. Popularize workers training and keep the training record contact 5. Workers understand and are familiar with H/S regulation and practice 6. Workers are fully aware of the consequences of breaking H/S rules 7. Workers are able to comply with H/S codes of conduct 8. Workers are able to perform their works based on standard operating procedure	
Category (HS2) environmental maintenance	
Inspection items 1. Materials are stacked and organized 2. Job sites are clean and have no water pool 3. Workers are familiar with operating environment 4. Completed construction moving path 5. Clear indication on job sites 6. Good lighting and construction moving path 7. Height over 1.5 m and with hoist device 8. Clean up waste on time 9. Completed safety equipment 10. Functional fire-fighting facilities	
Category (HS3) health and safety planning	
Inspection items 1. Clear H/S objective and feasible policies 2. Sufficient and reasonable H/S budget 3. H/S plans and SOP are developed completely	

(*continued*)

(*continued*)

Project:	Date of inspection:
4. Materials and construction methods are in compliance with the regulation	
5. Facilities meet H/S requirement	
6. Personal protective equipment meet H/S standard	
7. Risk assessments conducted before high risk operation	
8. Materials are in place, not causing problems while construction	
9. Exact planning of construction moving path	
10. Completed emergency response and medical care plan	
Category (HS4) health and safety management	
Inspection items	
1. H/S organization develops in accordance with H/S rules	
2. Site access control	
3. Auto check mechanism	
4. Regular workplace inspection	
5. Improvement and tracking data	
6. H/S management records	
7. Workers use helmets and protective equipment	
8. Construction scaffolds are set in right place	
9. A-type ladder meets the standard	
10. Protective measures are taken in open part space on the job site such as safety net	
11. Proper approach is taken to prevent objects from falling	
12. The pitch and strength of support frame meets the construction code	
13. Construction machinery has been inspected and meets the requirement	
14. Wires on wet ground are elevated	
15. Installation of leakage circuit breakers	
16. Reward and punishment system are developed	

References

Chang, Y.H.: Application of Bayesian networks to evaluation and prognosis of process performance—taking the cathode of CNT-BLU for example. Master's thesis, Institute of Industrial Engineering and Business Information, Donghai University, Taichung, Taiwan (2007)

Chen, C.T.: Transforming fault tree analysis into Bayesian network-case study: intake point of hydraulic power system. Master's thesis, Institute of Construction Engineering, National Taiwan University of Science and Technology, Taipei, Taiwan (2010)

Chen, F.M.: The risk analysis of the building retained excavation accident. Master's thesis, Institute of Civil Engineering and Disaster Prevention and Technology, National Kaohsiung University of Science and Technology, Kaohsiung, Taiwan (2005)

Chen, T.T., Leu, S.S.: Fall risk assessment of cantilever bridge projects using Bayesian network. Saf. Sci. **70**, 161–171 (2014)

Chen, T.T., Wang, C.H.: Fall risk assessment of bridge construction using Bayesian network transferring from fault tree analysis. J. Civ. Eng. Manage. **23**(2), 273–282 (2015)

Chong, H.G., Walley, W.J.: Rule-based versus probabilistic approaches to the diagnosis of faults in wastewater treatment processes. Artif. Intell. **1**, 265–273 (1996)

Doguc, O., Ramirez-Marquez, J.E.: An efficient fault diagnosis method for complex system reliability. In: Proceedings of the 7th Annual Conference on System Engineering Research 2009 (CSER 2009), Session 12—Systems Assessment & Assurance (2009)
Franke, U., Flores, W.R., Johnson, P.: Enterprise architecture dependency analysis using fault trees and Bayesian networks. In: Proceedings of the 42nd Annual Simulation Symposium (ANSS 2009), 22–27 Mar 2009, San Diego, CA, USA, pp. 209–216 (2009)
Graves, T.L., Hamada, M.S., Klamann, R., Koehler, A., Martz, H.F.: A fully Bayesian approach for combining multi-level information in multi-state fault tree quantification. J. Reliab. Eng. Syst. Saf. **92**(10), 1476–1483 (2007)
Griffin, C.W.: Fault tree as a safety optimization design tool. IEEE Trans. Electron Devices, pp. 64–83 (1973)
Hadipriono, F.C., Wang, H.K.: Analysis of causes of falsework failures in concrete structures. J. Constr. Eng. ASCE **112**(4), 120–131 (1986)
Heckerman, D., Breese, J., Rommelse, K.: Decision-theoretic troubleshooting. Commun. ACM **38**, 49–56 (1995)
Huang, C.R.: Application of Bayesian network in establishing fall type accident risk assessment model. Master's thesis, Institute of Construction Engineering, National Taiwan University of Science and Technology, Taipei, Taiwan (2009)
Khakzad, N., Khan, F., Amyotte, P.: Safety analysis in process facilities: comparison of fault tree and Bayesian network approaches. Reliab. Eng. Syst. Saf. **96**(8), 925–932 (2011)
Li, J.C., Tao, J.Y., Hu, J.Q., Wen, X.S.: Intelligent fault diagnosis method based on Bayesian network. J. Chin. Inertial Technol. **10**(4), 32–40 (2002)
Li, J.C., Tao, J.Y., Hu, J.Q., Wen, X.S.: Bayesian network construction method and fault diagnosis application based on fault tree. Comput. Eng. Appl. **39**(24), 13–22 (2003)
Millet, R.A.: Failures: how to avoid them. J. Manage. Eng. **15**(5), 32–36 (1999)
MOL: Yearbook of Labor Statistics. Institute of Labor, Occupation Safety and Health, Ministry of Labor, Taiwan. http://www.iosh.gov.tw/IAKP/ReportIndex.aspx?NID=575 (2013)
Morcos, M.S., Singh, G.: Decision support system for reliability assessment of management structures of organizations in the construction industry. Congr. Comput. Civ. Eng. Proc. **2**, 1577–1583 (1995)
Ogunyemi, O.I.: Assessing penetrating injury penetrating injury with abductive and geometric reasoning. TranmaSCAN, Ph.D. thesis, University of Pennsylvania, Philadelphia (1999)
Pan, N.F., Hadipriono, F.C., Duane, J.W.: Assessing reliability of construction systems using failure possibilities. In: Proceedings of the 4th International Conference on Engineering Computational Technology, Lisbon, Portugal (2004)
Powers, G.J., Tompkins, F.C.: Fault tree synthesis for chemical processes. AIChE J. **20**(2), 376–387 (1974)
Rodak, C., Silliman, S.: Probabilistic risk analysis and fault trees: Initial discussion of application to identification of risk at a wellhead. Adv. Water Resour. **36**, 133–145 (2012)
Xiao, L., Haijun, L., Lin, L.: Building method of diagnostic model of Bayesian networks based on fault tree. In: Proceedings of the International Society for Optical Engineering 71272C, 13 Oct 2008, Beijing, China. http://dx.doi.org//10.1117/12.806736 (2008)

Experimental Investigation for Determining Explosive Consumption and Researching Effect of Rock Fragmentation by Blasting During Shield Tunneling

Qingbin Zhang[1(✉)], Junsheng Yang[2], Congshi Wu[1], Xuemin Zhang[2], and Bailing Zhang[1]

[1] School of Civil Engineering and Architecture, Changsha University of Science and Technology, Changsha, China
cslgzqb@126.com

[2] School of Civil Engineering, Central South University, Changsha, China

Abstract. Blasting pretreatment is commonly used to break boulders before shielding to ensure shield machine tunneling safely. Considering that rock fragments after blasting should be met the special dimensional requirement of shield machine's mucking system, so the explosive consumption is usually larger than explosive charge of conventional blasting and should be determined accurately. However, explosive charge is usually determined based on empirical formula and regulation of rock fragment distribution is unknown, resulting in that the blasting effect is unsatisfactory. In order to solve these problems, model test of blasting is carried out and the following contents are studied. The contents conclude: the explosive consumption corresponding to the average size of rock fragment; the relation between explosive charge and the overburden depth; the relation between explosive charge and regulation of rock fragment distribution. There results show that: the explosive consumption of blasting at land is 5.4–6.5 times that of conventional blasting when the size of rock fragmentation after blast is less than 30 cm. The explosive consumption increase linearly with depth of overlying stratum in double logarithmic coordinates and the relation between average size of rock fragment and depth of overburden is quadratic polynomial. The influences of explosive charge and depth of overburden on average size of rock fragment after blast is different. Explosive charge is the main influence factors for average block degree when the boulder is buried in a shallower depth, but thickness of overlying stratum's influence on average block degree will exceed that of explosive consumption when the boulder' buried depth exceed a certain value.

1 Introduction

Blasting pretreatment is commonly used to break boulders during the tunneling construction in hard granite strata utilizing shield machine and there are some successful projects. For example: Shenzhen metro line 5 [1]; Fuzhou metro line 1 [2] Guangzhou metro line 6 [3], waterway engineering in Hitachi of Japan [4] and the intake water

W.-C. Cheng et al. (eds.), *Tunneling in Soft Ground, Ground Conditioning and Modification Techniques*, Sustainable Civil Infrastructures,
https://doi.org/10.1007/978-3-319-95783-8_10

tunnel of Taishan Nuclear Power Station [5], etc. The explosive charge must be accurate because the rock fragments after blasting must meet the special requirement of shield machine's mucking system. Too little charge will cause too large rock fragments and will affect badly and even destroy the mucking system. In contrast, overcharge will not only increase blasting cost but also increase grouting quantity after blasting. So the explosive consumption is fatal in blasting pretreatment. But the explosive consumption is usually determined based on empirical formula. In recent years, this problem is paid more attention by more and more scholars. Liu [6] has obtained the calculation formula of explosive consumption by theory of blasting in pretreatment of bedrock/boulders without free surface based on the intake tunnel project of Taishan nuclear station. Zheng [7] has studied the relationship between fragmentation distribution characteristic and explosive consumption by CDEM (Continuum Dis-continuum Element Method) and five factors that included the average fragmentation, ultimate fragmentation size, block non-uniform coefficient, system fracture degree and large block ratio. Liu [8] has analyzed the increment of explosive consumption between the 25–30 m water depth by blasting 10 concrete specimens at land and under water. Liu [9] has deduced that the explosive specific charge has uniform exponent relation to the blasting fragmentation distribution from theoretic. Chen [10] has obtained that the explosive specific charge will reduce double in case of the average blasting fragmentation double.

Above the research results are emphasized on the underwater blasting and obtained by experiment, so the results have its limitation and could not be used directly in the condition of boulder blasting with overburden. There are two reasons. First, the medium of overburden is different and the medium is water and rock respectively. Second, the requirement for rock fragmentation after blasting is different. The size of rock fragmentation must be less than 30 cm in order to meet the mucking system of shield machine in Taishan intake tunnel. The relationship between explosive consumption and blasting fragmentation distribution will be researched for the specific situation.

Based on project of the intake tunnel of Taishan Nuclear Power Station, model test of blasting is carried out and the following contents are studied. The contents conclude: the explosive charge corresponding to the average size of rock fragment; the relation between explosive consumption and the depth of overburden; the relation between explosive charge and regulation of rock fragment distribution.

2 Engineering Situation

The location of the intake of Taishan Nuclear Power Station is in the sea area between Yaoguzui and Dajin island. The length of tunnel is 4330.6 m and the diameter of tunnel is 9.03 m. The buried depth of tunnel is between 11 and 29 m and the span between two holes is 29.2 m. The tunnel is built by the slurry balance shield machine. The geological survey show that residual body of granite and bedrock will appear in the tunnel construction and the average strength is 100 MPa.

3 Test Parameters

3.1 Setup the Similarity of Criteria

In model test of blasting, the similarity criteria and the chosen parameters are mainly relevant to the explosive consumption and blasting rock fragmentation. The similarity criteria are setup based on material similarity, geometric similarity and blasting dynamic similarity at first and then the similarity criteria are determined by π theorem and dimensionless method. The dimensions and value of the site physical quantities in model test are shown in Table 1.

Table 1. Dimensions and value of the site physical quantities in model test

Number	Symbol	Name	Unit	Dimension	Value
1	Q	Explosive	kg	M	–
2	v	Velocity of explosive	m/s		4000
3	ρ	Density of charge	kg/m^3	ML^{-3}	1200
4	R	Charge radius	m	L	0.045
5	h	Charge loaded length	m	L	0.90
6	ρ_r	Density of rock	kg/m^3	ML^{-3}	2800
7	v_r	P-waves of rock	m/s	LT^{-1}	4000
8	σ_r	Strength of rock	N/m^2	$M/(LT^2)$	1×10^6
9	L	Size of rock(cube)	m	L	1.00
10	D	Average size of rock fragmentation	m	L	0.30
11	H	Depth of rock	m	L	–
12	ρ_c	Density of overburden	kg/m^3	ML^{-3}	2600
13	v_c	P-waves of overburden	m/s	LT^{-1}	2400

Explosive charge is influenced by Eq. 1:

$$\mathrm{Q} = \mathrm{f}(\mathrm{v}, \rho, \mathrm{R}, \mathrm{h}, \rho_r, c_r, \sigma_r, L, D, H, \sigma_c, v_c) \tag{1}$$

where Q is dependent variable and others is independent variable. The fundamental dimensions are three: v, ρ and R. Criterion of similarity is determined when the value of following dimension is 1 by π theorem and dimensionless method.

$$\pi_1 = \frac{R}{h}, \pi_2 = \frac{R}{L}, \pi_3 = \frac{R}{D}, \pi_4 = \frac{L}{H}, \pi_5 = \frac{\sigma_r}{\rho v^2}, \pi_6 = \frac{\rho v}{\rho_r v_r}, \pi_7 = \frac{\rho_r v_r}{\rho_c v_c}, \pi_8 = \frac{Q}{\rho R^3} \tag{2}$$

The equation of similarity is obtained:

$$\mathrm{f} = (\pi_1, \pi_2, \pi_3, \pi_4, \pi_5, \pi_6, \pi_7, \pi_8) \tag{3}$$

The value and meaning of Eq. 3 are shown in Table 2.

Table 2. Parameters and value of the similarity law

Parameters	Value	Meaning
π_1	0.05	Similarity ratio of charge dimension between model and practical blasting
π_2	0.045	Similarity ratio of dimension between model and actual blasting
π_3	0.15	Similarity ratio of size of rock fragmentation between model and actual blasting
π_4	0.143	Similarity ratio of geometric between model and prototype
π_5	0.0052	Similarity ratio between detonation pressure and rock strength
π_6	0.43	Similarity ratio of wave impedance between explosive and rock
π_7	1.8	Similarity ratio of wave impedance between overburden and rock
π_8	62.8	Similarity ratio of between charge weight and charge radium

3.2 Size and Material of Model

The ratio of geometric similarity is considered as 1:7, so the length of model (cube) is 15 cm. The other parameters of model are calculated as following: charge radius 0.68 cm, charge length 13.5 cm, density of explosive 1000 kg/m^3, detonation velocity of explosive 3200 m/s; test block strength 53.2 MPa, density of test block 2200 kg/m^3, P-wave speed of test block 3382 m/s; wave impedance of overburden material 4.1×10^6 kg/(m^2 s), density of overburden material 1600 kg/m^3, P-wave speed of overburden material 2562 m/s; size of blasting rock fragmentation 4.5 cm.

4 Blasting Model Test

4.1 Test Sample' Preparing and Parameters

Test sample are made referring to C50 concrete block and mix proportion must be adjusted in order to satisfy the test requirement. The mix proportion of C50 concrete block is listed in Table 3 and the distribution range of test samples parameters are shown as follow: density 2200–2300 kg/m^3, P-wave speed 3300–3500 m/s, strength 52–55 MPa.

Table 3. Mix proportion of C50 concrete sample (kg/m^3)

Cement	Sand	Gravel 5–10 mm	Gravel 10–20 mm	Fly ash	High efficiency water reduce agent	Water
376	659	468	703	94	4.7	145

4.2 Model Test

According to the different depth of overburden, there are 6 test situations in model test. The depth of overburden is 0, 30, 50, 70, 100 and 150 cm separately. Experimental phenomena and data are listed in Table 4.

Table 4. Charge weight of samples and fragmentation statistics

Sample number	Q (g)	D (cm)	Q (kg m^{-3})	H (cm)	Test site	Q_b (kg m^{-3})	Experimental phenomena
1–1	20	2.00	5.93	0	Experimental tank	3.26	–
1–2	11	4.50	3.26				
1–3	8	5.54	2.37				
2–1	12	4.63	3.56	30	Experimental tank	3.85	Collapsed cavity with a diameter of 30 cm emerged on the surface of overburden
2–2	13	4.54	3.85				
2–3	14	4.40	4.15				
3–1	14	4.67	4.15	50	Experimental tank	4.44	Annular crack with a diameter of 45 cm emerged on the surface of overburden and cavity appeared on the top of sample block
3–2	13	5.33	3.85				
3–3	15	4.48	4.44				
4–1	16	4.46	4.74	70	Stock ground	4.74	Cavity emerged around the sample block
4–2	17	4.28	5.04				
4–3	18	3.98	5.33				
5–1	15	4.85	4.44	100	Stock ground	5.04	Cavity emerged around the sample block
5–2	18	4.09	5.33				
5–3	17	4.45	5.04				
6–1	22	3.78	6.52	150	Stock ground	5.93	Cavity emerged around the sample block
6–2	20	4.65	5.93				
6–3	19	4.31	5.63				

The diameter of slurry pipeline in TBM for Taishan Water Tunnel is 40 cm, so the average blasting fragmentation shall be controlled less than 30 cm in order to meet the requirement of mucking system. According to the similarity criterion of blasting test, 30 cm in actual is corresponded to 4.5 cm in model test. The average size of fragmentation close to 4.5 cm in Table 4 is selected, and the charge weight corresponding to it is the optimum explosive charge (Table 5).

Table 5. Optimum explosive charge and average fragmentation degree under different depth of overburden

Depth of overburden (cm)	0	30	50	70	100	150
Optimum explosive charge	3.26	3.85	4.44	4.74	5.04	5.93
Average fragmentation degree	4.50	4.64	4.48	4.46	4.45	4.65

4.3 Experimental Phenomena Analysis

The conclusions are obtained from the blasting model test:

(1) The optimum explosive charge which meet the special fragmentation requirement is 3.26 kg/m^3 when the depth of overburden is 0, which is 5.4–6.5 times explosive charge on open fields.
(2) Blasting crater became collapse cavity on the surface of overburden when the overburden depth increase from 0 to two times the test tube size. When the overburden depth is the 3.3 times the test tube size, collapse cavity will disappear slowly and the circle crack on the surface of overburden will appear.
(3) When the overburden depth is 4.7 times the test cube size, blasting cannot influence the surface of overburden. But the cavity around the test cube will be produced because of the consolidated sand by explosion pressure. So in practical engineering, if the cavity are not treated or not treated well, the pressure jump will appear and cause difficulty to dwell, even will destroy stability of the tunnel face.

5 Data Analysis

5.1 Relation Between Depth of Overburden and Optimum Explosive Charge

According to the data in Table 5, the relation between the two factors is obtained and the fitting formula is shown as Formula 4.

$$y = 1.7368x + 3.3855 \tag{4}$$

where y is the optimum explosive charge, x is the depth of overburden.

From Formula 4, the result is that the relation between the two factors was linear.

5.2 Relation Between Explosive Charge and Average Fragmentation Size

Extracting the data of optimum explosive charge and average fragmentation size in Table 5, scatter diagram is drafted as Fig. 1. The data excluding the minimum and the maximum value of explosive charge is almost linear, and the preserved data is fitted in double logarithm coordinate (Fig. 2).

The fitting formula between optimum explosive charge and average fragmentation size can be shown as Formula 5, and the relative coefficient is 0.987.

$$d = 10^{0.70149} \times q^{-0.077} \tag{5}$$

where d is the average fragmentation size, q is the optimum explosive charge. The formula shows that the relation between optimum explosive charge and average fragmentation size is decaying power function. In other words the average fragmentation size will decrease with the increase of explosive charge. For example, the average fragmentation size will decrease 0.8 cm with the explosive charge increase 1 kg/m^3.

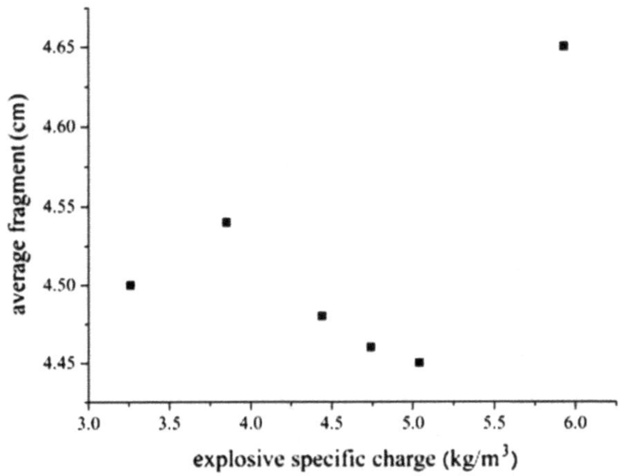

Fig. 1. Scatter diagram between optimum explosive charge and average fragmentation size

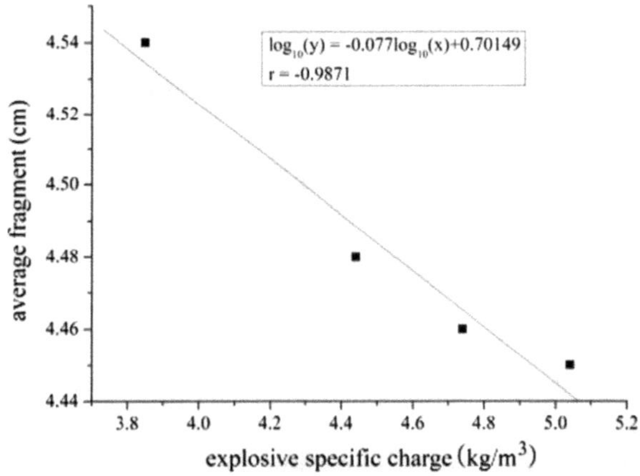

Fig. 2. Fitting curve between optimum explosive charge and average fragmentation size in double logarithm coordinate

The explosive charge data corresponding to the minimum and maximum depth of overburden is rejected because the effect of depth of overburden. According to the blasting model test, the explosive charge is more than 3 times that on open field, in other words, the explosive charge is less on open field than the blasting under overburden in order to meet the requirement of special fragmentation size. When the depth of overburden is more than 150 cm, the average fragmentation size is not decrease with the increase of explosive charge because of the clamping action and energy dissipation caused by overburden.

5.3 Relation Between Depth of Overburden and Average Fragmentation Size

The depth of overburden can influence the average fragmentation size, so the relation between the two factors is analyzed excluding the depth of overburden is zero and 150 cm (Fig. 3).

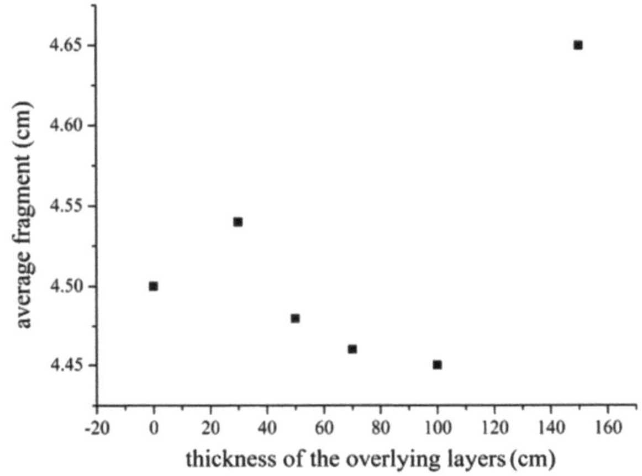

Fig. 3. Scatter diagram between depth of overburden and average fragmentation size

The fitting formula is shown when the overburden depth is 30–100 cm.

$$y = 4.65779 - 0.00482x + 2.75168E^{-5}x^2 \quad (6)$$

where y is the average fragmentation size, x is the overburden depth, the relative coefficient is 0.9939. According to the fitting formula and the fitting curve, the conclusion can be obtained that the average fragmentation size decrease with increase of the depth overburden when the overburden depth is less than 90 cm, but the average fragmentation size increase with increase of the overburden depth when overburden depth is more than 90 cm (Fig. 4).

The average fragmentation size is influenced mostly by explosive charge than that of overburden depth when the overburden depth is between 30 and 90 cm. But the influence degree is reducing with increase of the overburden depth. The variable quantity of average fragmentation size corresponding to the overburden depth is between 30 and 50 cm is 1.725 times that corresponding to the depth overburden is between 50 and 70 cm, Meanwhile, the variable quantity of average fragmentation size corresponding to the overburden depth is between 50 and 70 cm is 3.638 times that corresponding to the overburden depth is between 70 and 90 cm. When the overburden depth is more than 90 cm, the average fragmentation cannot meet test requirement if the explosive charge is increased only because of the influence of overburden.

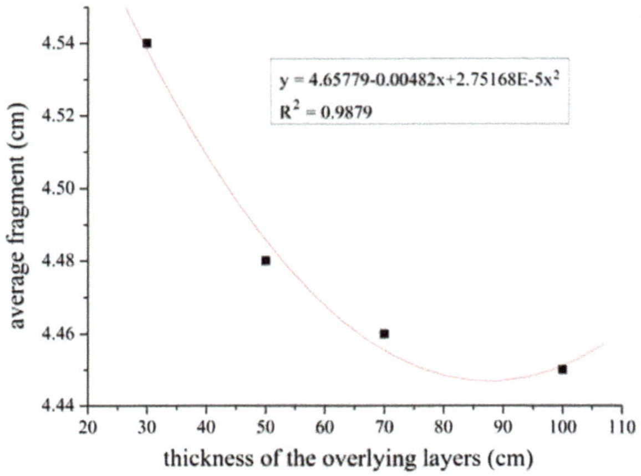

Fig. 4. Fitting curve between depth of overburden and average fragmentation size

5.4 Relation Among Explosive Charge, Overburden Depth and Average Fragmentation Size

The average fragmentation size is influenced differently by explosive charge and overburden depth, meanwhile the depth of overburden and the explosive charge interacted. So it is necessary to analysis the relation among the three factors. The data

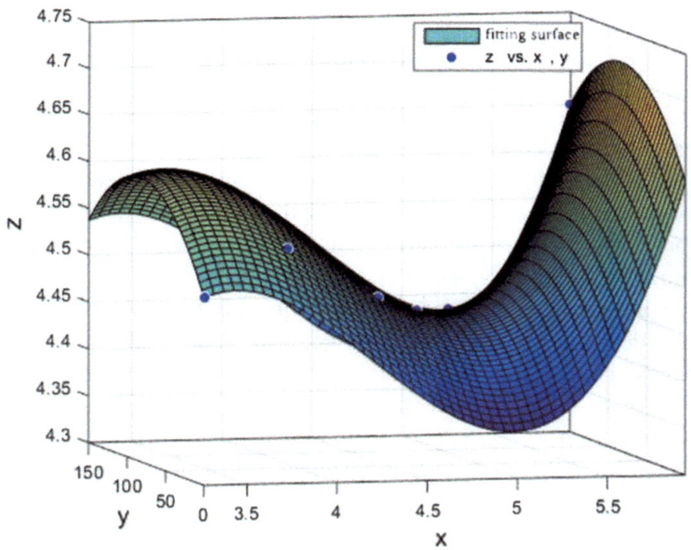

Fig. 5. Relation among explosive charge, overburden depth and average fragmentation size

in Table 5 is fitted by MATLAB and the 3-d fitting chart and fitting formula can be obtained (Fig. 5).

$$z_1 = -1.434 + 4.491x_1 - 1.105x_1^2 + 0.0876x_1^3 + 0.0025y_1 - 1.483E^{-5}y_1^2 \quad (7)$$

where x is the explosive charge, y is the overburden depth, z is the average fragmentation size. The relative coefficient of formula is 0.9989.

The data of explosive charge and depth of overburden is distributing near the intersection of the diagonal surface and the fitted surface because of the requirement of optimum average fragmentation size. The explosive charge with overburden is more than that without overburden in order to meet the requirement of special fragmentation size. With the increase of explosive charge and overburden depth, the trend of average fragmentation size approaching the optimum is mitigating, which means that the average fragmentation is affected more seriously than that of the overburden depth. But when the overburden depth is more than a certain value, the trend of average fragmentation size away from the optimum value (4.5 cm) is increasing although the explosive charge is increasing too, which means that overburden depth affects the average fragmentation size more seriously than that of explosive charge. The result is uniform to the aforementioned conclusion. That is to say, in practical engineering the explosive charge cannot be increased only to meet the special fragment requirement of mucking system of TBM when the overburden depth is more than a certain value, and measures including change the type of explosive or adjusting spacing pattern parameter shall be considered.

5.5 The Formula of Explosive Charge in Practice

The explosive charge is usually determined based on empirical formula or referring to empirical formula of explosive consumption in underwater blasting. The explosive charge empirical formula of Swede is the most commonly used (Formula 8).

$$q = q_1 + q_2 + q_3 + q_4 \quad (8)$$

where q_1 is the basic explosive consumption, and usually is two times that of open field blasting, q_2 is the increments of water pressure above blasting area ($q_2 = 0.01h_2$, h_2 is the water depth), q_3 is the increments of overburden depth ($q_3 = 0.02h_3, h_3$ is the overburden depth), q_4 is the increments of rock expansion ($q_4 = 0.03h$, h is the gradient section).

The effect of water depth and overburden is taken in count, but the special requirement of fragmentation size is not consider. So it is not used directly in practical engineering.

According to Formulas 4 and 7, the relation among the optimum explosive charge, the average fragmentation size and the overburden depth is assumed as Formula 9.

$$z = ax + by^{-\frac{5}{4}} + c \quad (9)$$

Combining the data from Table 5 and Formula 9, the Formula 9 is turned as Formula 10.

$$z = 1.76x + 13.83y^{-\frac{5}{4}} + 1.276 \tag{10}$$

where, z is the optimum explosive charge (kg/m^3), x is the overburden depth (m), y is the average fragmentation size (cm), and the relative coefficient is 0.9919.

Based on the test data and similarity criterion, the explosive charge can be calculated in practical engineering (Table 6) and the relation among overburden depth, average fragmentation size and explosive charge in practical engineering can be derived.

$$z = 0.2631x + 1.885y^{-\frac{3}{2}} + 3.356 \tag{11}$$

where meaning of x, y and z is same as Formula 10. The relative coefficient is 0.9919.

Table 6. Explosive charges of model test and in practical engineering separately

Overburden depth of model test (m)	0.0	0.3	0.5	0.7	1.0	1.5
Overburden depth in practical engineering (m)	0.0	2.1	3.5	4.9	7.0	10.5
Average fragmentation size of model test (cm)	4.5	4.64	4.48	4.46	4.45	4.65
Average fragmentation size in practical engineering (cm)	31.50	32.48	31.36	31.22	31.15	32.55
Optimum explosive charge (kg/m^3)	3.26	3.85	4.44	4.74	5.04	5.93

5.6 Comparison and Verify

In order to verify whether the Formula 11 is right, the formula is contrasted with the Formula 4. In the same time, the empirical formula of Swede is considered in the comparison too and the results are shown in Table 7.

Table 7. Explosive charge's comparison among the Swede empirical formula, practical engineering and fitting Formula (11)

Overburden depth		0.0	5.0	10.0	15.0	20.0	25.0
Explosive charge (kg/m^3)	Practical engineering	3.39	4.63	5.87	7.11	8.25	9.59
	Fitting formula	3.52	4.77	6.03	7.29	8.54	9.80
	Swede empirical formula	1.10	1.37	1.52	1.67	1.82	1.97
Difference between the fitting formula and practical engineering		1.04	1.03	1.03	1.03	1.04	1.02
Ratio of the fitting formula and Swede empirical formula		3.20	3.48	3.97	4.37	4.69	4.97

The explosive charge's difference between practical engineering and fitting formula is small, which verified the fitting formula is credible. Another, the explosive charge of fitting formula is larger than that of Swede empirical formula and the former are 3.2–4.97 times the latter. The reason is the requirement of fragment size in Swede empirical formula is taking mining and transportation as standard, which is larger than the special requirement of fragmentation size (30 cm) in practical engineering.

6 Conclusions

In this paper, blasting model test is carried out based on the special requirement of rock fragment in boulder's blasting under overburden, and the relation among the explosive charge, average fragmentation size and overburden depth is studied. The conclusions are obtained as following.

(1) The explosive charge at land is 5.4–6.5 times that of conventional blasting in order to meet the special requirement of rock fragment that the size must be less than 30 cm.
(2) The relation between the explosive charge and the average fragmentation size is liner in double logarithm coordinate. The relation between the overburden depth and the average fragmentation size is quadratic polynomial.
(3) The influence of explosive charge and overburden depth to average fragmentation size is different. When the overburden depth is small, the average fragmentation size is influenced mostly by the explosive charge, but when the overburden depth exceeds a certain value, the average fragmentation size is influenced mainly by the overburden depth. So in practical engineering, measure shall be considered including change the type of explosive and adjusting the spacing pattern parameter when the overburden depth is exceed a certain value.
(4) The relation among the explosive charge, the average fragmentation size and the overburden depth can be referenced by other similar projects.

References

1. Liu, J.: Research and Practice on Shield Tunnel of Shenzhen Metra. China Communications Press, Beijing (2011)
2. Liu, H., et al.: Application of microtremor method to boulders detection in urban metro shield construction: case study of Fuzhou Metro line no. 1. Tunn. Constr. (2016). https://doi.org/10.3973/j.issn.1672-741X.2016.12.014
3. Zhong, C., et al.: On construction risks of a shield machine operated in weathered granite stratum and its countermeasures. Mod. Tunn. Technol. (2013). https://doi.org/10.3969/j.issn.1009-6582.2013.03.003
4. Li, D.: (Trans.).: Blasting rock without free surface. Blasting 4, 66–69 (1990)
5. Zhang, Q., et al.: Experimental investigation for determining explosive consumption by blasting to pretreating boulders in the ground. Explos. Shock Waves (2016). https://doi.org/10.11883/1001-1455(2016)05-0695-08

6. Liu, H.: Study on determination of unit consumption of explosives in pretreatment of bedrock/boulders along alignment of shield-bored tunnel without free surface. Tunn. Constr. (2015). https://doi.org/10.3973/j.issn.1672-741X.2015.03.004
7. Zheng, B., et al.: Numerical study on relationship between specific charge and fragmentation distribution of hematite. Blasting (2015). https://doi.org/10.3963/j.issn.1001-487X.2015.03.011
8. Liu, M., et al.: Experimental analysis on specific charge of underwater explosion of concrete. Blasting (2007). https://doi.org/10.3963/j.issn.1001-487X.2007.01.003
9. Liu, H., Feng, S.: Theoretical research of the effect on the blasting fragmentation distribution from the explosive specific charge. Explos. Shock Waves **17**(4), 359–362 (1997)
10. Chen, Y.: Effect on the explosive specific charge considering blasting fragmentation. Blasting **13**(3), 19–22 (1996)
11. Herrenknecht.: The Operating Instructions of TBM for Taishan Water Tunnel. (2010)

A Web-Based Ground Settlement Prediction System for Subway Construction

Bo Liu[1,2(✉)], Weihong Yang[1], Lei Gao[1], and Yixin Wang[1]

[1] School of Mechanics and Civil Engineering, China University of Mining and Technology (Beijing), Beijing 100083, China
dr_boliu@163.com

[2] State Key Laboratory of Geomechanics and Deep Underground Engineering, Beijing 100083, China

Abstract. The development of a web-based ground settlement prediction system for subway construction is presented. The system is developed based on STEAD, and it is named STEAD-Web. STEAD is a ground settlement prediction system developed in China earlier. It is short for Subway Tunneling-induced ground-Environment-damage Assessment and control Design system. STEAD works well in predicting ground settlement. Unfortunately, it works off-line and it is not so convenient to deal with massive measured data every day. STEAD-Web is an upgraded version of STEAD, which runs in Browser/Server mode. The server automatically analyzes the monitoring data and makes a prediction by a back-analysis method using the real-time measured data. Users can access the results through web browsers from anywhere after authorization. Finally, the paper presents two case applications of the new system in Beijing subway construction.

1 Introduction

Subway construction may cause serious damage to existing structures. Therefore, one of the most important issues of subway construction in urban areas is the estimation of ground movements, so that the possible damage level of adjacent buildings due to subway construction can be assessed.

The safety evaluation of subway construction in China is usually made by the accumulated ground movements and the rate of ground movements. A large number of observation points are being monitored every day. In order to deal with the massive measured data, many attempts have been made to build an information system (Ding and Zhou 2013; Li and Zhu 2013). However, most of them are aimed at inspecting and manage the measured data. In this paper, the development of a ground settlement prediction system for subway construction is presented, which is an upgraded version of STEAD (Liu and Han 2006; Liu and Li 2011) and is web-based.

W.-C. Cheng et al. (eds.), *Tunneling in Soft Ground, Ground Conditioning and Modification Techniques*, Sustainable Civil Infrastructures,
https://doi.org/10.1007/978-3-319-95783-8_11

2 Settlement Predicting Methods

2.1 Peck Formula for Settlement Prediction (Empirical Prediction Method)

In engineering practice, ground surface settlements are often described by empirical formulae. The transverse settlement trough can be described by the normal probability density function (Peck 1969). The longitudinal settlement trough can be described by the normal cumulative distribution function, so that the three-dimensional ground movements can be estimated using Eqs. (1)–(5) (Attewell and Woodman 1982).

$$w(x,y,z) = \frac{V_{gl}}{i_z\sqrt{2\pi}}\exp\left(-\frac{x^2}{2i_z^2}\right)\left[\Phi\left(\frac{y-y_i}{i_z}\right) - \Phi\left(\frac{y-y_f}{i_z}\right)\right] \tag{1}$$

$$v(x,y,z) = \frac{-n}{z_0 - z}x \cdot w(x,y,z) \tag{2}$$

$$u(x,y,z) = \frac{nV_{gl}}{2\pi(z_0 - z)}\exp\left(-\frac{x^2}{2i_z^2}\right)\left[\exp\left(-\frac{(y-y_i)^2}{2i_z^2}\right) - \exp\left(-\frac{(y-y_f)^2}{2i_z^2}\right)\right] \tag{3}$$

$$\Phi(x) = \frac{1}{\sqrt{2\pi}}\int_{-\infty}^{x}\exp\left(-\frac{\alpha^2}{2}\right)d\alpha \tag{4}$$

$$i_z = K(z_0 - z)^n \tag{5}$$

In the above equations, w is the vertical ground displacement or the ground settlement, v and u are the horizontal displacements. V_{gl} is the ground loss per unit length. Φ is cumulative function of the standard normal distribution. i_{z} is transverse horizontal distance between the points of maximum settlement and inflexion at depth z. z_0 is the depth of the tunnel center line. K and n are empirical parameters. y_{i} is the y-coordinate of the starting points. y_{f} is the y-coordinate of the current points.

2.2 Stochastic Medium Theory for Settlement Prediction

Stochastic medium theory was firstly introduced by Polish scholar (Litwiniszyn 1957), and then was improved by Chinese scholar (Liu 1993). It is assumed that the underground excavation can be divided into infinitesimal excavation elements, and the ground movement induced by the excavation equals the sum of all the elemental excavations. The three-dimensional ground movements can be estimated using Eqs. (6) and (7).

$$
\begin{aligned}
W(x,y) = & \int_{y_i}^{y_f} \int_{H-a}^{H+a} \int_{-\sqrt{a^2-(H-\eta)^2}}^{\sqrt{a^2-(H-\eta)^2}} We(x,y) d\varsigma d\eta d\xi \\
& - \int_{y_i}^{y_f} \int_{H-b}^{H+b} \int_{-\sqrt{b^2-(H-\eta)^2}}^{\sqrt{b^2-(H-\eta)^2}} We(x,y) d\varsigma d\eta d\xi
\end{aligned}
\tag{6}
$$

$$
We(x,y) = \frac{\tan^2\beta}{\eta^2} \exp\left\{ -\frac{\pi \tan^2\beta}{\eta^2} \left[(x-\xi)^2 + (y-\varsigma)^2 \right] \right\} \tag{7}
$$

In the above equations, W is ground surface settlement, We is settlement of elemental excavation. a and b are the initial radius of the tunnel excavation and convergence radius after the effective support; H represents the tunnel depth; β is the angle of main influence sphere.

3 Development of STEAD-Web

3.1 System Architecture

STEAD-Web is developed based by STEAD (Liu and Han 2006) and STEAD-Risk (Liu and Li 2011). STEAD-Web is programmed in C# and Javascript, which is a web based application. The system can be divided into three parts: the monitoring side, the server side and the user side (Fig. 1).

The monitoring side includes automatic data acquisition service and data upload service. Both automatic monitoring data and manual monitoring data can be uploaded to the platform.

The server side includes map service and data analysis service. The real-time monitoring data can be displayed on a map. The system adopts the Baidu Map API as the map service engine, which is light-weighted and more flexible than Geographic Information System (GIS). The system reads off data from the database and the prediction is made by a back-analysis procedure. Both empirical method and stochastic medium theory method can be employed in the back-analysis procedure.

The system runs in Browser/Server mode. On the user side, authorized users can access the data through web browsers without installation of any apps, or they can also install the user client to get a better experience.

3.2 A Data Model for Subway Construction

The system needs a database to save and manage data. The database contains several tables to store various information (Fig. 2). The table Lines is used to the subway line names and the line ID. Every subway line contains many projects. The table Projects is used to store the project name and which subway line it belongs to. The table Points contains the basic information and the current status of the observation points. The

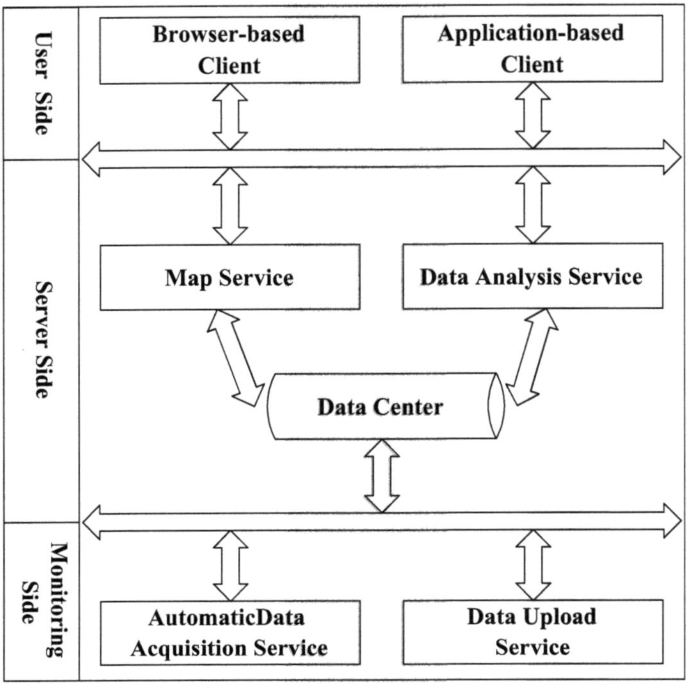

Fig. 1. Architecture for the web-based ground settlement prediction system

observation points are of different kinds, such as ground surface settlement, inclination of building, crown settlement etc. The table Types is used to store types of observation points. The table Status stores the security classification of the observation points. The table Records stores every single record of the observation points.

4 Implementation of STEAD-Web

4.1 Application of Hydrostatic Level Gauges

The monitoring data are obtained both by electronic sensor and manual work. The hydrostatic level gauges are applied on the worksite of Liaogongzhuang Station of Beijing Subway Line 6 (Fig. 3). The hydrostatic level gauges are connected to data acquisition box. The data are acquired by STEAD Data Acquisition Client and then uploaded to the web platform (Fig. 4). A backup copy of the measured data will be stored in the worksite host, in case the network is not well.

Liaogongzhang station is excavated by PBA method. These automatic observation points are installed right over the station. The ground surface is stable after the top arch construction (Fig. 5b).

The installation of the automatic observation points is more complicated than the manual observation points, so that the automatic monitoring is currently applied where

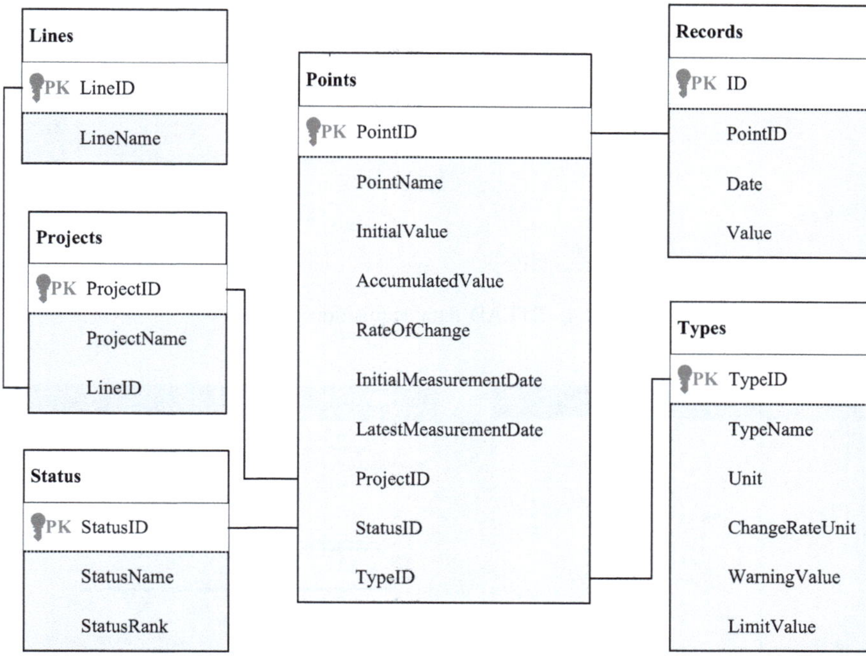

Fig. 2. Database of the web-based ground settlement prediction system

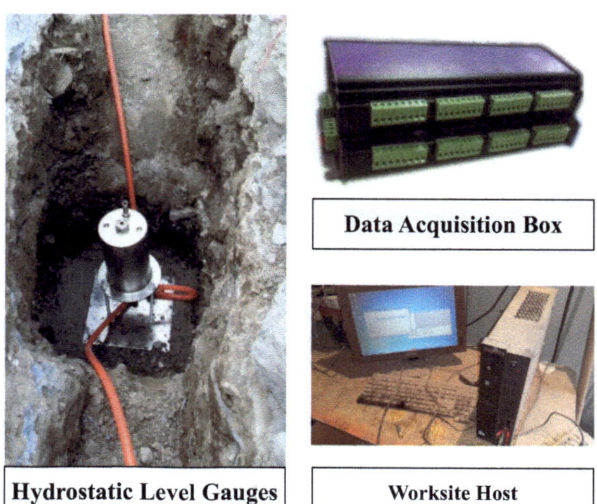

Fig. 3. Automatic data acquisition system

long-time monitoring is needed, such as subway stations and vertical shafts. As for the intersectional tunnel, the construction is going very fast and the automatic monitoring method is not practical.

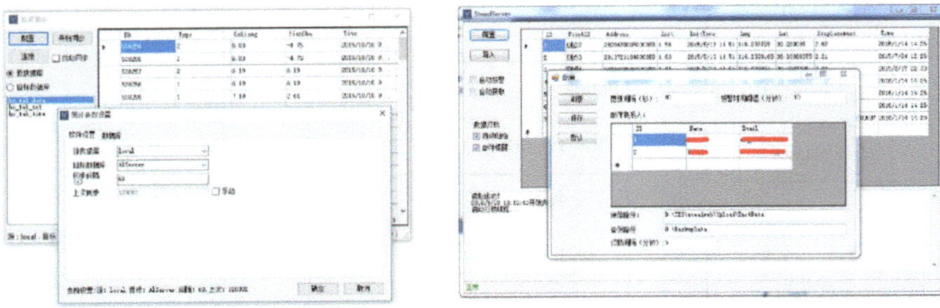

Fig. 4. STEAD data acquisition client

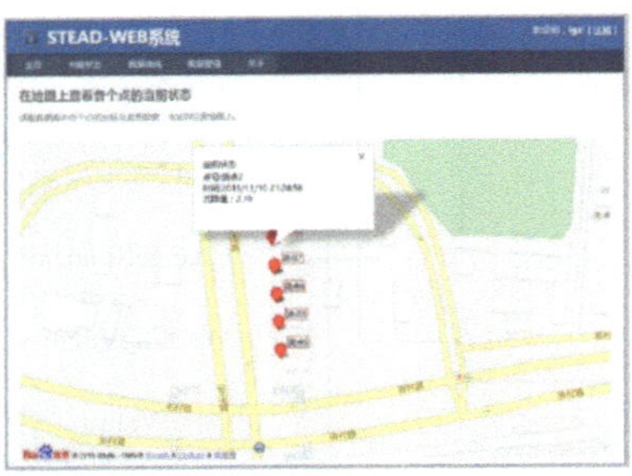

(a) Status of the observation points

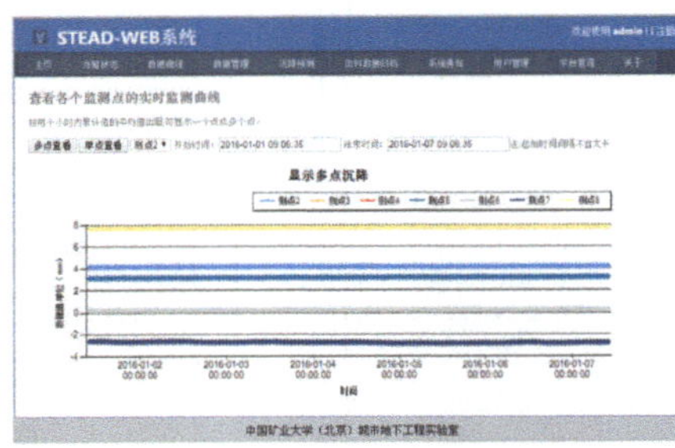

(b) Ground surface settlement vs time

(c) Measurement data upload interface

(d) Historical measurement data

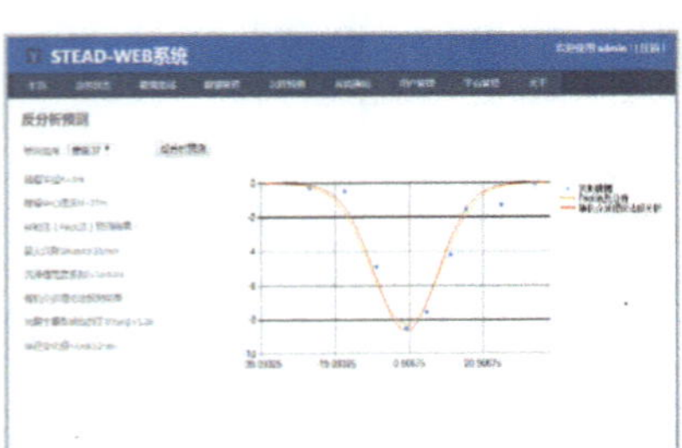

(e) Back-analysis results

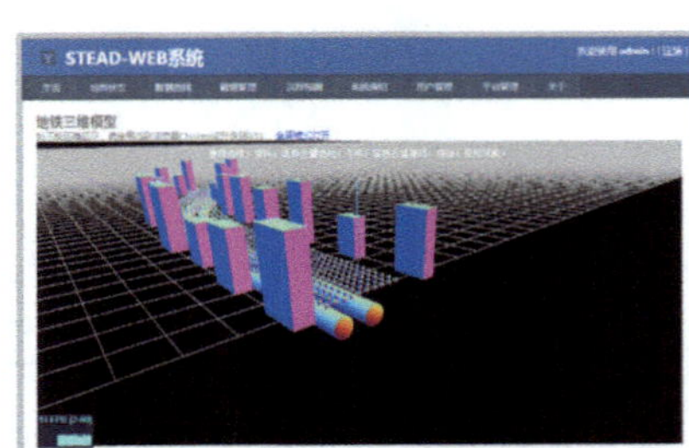

(f) 3D view of the subway model

Fig. 5. Graphical user interface of STEAD-web

4.2 Automatic Back Analysis

The Tianqiao-Yongdingmenwai Project of Beijng Subway Line 8 is constructed by shield tunneling method. The monitoring data of several cross sections of the tunnel are specified as the back-analysis data sample. A server program is running on the server host, which automatically reads and analyzes the monitoring data. The results of the back-analysis work are stored in the database. Authorized user can access the monitoring data and the back-analysis results through the web platform. The predicted data by back- analysis method is compared with the measured data (Fig. 5e).

5 Conclusions

The development of a web-based ground settlement prediction system for subway construction is presented in this study. The architecture and the database model are described in the paper. The system is successfully applied to Beijing Subway Line 6 and Line 8. The monitoring data and predicted data are compared on the web platform, which verified the rationality and applicability of the prediction system.

Acknowledgements. This research was supported by National Natural Science Foundation of China (No. 41472259, No. 51374209), 13th Five Year National Science and Technology Major Project (No. 2016YFC0802505)

References

Attewell, P.B., Woodman, J.P.: Predicting the dynamics of ground settlement and its derivitives caused by tunnelling in soil. Ground Eng. **15**(8) (1982)

Ding, L.Y., Zhou, C.: Real-time safety early warning system for cross passage construction in Yangtze Riverbed Metro Tunnel based on the internet of things. Autom. Constr. **36**, 25–37 (2013). https://doi.org/10.1016/j.autcon.2013.08.017

Li, X., Zhu, H.: Development of a web-based information system for shield tunnel construction projects. Tunn. Undergr. Space Technol. **37**, 146–156 (2013). https://doi.org/10.1016/j.tust.2013.04.002

Litwiniszyn, J.: Fundamental principles of the mechanics of stochastic medium. In: Proceedings of 3rd Conference on Theoretical and Applied Mechanics, Bongalore, India (1957)

Liu, B.C.: Ground surface movement due to underground excavation in the PR China. Compr. Rock Eng. **4**, 780–816 (1993). https://doi.org/10.1016/0148-9062(94)91319-6

Liu, B., Han, Y.: A FLAC3D-based subway tunneling-induced ground settlement prediction system developed in China. In: Proceedings of the 4th International FLAC Symposium on Numerical Modeling in Geomechanics. Hart & Varona, Madrid (2006). ISBN0-9767577-02

Liu, B., Li, Y.: A tunneling-induced stratum settlement assessment and prediction system: STEAD-RISK, Hunan, China. Am. Soc. Civil Eng. (ASCE) (2011). https://doi.org/10.1061/47632(411)1

Peck, R. B.: Deep excavations and tunnelling in soft ground. In: Proceedings of 7th International Conference on SMFE, pp. 225–290

Implementation of Highly Flowable Strain Hardening Fiber Reinforced Concrete (HF-SHFRC) to New RC Bridge Columns for Sustainability Development

Wen-Cheng Liao(✉) and Chih-Chiang Yeh

Department of Civil Engineering, National Taiwan University, Taipei, Taiwan
wcliao@ntu.edu.tw

Abstract. A remarkable development of high strength concrete and reinforcement has been achieved nowadays. The purpose of New RC project is aimed to reduce member section size by using high strength concrete ($f'_c > 70$ MPa) and high strength rebars ($f_y > 685$ MPa). Material consumption can be further reduced owing to the upgrade of strength. However, the nature of brittleness of high strength concrete may also cause early cover spalling and other ductility issues. Addition of steel fibers is an alternative as transverse reinforcement in New RC infrastructure systems. Highly flowable strain hardening fiber reinforced concrete (HF-SHFRC) has excellent workability in the fresh state and exhibits the strain-hardening and multiple cracking characteristics of high performance fiber reinforced cementitious composites (HPFRCC) in their hardened state. This study aims to investigate the cyclic behavior of New RC bridge columns made of HF-SHFRC. Five large scale bridge columns are subjected cyclic lateral loading to verify their responses and deformation capacity. The test results show that by adding 1.5% of high strength hooked steel fibers, great deformation capacity is developed either stirrups spacing is even increased to two times of that of control specimen or elimination of all tires in New RC bridge columns. Implementation of HF-SHFRC to New RC infrastructure offers opportunities to significantly simplify the design and construction of members for sustainable urbanization, while ensuring adequate ductility and damage tolerance.

1 Introduction

The purpose of New RC project is aimed to reduce member section size by using high strength concrete ($f'_c > 70$ MPa) and high strength rebars ($f_y > 685$ MPa). It can be expected that not only structure life is lengthened, but life cycle cost is reduced by implementation of New RC project in infrastructure systems. However, the excessive brittleness associated with high strength concrete has long been a primary concern.

In order to improve this weakness, many studies have suggested adding fibers to high strength concrete to increase its toughness and ductility. Once a crack has formed, fibers inhibit the growth of the crack and only in the extreme scenario that all fibers

W.-C. Cheng et al. (eds.), *Tunneling in Soft Ground, Ground Conditioning and Modification Techniques*, Sustainable Civil Infrastructures,
https://doi.org/10.1007/978-3-319-95783-8_12

have been pulled from the concrete does failure occur. This structural characteristic helps avoid sudden failure and minimize the early spalling of the concrete cover.

Decades of research have been dedicated to understand the behavior of fiber reinforced concrete (FRC) in response subjected to axial, flexural and shear forces. The addition of steel fibers enhances tensile response, shear strength and deformation capacity in tension of concrete. Additionally, concrete toughness, ductility and failure mechanisms are also improved. Moreover, the addition of steel fibers to concrete tends to reduce crack widths and propagation of cracks, which keeps cracks from rapidly spreading through the members. Resultantly, global failure caused by locally instability due to large area spalling and damage can be effectively delayed, particularly in high strength concrete.

Regarding the aforementioned benefits, steel fiber reinforced concrete has a high potential for application in structural members subjected to high shear forces. However, for steel fiber reinforced concrete to be practically used in typical engineering projects, appropriate design specifications must be developed. There are no equations to quantify the confinement efficiency and shear strength contribution for design purposes. Such specifications can only be developed from in depth experimental analysis and quantitative data about the physical behavior of the concrete. This paper presents the experimental study to verify design equations for confinement efficiency and shear strength substituted by fibers and verified by five large scale bridge columns for their adequacy.

2 Highly Flowable Strain Hardened Steel Fiber Concrete, HF-SHFRC

Self-consolidating concrete (SCC) has a high flowability and a moderate viscosity, and has no blocking by the reinforcement during flow. The concept of SCC was proposed first by Okamura in 1986 (Ozawa et al. 1989). The development of SCC can be deemed as one of the most important breakthrough in concrete technology in the decades. Not only SCC, considerable attention has also been paid to fiber reinforced concrete/cement composites (FRCC) in recent years. In order to classify FRCC based on their tensile performance, Naaman and Reinhardt (1996), Naaman (2003) proposed a new class of FRCC, referred to high performance fiber reinforced cement composites (HPFRCC). The idea behind this new classification of FRCC was to distinguish between the typical tensile performance obtained with traditional FRCC, characterized by a softened response after first cracking, and the tensile strain-hardening response with multiple cracking exhibited by HPFRCC. Typical tensile stress-strain responses of FRCC and HPFRCC are illustrated in Fig. 1.

Highly flowable strain hardening fiber reinforced concrete (HF-SHFRC) has good workability in the fresh state and exhibits the strain-hardening and multiple cracking characteristics of high performance fiber reinforced cementitious composites (HPFRCC) in the hardened state. More characteristics of HF-SHFRC, such as mix design, mixing procedure, fresh properties, tensile strain hardening behavior, resistance of crack propagation and structural applications, can be found in the literatures (Liao et al. 2010; 2015; Liao and Chao 2015).

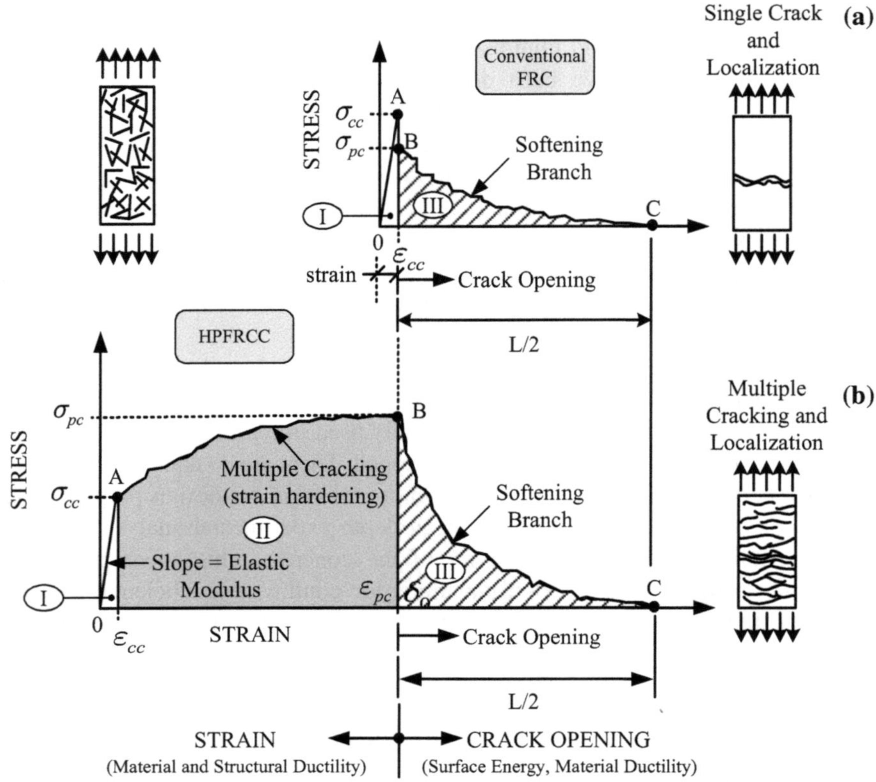

Fig. 1. Stress-strain response of conventional FRC and HPFRCC (Naaman 2003)

3 Experimental Program

3.1 Materials

The mixtures of HF-SHFRC was developed based on systematic mix procedures by using tensile strain hardening responses as performance criteria (Liao et al. 2017) to ensure its workability and mechanical properties. The cementitious materials used in this study were ASTM Type I Portland cement, ground granulated blast-furnace slag and class C fly ash. The coarse aggregate had a maximum size of 9.5 mm and consisted of solid crushed limestone from a local source, with a density of about 2.70 g/cm^3. A polycarboxylate-based superplasticizer was used in the mixture. Hooked steel fiber with circular cross-section was used, with normal tensile strength of 2300 MPa and aspect ratio of 79. The concrete compressive strengths were determined in accordance with ASTM C-39 standard compressive tests on at least six 100 × 200 mm cylinders. The concrete specified compressive strength was 70 MPa. The slump flow of HF-SHFRC used in this study is around 600 mm. Two different sizes of reinforcing bars were used: D25 (5.07 cm^2) with yield stress of 685 MPa, and D13 (1.27 cm^2) with yield stress of 785 MPa. The baseline tensile test for a single control reinforcing bar to determine yield strength, ultimate strength and bar strain at yield stress in accordance with ASTM E8/E8M and ASTM A370.

3.2 Specimen Design

The lateral cyclic test of five bridge columns was conducted in this study. Design parameters are listed in Table 1. Column BMCR1 is a baseline specimen made of conventional concrete and normal strength rebars to comply with current design code. Columns NEWRC1 and NEWRC2 consist of high strength concrete and steel reinforcement bars. Columns NEWRCSF_10 and NEWRCSF_20 were made of HF-SHFRC at their plastic hinge zone parts and identical with NEWRC1 and NEWRC2 elsewhere. The cross section of all five columns are squared. For the same nominal moment capacity with BMCR1, the cross-sectional area of NEWRC1 was reduced by 30% and the amounts of longitudinal and transverse reinforcement were also diminished by 25% owing to the upgrade of material strengths.

Table 1. Bridge column specimen design parameters

Parameters	Unit	BMRC1	NEWRC1	NEWRC2	NEWRCSF_10	NEWRCSF_20
Cross-sectional width	mm	600	500	500	500	500
Column height	m	3.6	3.6	3.6	3.6	3.6
f_c'	MPa	36.7	80.9	80.9	87.0	87.0
f_y	MPa	420	685	685	685	685
f_{yt}	MPa	280	785	785	785	785
Transverse reinforcement spacing	mm	100	100	200	100	200
Number of longitudinal bars	Pieces	16	12	12	12	12
ρ	%	2.3	2.4	2.4	2.4	2.4
ρ_s	%	2.38	2.34	1.17	1.17	1.17
V_f	%	0	0	0	1.50	1.50

The only difference of NEWRC1 and NEWRC2 is the spacing of transverse reinforcement at the plastic hinge zone, which is 100 and 200 mm, respectively. As mentioned, the plastic hinge zone parts were cast with HF-SHFRC in NEWRCSF_10 and NEWRCSF_20, so the transverse reinforcement can be further reduced. Compared to NEWRC1, the hoop spacing kept 100 mm but all ties were eliminated in NEWRCSF_10. In NEWRCSF_20, the spacing of transverse reinforcement with tie arrangement increased two times to 200 mm.

Regarding determination of steel fiber volume fraction, V_f, this study assumes that the toughness provided by the lateral ties can be substituted with steel fibers. Under this assumption, steel fibers are added to compensate for the decrease in confinement that results from increasing the spacing of transverse reinforcement from NEWRC1 to that of NEWRC2. The conversion of confinement efficiency provided by transverse reinforcement into that by steel fibers can be found in the literatures (Wisena et al. 2016). Thus, V_f of 1.5% is determined in this study and construction feasibility of pouring HF-SHFRC is also considered.

To summarize, through comparison of BMRC1 and NEWRC1, the possibility of reducing cross-section dimensions and material consumptions by using high strength materials is discussed. By comparing NEWRC1 and NEWRC2, the effect of transverse reinforcement spacing on the failure mechanism is examined. Also, the feasibility of HF-SHFRC to replace lateral ties in the plastic hinge zone is investigated by comparing NEWRC1 and NEWRCSF_10. In NEWRC2 and NEWRCSF_20, the effect of steel fiber on transverse reinforcement spacing is also assessed.

3.3 Test Setup

The test setup is shown in Fig. 2. Two vertical rods were secured at one end to the strong floor and at the other end to the steel cap beam. They were used to apply a constant force of 1335 kN to the specimen along the gravity direction. One actuator at one end anchored to the reaction wall and at the other end to the steel cap beam were employed to apply lateral cyclic loading to the specimen using displacement control to drift levels of 0.25, 0.375, 0.5, 0.75, 1.0, 1.5, 2.0, 3.0, 4.0, 5.0, 6.0, 7.0, 8.0, 9.0 and 10%. The drift is defined as the horizontal displacement of the loading point divided by the distance from the loading point to the base of the column, 3600 mm. Each drift level was repeated twice to investigate the strength degradation of the specimens.

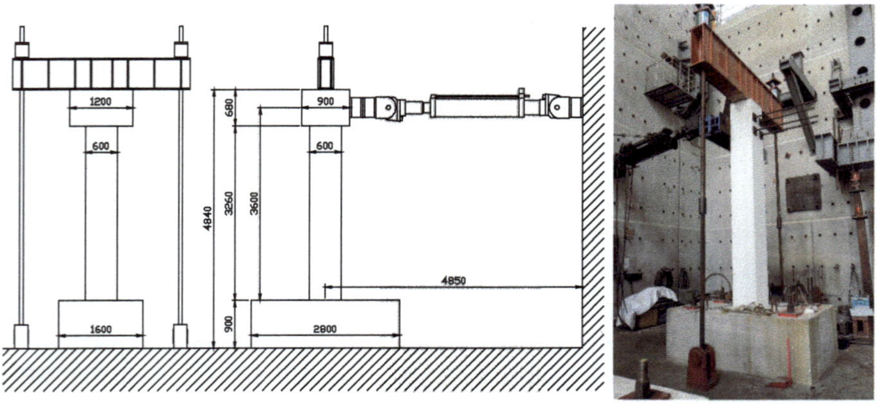

Fig. 2. Test setup

4 Test Results

4.1 Hysteretic Behavior

Hysteresis loops were generated from horizontal displacement and loading measurements recorded by the hydraulic jack system at the top of the column, as illustrated in Fig. 3. Full hysteresis loops of columns BMRC1, NEWRC1, NEWRC2, NEWRCSF_10 and NEWRCSF_20 specimens are shown and elastic-perfectly plastic (EPP) backbone curves owing to ideal flexural behaviors before 6% drift ratio are also observed. It is noticed that sudden shear failure occurred beyond 6% drift ratio in NEWRC2 due to its inadequate transverse reinforcement arrangement.

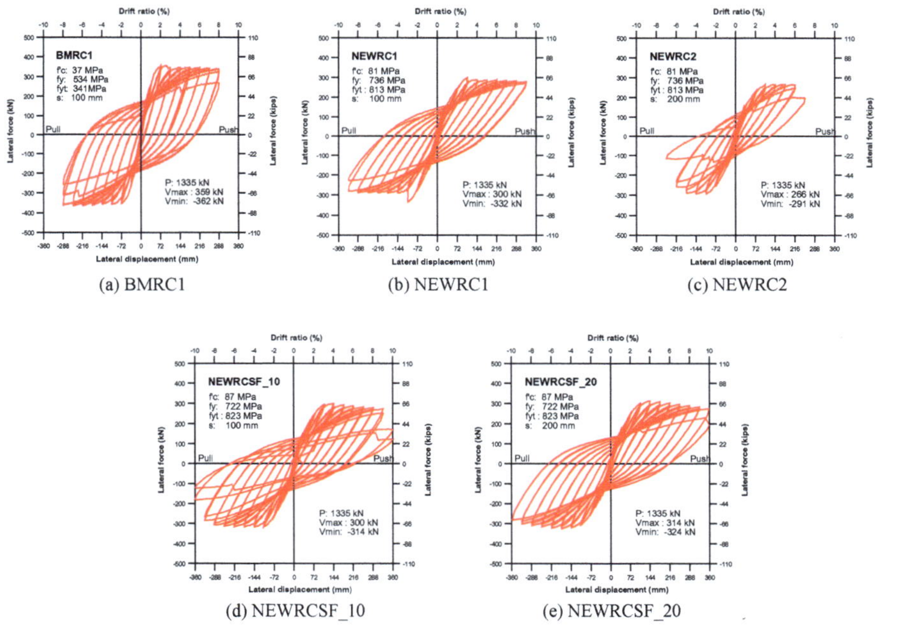

(a) BMRC1 (b) NEWRC1 (c) NEWRC2

(d) NEWRCSF_10 (e) NEWRCSF_20

Fig. 3. Hysteresis loops of specimens

Table 2 lists the yield strength of each specimen, P_y; yield displacement, Δ_y; ultimate displacement, Δ_u, and ductility, μ, where $\mu = \Delta_u/\Delta_y$. The ultimate drift is defined as the drift corresponding to a significant reduction in lateral strength capacity. It is worth mentioning that the ductility of all specimens are higher than 3.5. According to the public road bridge seismic design specifications, the ductility must be higher than or equal to 3.0 to be deemed as a qualified bridge column. This finding demonstrates that the ductility requirement can be assured using high strength materials in bridge columns.

Figure 4 shows the comparisons of hysteresis loops for each specimen. In Fig. 4a, larger transverse reinforcement spacing with less confinement of NEWRC2 did not provide enough strength to prevent shear failure at large displacement. From Fig. 4b, it

Table 2. Yield strength and displacement characteristics

Specimen ID	Yield strength P_y (kN)	Yield displacement Δ_y (mm)	Ultimate displacement Δ_u (mm)	Ductility, μ
BMRC1	352	50.0	288	5.8
NEWRC1	290	59.0	324	5.5
NEWRC2	267	58.7	216	3.7
NEWRCSF_10	298	69.0	324	4.7
NEWRCSF_20	303	75.3	360	4.8

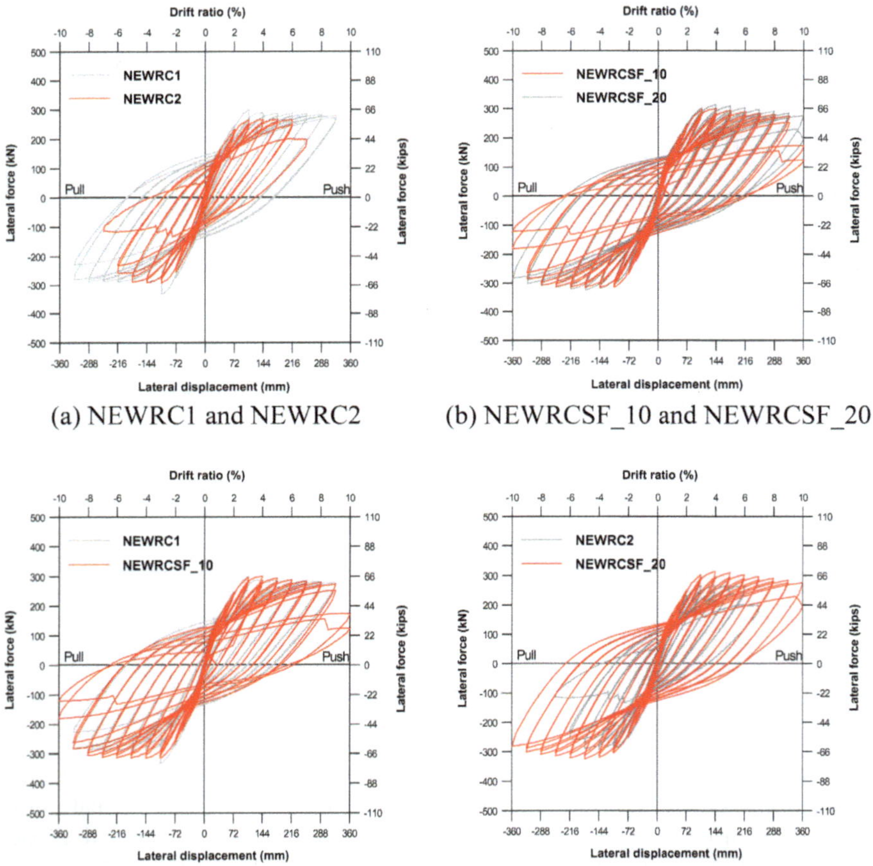

(a) NEWRC1 and NEWRC2

(b) NEWRCSF_10 and NEWRCSF_20

(c) NEWRC1 and NEWRCSF_10

(d) NEWRC2 and NEWRCSF_20

Fig. 4. Comparisons of hysteresis loops

is clear that columns NEWRCSF_10 and NEWRCSF_20 have nearly identical performance and it implies either increasing transverse reinforcement spacing or eliminating ties is acceptable with substitution of HF-SHFRC. Figure 4c shows even the transverse reinforcement of NEWRCSF_10 is 50% less than that of NEWRC1, the presence of HF-SHFRC can be an effective alternative to the transverse reinforcement. The ultimate drift ratio of NEWRCSF_10 was able to exceed 9% with a ductility of 4.7. From Fig. 4d, the addition of steel fibers not only increase confinement, the lack of shear strength observed in NEWRC2 relative to NEWRC1 can be compensated by using steel fibers. Shear failure in NEWRC2 was avoided in NEWRCSF_20.

4.2 Crack Development and Pattern

Figure 5 shows the crack pattern of each specimen at the end of experiment. In general, flexural cracks paralleling the transverse reinforcement formed on the specimen surface at 0.5% drift ratio. From 0.75 to 2%, the flexural cracks began to expand towards the outside edges and the neutral axis of the column. At 2%, the flexural shear cracks propagated and crack angle changed to 45°. The concrete cover of concrete on the

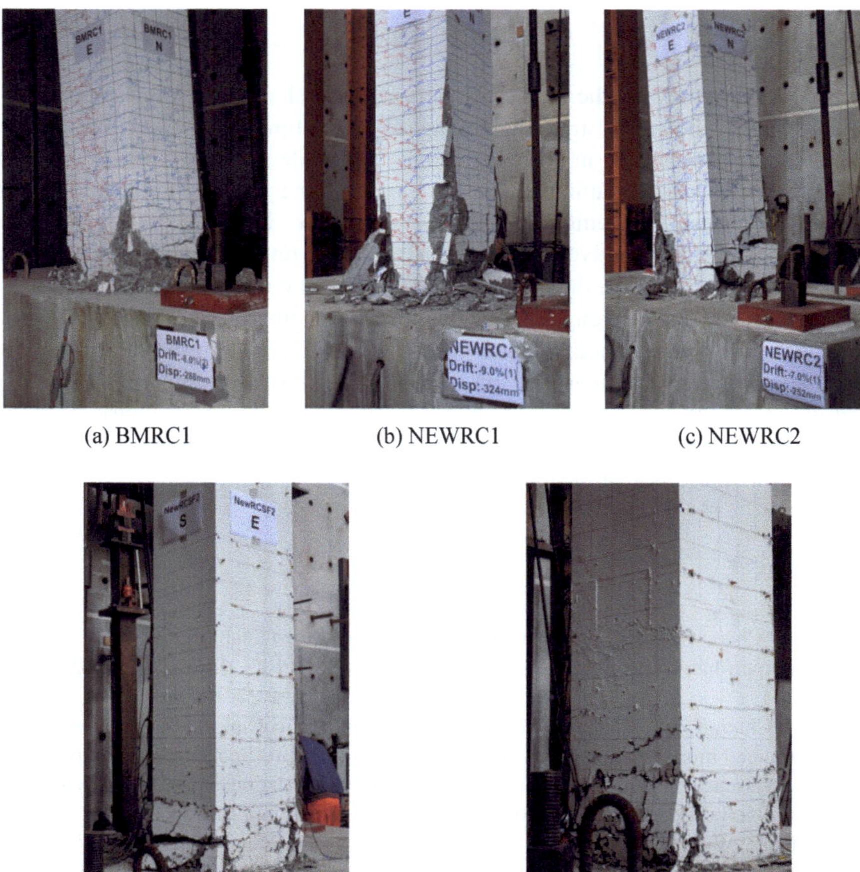

(a) BMRC1 (b) NEWRC1 (c) NEWRC2

(d) NEWRCSF_10 (e) NEWRCSF_20

Fig. 5. Appearance of cracks and damage at the end of each experiment corner of the column began to spall off at 4%. At 5%, the entire surface of the column began to fall off, and at 6%, the cracks on the surface of the column began to widen. Except NEWRC2, the experiment concluded when buckling and reinforcement fracture was observed at more than 8% drift.

As mentioned, the pinched characteristics was observed in NEWRC2 at drift ratio higher than 7%. In terms of crack development, shear cracks of NEWRC2 were more prevalent than those observed in NEWRC1 and this resulted in an extensive area of the concrete had spalled off the column, ending the experiment at 7% drift.

In NEWRCSF_10 and NEWRCSF_20, very fine flexural cracks were observed until 3%. Little spalled concrete began to fall from the column at 8%. At 10% drift, the column base appeared to be pulverized and the reinforcing bars were fractured. Notably, within the area of the plastic hinge zone made of HF-SHFRC, cracks were mostly horizontal and lacked the prevalence of shear cracks which was characteristic of the rest of the column surface.

5 Conclusions

An experimental study on the cyclic behavior of normal strength, high strength concrete and highly flowable strain hardening fiber reinforced concrete, HF-SHFRC, bridge columns is presented in this paper. Five large-scale specimens were tested. The primary objective was to examine the plastic hinge and the potential for steel fibers as a substitute to transverse reinforcement in the plastic hinge. Results show that HF-SHFRC can be an effective alternative to transverse reinforcement. Compared to a conventional bridge column, the high strength reinforced concrete columns used in this study have considerable strength capacity. Once steel fibers are added, the column exhibits flexural failure without sudden shear failure at large drift; the steel fibers both increase confinement and shear strength of high strength concrete which in turn helps inhibit the formation and widening of cracks. In summary, this study contributes to the development of new material and construction technique research and should be considered for future bridge designs.

References

Liao, W.-C., Chao, S.-H.: Crack opening evaluation and sustainability potential of highly flowable strain-hardening fiber-reinforced concrete (HF-SHFRC). J. Test. Eval. **43**(2), 326–335 (2015)

Liao, W.-C., Chao, S.-H., Naaman, A.E.: Experience with self-consolidating high performance fiber reinforced mortar and concrete. ACI J. **247**, 79–94 (2010) (Special Publication No. 247)

Liao, W.-C., Wisena, P., Liu, E.-J.: Compressive stress-strain relationship of high strength steel fiber reinforced concrete. J. Adv. Concr. Technol. **13**, 379–392 (2015)

Liao, W-Cheng, Wisena, P., Yu, L.-C.: Systematic mix procedures for highly flowable-strain hardening fiber reinforced concrete (HF-SHFRC) by using tensile strain hardening responses as performance criteria. Sci. Adv. Mater. **9**(7), 1157–1168 (2017)

Naaman, A.E.: Strain hardening and deflection hardening fiber reinforced cement composites. In: Naaman, A.E., Reinhardt, H.W. (eds.) High performance fiber reinforced cement composites (HPFRCC-4), pp. 95–113. RILEM Publications, Bagneux (2003)

Naaman AE, Reinhardt HW (1996) High performance fiber reinforced cement composites: HPFRCC 2, RILEM, No. 31, E. & FN Spon, London, 505 pages

Ozawa, K., Maekawa, K., Kunishima, M., Okamura H.: Development of high performance concrete based on the durability design of concrete structures. In: Proceedings of the second East-Asia and Pacific Conference on Structural Engineering and Construction (EASEC-2), vol. 1, pp. 445–450 (1989)

Wisena, P., Liao, W.-C., Wang, Y.: High strength concrete columns under axial compression load: hybrid confinement efficiency of high strength transverse reinforcement and steel fibers. Materials **9**(4), 264 (2016)

Design and Numerical Analysis of an Externally Heated Geothermal Bridge Deck

Gang Lei, Xinbao Yu(✉), and Teng Li

Department of Civil Engineering, University of Texas at Arlington, Arlington, TX 76019, USA
{gang.lei,teng.li}@mavs.uta.edu, xinbao@uta.edu

Abstract. In winter, ice and snow on pavement slabs and bridge decks cause serious driving conditions to motorist both in safety and ability to accelerate and clime grade. Geothermal energy has gradually emerged as a new heating source for heating bridge decks and pavements to keep bridges and roads free of snow and ice in winter. However, current geothermal bridge deicing design and studies mostly focus on new bridges that hydronic loops are embedded in concrete deck during construction phase, which is considered as internal heating. This paper presents a conceptual design and numerical analyses of external geothermal bridge deck heating that the hydronic loops are attached to the bottom of bridge deck and encapsulated in a layer of geofoam. A series of parametric analyses is performed to investigate the principles of external heating process. The controlling factors, such as ambient temperature, inlet temperature, wind speed, flow velocity, and foam thickness are considered. The results show that the temperature of the proposed external heated bridge deck can rise above freezing in mild winters through heating for at least 3 h. Limitations of the design are also discussed that the proposed heating system would be unfeasible in the area where the ambient temperature and inlet temperature are lower than −2 and 12 °C, respectively, and wind speed is higher than 6 m/s. It is also found that foam thickness slightly affects the heating process.

1 Introduction

Adverse weather conditions have a detrimental effect on the safety and operation of bridge decks and roads. The snow and ice on bridge deck in winter are potential problems which generate harsh driving conditions for motorists. Existing bridge deck deicing methods, such as using salts and chemicals as deicing materials, results in energy intensity, bridge structure corrosion, and dangers to the environment. For example, salts like NaCl and $CaCl_2$ accelerate corrosion of the bridge structure (Baboian 1995; Cady and Weyers 1983). Hence, alternative bridge de-icing techniques utilizing solar energy and geothermal energy are increasingly developed, which actively promote transportation safety and energy conservation and environmental protection.

Geothermal energy is increasingly popular due to its sustainable, renewable and non-polluted features and is used in bridge deck and pavement deicing through

W.-C. Cheng et al. (eds.), *Tunneling in Soft Ground, Ground Conditioning and Modification Techniques*, Sustainable Civil Infrastructures,
https://doi.org/10.1007/978-3-319-95783-8_13

ground-source heat pumps (GSHPs), geothermal energy piles (GEPs), and borehole thermal energy storage (BTES) (Zhang et al. 2017). Geothermally heating on outside surfaces typically relies on hydronic heat exchangers installed in the bridge deck and pavement which are related to the climatic conditions and the system specifications (Eugster 2007). The SERSO pilot plant in central Switzerland aiming to prevent ice-formation on a high-way bridge surface is one of the well documented geothermal installations (Eugster and Schatzmann 2002).

The current geothermal bridge deicing design and studies mostly focus on new bridges that hydronic loops are embedded in concrete deck during construction phase, which is considered as internal heating. A two-dimensional numerical model was performed to analyze effects of the snow melting process on a pavement snow melting system (Rees et al. 2002). This model was improved by Liu et al. (2003) to simulate a bridge deck heated hydronically over a lifetime as opposed to singular storm events incorporating a ground source heating pump. Chen et al. (2011) used a two-dimensional finite element model built by Wang et al. (2009) to analyze de-icing to design a pavement hydronic de-icing system. A series of parametric analyses was conducted by Bowers and Olgun (2014) to analyze the operational principles of bridge deck heating. Their results presented that bridge deck surface temperature can reach above freezing while heated by energy piles alone. Recently, several models were performed by Abraham et al. (2016) to analyze pavement heating by burying loops in the aggregate base layer of pavements. Han and Yu (2017) modeled a bridge deck to investigate the feasibility of geothermal heat exchanger pile based snow melting system for several cities in the United States.

The existing geothermal bridge/pavement de-icing researches focus on internal heating systems. This paper presents a conceptual design of external bridge deck heating that the hydronic loops are attached to the bottom of bridge deck and encapsulated in a layer of geofoam. Base on the design, finite element models of external heating bridge deck heating were built through COMSOL Multiphysics (COMSOL 2016) to investigate the feasibility of the external heating process. A series of parametric analyses was performed to analyze the effects of parameters.

2 Conceptual Design of External Heating Bridge Deck

TxDOT owns a huge number of overpass bridges that are in critical need for bridge deicing. Approximately 85% of the bridges in Texas were constructed using CIP-PCP bridge decks which consist of precast, prestressed concrete panels (PCPs) and a cast-in place (CIP) concrete deck (Merrill 2002). Therefore, the heated bridge deck design is intended to be used on CIP-PCP bridge decks. Deicing of bridge deck for existing bridges requires external heating source which is made possible by supplying heat from two sides of the bridge deck. Considering the installation requirement and minimizing disturbance to the traffic, heat supply from bottom of bridge deck is a reasonable choice although it is not the most heating efficient. A schematic of the concept is presented in Fig. 1 where the hydronic heating loop which consists of hydronic pipes and foam insulation materials are attached to the bottom of the bridge deck through mental fixtures. Insulation foam (polyurethane or geofoam) is installed to encapsulate the heat pipes to minimize the heat loss from the bottom. Warm fluids (water and antifreeze mixtures) supplied from underground loops provides heat source to warm the bridge deck surface above due to high thermal resistance of the underneath insulation.

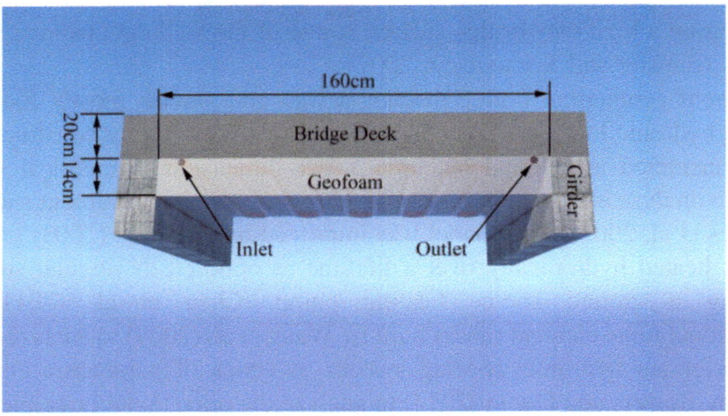

Fig. 1. Schematic of heated bridge deck for existing bridge

3 Numerical Modeling

This section presents (a) the features of finite-element model which was extracted from the conceptual design to evaluate the efficiency of external heating and heat transfer of bridge deck based on different scenarios; (b) boundary conditions, material properties and heat transfer properties that were employed for the numerical simulations.

3.1 Finite-Element Modeling

A series of 3-D finite-element models is developed using software COMSOL Multiphysics (COMSOL 2016) to analyze the bridge deck heating process under different winter scenarios. The geometry of the 3-D model is shown in Fig. 2. The dimensions of the bridge deck slab and geofoam are 1.6 m × 1.2 m × 0.2 m and 1.6 m × 1.2 m 0.14 m, respectively. The total bridge deck is regarded as convective interface between concrete and air. The rebar in bridge deck is not modeled due to its small volume and high thermal conductivity in comparison to the concrete slab. The selected polyethylene (PEX) pipe with 2.4 cm outer diameter and 0.2 cm wall thickness is attached to the bottom surface of the bridge deck. The pipe centerline spacing is 16 cm. Water with 25% propylene glycol is utilized as heat carried fluid circulating through tubes at a constant flow velocity.

3.2 Boundary Conditions and Material Properties

Based on the heat transfer module in COMOSOL, the initial temperatures in concrete slab, pipe, geofoam, and air were set to equal to the ambient temperature at the beginning of all numerical runs. The initial velocity and pressure of circulating fluid

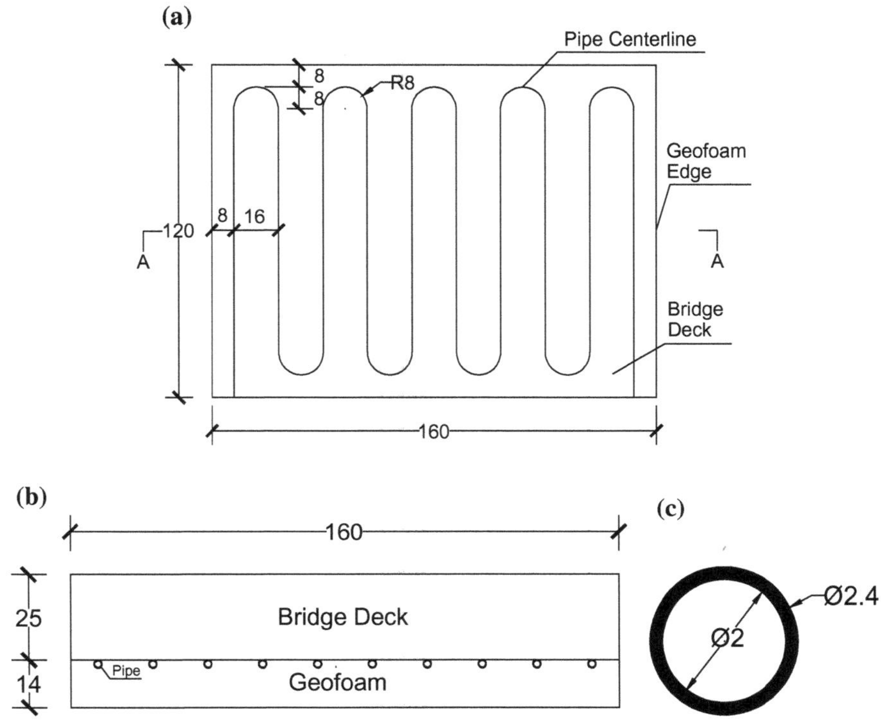

Fig. 2. Bridge deck slab used in simulations and layout of the heating tube; **a** plane view; **b** A-A cross-section view; **c** cross-section view of tube (unit: cm)

were set to zero. Adiabatic conditions are imposed on the two sides of the concrete deck along length (perfect thermal insulation). During the simulations, the inlet temperatures and fluid velocities shown in Table 2 are kept constant and are fixed at the pipe inlet face considered as the external heat source. The interfaces between carrier fluid and pipe, pipe and concrete slab, pipe and geofoam, and geofoam and concrete slab are considered as perfect heat contacts. Heat resistances at the pipe-fluid interface, the pipe-geofoam interface, and the pipe-concrete interface are not modeled. The ice or snow melting process is not considered.

Properties of the materials considered for the numerical analyses are summarized in Table 1. Regarding the thermal properties of concrete, air, and circulating fluid, the conductivities, heat capacities, and surface emissivities have been adopted based on the indications provided by Bowers and Olgun (2014).

4 Parametric Analyses

This section shows the parametric analysis plan and results regarding bridge deck heating simulations. The plan includes proposed factors of the slab heating including wind speed, inlet fluid temperature, ambient temperature, flow velocity, and foam

Table 1. Summary of the material properties used in numerical simulations

Material	Property	Value
Carrier fluid	Dynamic viscosity	0.00273 Pa s
	Heat capacity	3691 J/(kg K)
	Density	1000 kg/m^3
	Thermal conductivity	0.61 W/(m K)
Concrete	Density	2300 kg/m^3
	Thermal conductivity	1.8 W/(m K)
	Heat capacity	880 J/(kg K)
	Surface emissivity	0.91
Air	Density	1.23 kg/m^3
	Heat capacity	1006 J/(kg K)
	Thermal conductivity	0.0239 W/(m K)
	Kinematic viscosity	1.315×10^{-5} m^2/s
Geofoam	Density	20 kg/m^3
	Thermal conductivity	0.38 W/(m K)
	Heat capacity	1300 J/(kg K)
	Surface emissivity	0.60
Polyethylene pipe	Density	1360 kg/m^3
	Thermal conductivity	0.46 W/(m K)
	Heat capacity	950 J/(kg K)

thickness. Additionally, the numerical results contain temperature variations along vertical section in between tubes, influences on average deck top surface temperature and heating time induced by different ambient temperatures and inlet temperatures, wind speeds, and foam thickness.

4.1 Parametric Analysis Plan

Parametric analyses were performed to find and investigate the effect of proposed factors on external bridge deck heating process. These factors include wind speed, inlet fluid temperature, ambient temperature, flow velocity, and foam thickness. Each factor is systematically varied and analyzed coupled with other varied factors. A total of 88 model runs were analyzed for each process where different model parameters are summarized in Table 2.

For case 1 (baseline case), the initial temperature of deck slab, pipe, carrier fluid, and foam is set to −2 °C and no wind is considered. The center-to-center spacing of tubes is 16 cm. Fluid circulated at a flow velocity of 0.6 m/s has a 12 °C inlet temperature. Case 2–3 studied the effect of different ambient temperatures and wind speeds coupled with disparate inlet fluid temperatures, respectively. Case 4–5 presents the impact of different inlet temperatures and wind speeds with varied foam thickness, respectively. The sixth case was performed to investigate the effect of fluid flow velocity.

Table 2. Model parameters used in the simulation of winter scenarios

Case	Wind speed (m/s)	Inlet fluid temperature (°C)	Ambient temperature (° C)	Flow velocity (m/s)	Foam Thickness (cm)	Number of runs
1	0	12	−2	0.6	14	Base Case
2	0	8, 12, 16, 20	−4, −3, −2, −1	0.6	14	15
3	0, 1, 2, 4, 6	8, 12, 16, 20	−2	0.6	14	15
4	0	8, 12, 16, 20	−2	0.6	6, 8, 10, 12,14, 16,18	27
5	0, 1, 2, 4, 6	12	−2	0.6	6, 8, 10, 12, 14, 16, 18	27
6	0	12	−2	0.6, 0.9, 1.2, 1.5	14	3

4.2 Results

4.2.1 Temperature Variations Along the Vertical Section in Between Circulation Tubes

The temperature distribution at steady state for baseline case is presented in Fig. 3a. It shows the external heating process. The temperatures at the bottom and two sides of the geofoam equal to ambient temperature indicating the insulation function works well.

Figure 3b presents the temperature gradient along the vertical line of symmetry of the circulation tubes. The figures show that the domain area near loops get warmer than other parts of the bridge deck and geofoam. It is also seen that the top 10 cm of the geofoam and bottom 4 cm of the slab reach to a temperature above 0 °C after roughly

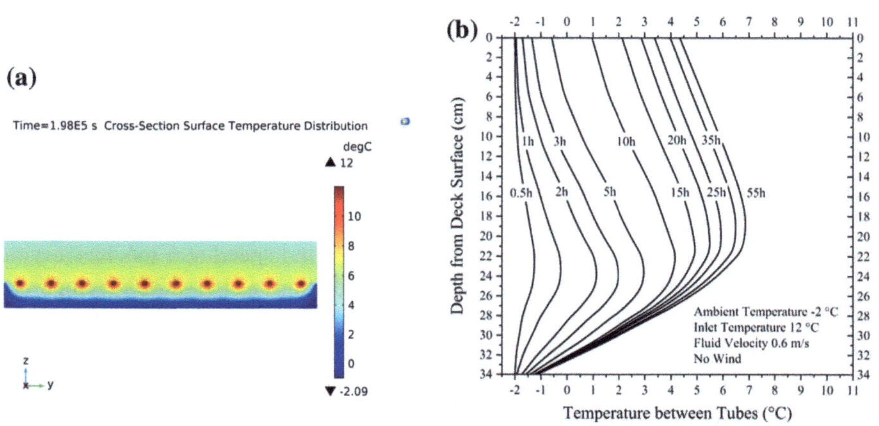

Fig. 3. Temperature variations at steady state for baseline case; **a** temperature distribution on the middle bridge deck cross-section along the length; **b** temperature profiles along the vertical symmetry axis of the cross section

2 hours' heating with 12 °C fluid flow. The bottom of geofoam keeps at a constant temperature after 10 hours' heating, showing that geofoam negatively affects heat transfer. It takes roughly 6 h to get the temperature at bridge deck top surface larger than 0 °C and nearly 55 h for the heating process to reach steady state.

4.2.2 Impacts of Ambient Temperature and Wind Speeds

The results shown in Fig. 4 indicate that ambient temperature and wind speed significantly affect heating process coupled with different inlet temperatures. Figure 4a shows the heating times required to raise the average deck top surface temperature up to 0 °C at different ambient temperatures. More heating time is needed for lower ambient temperature. When the inlet temperature is 8 °C, the heating time to reach top concrete surface temperature to 0 °C increases 41–74%. For the same ambient temperature, the inlet temperature drastically decreases heating time, especially when the ambient temperature is 8 °C.

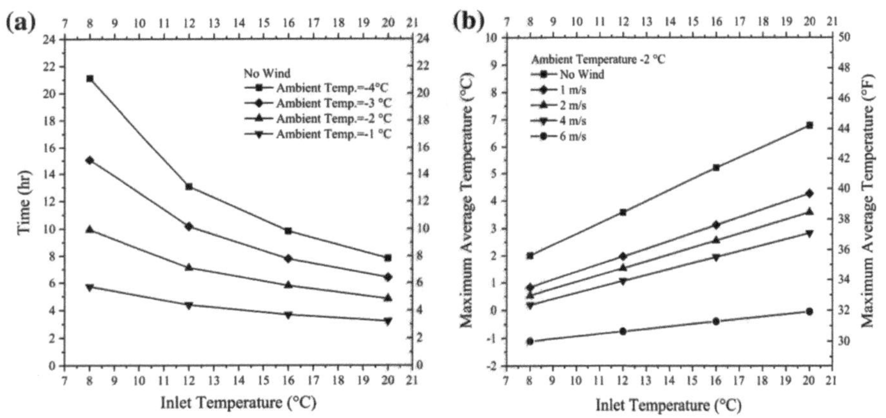

Fig. 4. Heating time required to raise the average deck top surface temperature to 0 °C for **a** different ambient temperatures; and maximum deck top surface temperature for **b** different wind speeds

The effect of wind speed can be also seen from Fig. 4b that higher wind speeds result in lower average bridge deck top surface temperatures. In this analysis, the ambient temperature and inlet temperature are −2 and 12 °C, respectively, the flow velocity 0.6 m/s, and the foam thickness 14 cm. The wind speed at 6 m/s has a critical impact on average top surface temperature, introducing a 101–122% temperature decrease with respect to no wind scenario. Additionally, at this wind speed the average top surface temperatures are below 0 °C even with inlet temperatures of 20 °C. Other wind speeds at 1, 2, and 4 m/s generate 37–59%, 47–74%, and 58–91% temperature decrease, respectively, compared with no wind cases.

4.2.3 Impact of Foam Thickness

The effects of foam thicknesses are shown in Fig. 5 coupled with different inlet temperatures and wind speeds. In these cases, the ambient temperature is −2°C and the flow velocity 0.6 m/s. The results indicate that foam thickness does not significantly affect average deck top surface temperature for a given ambient temperature. Average temperature reduces by 2–7% shown in Fig. 5a in comparison with the results in Fig. 4. A similar temperature response is found for the case where foam thickness changes with different wind speeds shown in Fig. 5b. As shown, the temperature has a 2–10% decrease.

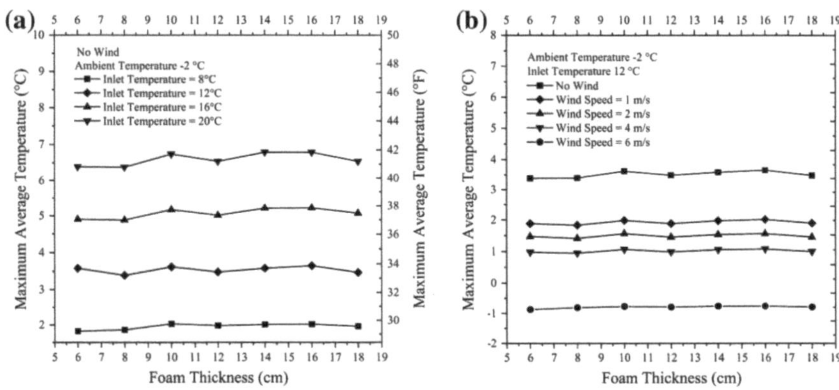

Fig. 5. The average deck top surface temperature at steady state for different foam thicknesses with **a** different inlet temperatures and **b** different wind speeds

The effect of the flow velocity on heating process is small and the results are not shown here. A series of simulations were performed for fluid velocities at 0.9, 1.2, and 1.5 m/s in comparison with that at 0.6 m/s in baseline case. For these simulations, the ambient temperature and inlet temperature are −2 and 12 °C, respectively. No wind is considered. It is seen that slow velocity induces long heating time. A flow velocity at 0.9 m/s generates a 5% heating time reduction with respect to base case.

5 Conclusions

Geothermal bridge deicing is an innovative renewable technology designed to utilize the relative constant temperature of the ground for efficient heating bridge decks and pavements during winter snow events. Current studies mostly focus on heating process induced by hydronic loops embedded in bridge deck. This paper presents conceptual design and feasibility of external heating bridge deck under mild winter conditions. A series of numerical simulations regarding different parameters were performed to investigate the key factors of the external heating process. Considered parameters contain ambient temperatures, inlet fluid temperatures, flow velocities, wind speeds, and foam thicknesses.

It is summarized that the proposed external heated bridge deck can reach above freezing temperatures in mild winters through heating for at least 3 h. It is also found that in cold areas where temperatures are lower than −2 °C and wind speeds higher than 6 m/s, the proposed heating system would be unfeasible if the inlet temperature is lower than 20 °C. Additionally, foam thickness has a minor effect on the heating process when the thickness is greater than 6 cm.

This study sheds light on the principles of external heating process and highlights the need for more detailed investigation. Future studies on the external bridge deck heating system will focus on size effect analysis, mesh sensitivity analysis, and the pipe-geofoam-concrete interactions investigation. Experimental studies of the proposed external heating system are also needed to analyze the efficiency of the system in comparison with numerical results.

Acknowledgements. The authors appreciate the financial support of this research from Texas Department of Transportation, USA.

References

Abraham, S.P., Abdelaziz, S.L., Longtin, J.: Heat Exchangers for pavement surface de-icing. In: Geo-Chicago 2016 Technical Papers, ASCE (2016). https://doi.org/10.1061/9780784480137.060

Baboian, R.: Environmental conditions affecting transport infrastructure, Materials Performance, NACE (1995)

Bowers, G.A. Jr., Olgun, C.: Ground-source bridge deck deicing systems using energy foundations. In: Geo-Congress 2014 Technical Papers, ASCE (2014). https://doi.org/10.1061/9780784413272.261

Cady, P.D., Weyers, R.E.: Chloride penetration and deterioration of concrete bridge decks. In: Cement, Concrete and Aggregates. ASTM (1983). https://doi.org/10.1520/cca10258j

Chen, M., Wu, S., Wang, H., Zhang, J.: Study of ice and snow melting process on conductive asphalt solar collector. Solar Energy Mater. Solar Cells (2011). https://doi.org/10.1016/j.solmat.2011.07.013

COMSOL.: COMSOL Multiphysics version 5.2a: user's guide and reference manual. COMSOL, Burlington (2016)

Eugster, W.: Road and bridge heating using geothermal energy. overview and examples. In: Proceedings European Geothermal Congress (2007)

Eugster, W., Schatzmann, J.: Harnessing solar energy for winter road clearing on heavily loaded expressways. In: Proceedings of XI'th PIARC International Winter Road Congress (2002)

Han, C.J., Yu, X.: Feasibility of geothermal heat exchanger pile-based bridge deck snow melting system: a simulation based analysis. Renew. Energy (2017). https://doi.org/10.1016/j.renene.2016.08.062

Liu, X., Rees, S., Spitler, J.: Simulation of a geothermal bridge deck anti-icing system and experimental validation. In: Proceedings of Transportation Research Board 82nd Annual Meeting. TRB (2003)

Merrill, B. D.: Texas' use of precast concrete stay-in-place forms for bridge decks. In: Proceedings of Concrete Bridge Conference, National Concrete Bridge Council (2002)

Rees, S.J., Spitler, J.D., Xiao, X.: Transient analysis of snow-melting system performance. In: Proceedings of ASHRAE Transactions 2002. ASHRAE (2002)

Wang, H., Wu, S., Chen, M., Zhang, Y.: Numerical simulation on the thermal response of heat-conducting asphalt pavements. In: Proceedings of 3rd International Symposium on Functional Materials 2009. IOP Publishing (2009). https://doi.org/10.1088/0031-8949

Zhang, N., Yu, X.B., Li, T.: Numerical simulation of geothermal heated bridge deck. In: Proceedings of 2017 International Conference on Transportation Infrastructure and Materials, DEStech Publications (2017). https://doi.org/10.12783/dtmse/ictim2017/9952

Influence of the Canopy Filling on Shallow Tunnels

Vinícius Resende Domingues[1(✉)], Bernardo Cascão Pires e Albuquerque[1], and André Pacheco de Assis[2]

[1] PiezoGreen Solutions, Brasília, DF, Brazil
vinicius.rdomingues@gmail.com
[2] Department of Civil and Environmental Engineering, University of Brasília, Brasília, DF, Brazil

Abstract. The construction of shallow tunnels using a pre-support system based on pipe canopies may require important decisions during the planning stage. Once the decision of using pipe canopies has been made, the engineer has to choose between steel cylinders or steel tubes for the canopy. If the later ones are chosen, it is necessary to decide whether to keep the soil inside, remove the soil or remove the soil and fill the tube with concrete. In order to clarify the benefits and harms that guide this interactive solution, a study about the influence of the filling material in the bending behavior of the canopy was conducted. The results were obtained for steel behaving as an elastic-perfectly plastic material and concrete being represented by the parabola-rectangle diagram, however the methodology may be applied for different stress-strain curves. A case study, based on the site conditions proposed in Domingues et al. (Society for Mining, Metallurgy and Exploration, The World Tunnel Congress 2016, San Francisco, California, United States, 2016) exemplify the usefulness of the methodology proposed. As a result, it was shown that it may be favorable to keep the soil inside the beam instead of replacing it by concrete or use a massive structural element. In other words, as any other structural system, there is an optimal solution that leads to an economic and safe construction. Hence, this paper provides important tools for decision-making as it expands the possibilities of using the pre-supported pipe canopy technology in tunnel construction.

1 Introduction

The increasing demand for underground works in urban centers requires constant evolution in building techniques in order to achieve success and safety. The geomechanical conditions of the soil mass, usually unfavorable, added to the low distance between tunnel's ceiling and the surface requires intense treatment of the soil massif.

The soil excavated for the construction of undercrossing tunnels are usually composed by layers of fill and/or soft soil or weak rock, leading to stability problems when subjected to tunneling operations. The combination of low coverage and week ground is usually solved by soil improvement techniques and strengthening of the excavated cross-section with further advance with the well-known New Austrian Tunneling Method (NATM).

W.-C. Cheng et al. (eds.), *Tunneling in Soft Ground, Ground Conditioning and Modification Techniques*, Sustainable Civil Infrastructures,
https://doi.org/10.1007/978-3-319-95783-8_14

However, as shown in Domingues et al. (2016), another option is to build structural frames that support the surface. This technique consists of a pipe canopy supported by two porticoes on each end of the tunnel, as can be seen in Fig. 1. The construction process of the pipe canopy involves driving the tubes from one side of the tunnel to the other. After this procedure, the inside of the pipes can be left with soil or replaced by concrete. This research presents the influence of the pipe canopy filling.

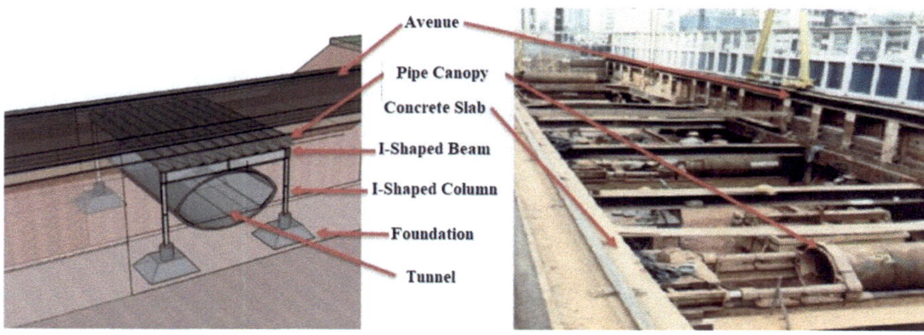

Fig. 1. Schematic design of the construction technique (Domingues et al. 2016)

The specifications of the beam filling material should be in accordance with technical standards and may also take into account economic factors. Although the soil replacement by concrete could improve the inertia, it is not necessarily the proper solution for this construction technique. Hence, using analytical solutions, different scenarios were studied in order to clarify which filling material might be suitable for a generic case.

2 Pipe Canopy Function

Domingues et al. (2016) and Domingues (2016) exposed a technical solution to make feasible tunneling operations in urban areas. This technique, named as portico composed of pipe canopy, is presented in Figs. 1 and 2.

The pipe canopy, structural element under analysis, represents the longitudinal cross section of this pre-support system. Each pipe works as a simply supported beam that retransmits the loads to the porticoes. Therefore, these elements must be designed to support the overburden, providing eligible settlements. Domingues et al. (2016) exposed, using numerical tools, that the soil mass surrounding the tunnel may resist a part of the imposed loading. This fact is not taken into account in the analytical solution, hence an extra factor of safety is implicit. However, the tools presented in this paper are important for a pre-design stage and for decision-making.

Fig. 2. Schematic view of the excavation

3 Methodology

According to the instructions given in Domingues et al. (2016), in order to verify the cross-section dimensions that lead to small settlements, a 3-step procedure was implemented in the software Mathematica® (Appendix). First, the cross-section is divided into 4 different parts, as shown in Fig. 3. This step aims to separate parts that work similarly. Second, the position of the neutral axis is determined. This is calculated by an equilibrium of stresses in the cross section, Eq. 1. However, to perform this calculation one needs to know the strains that occur in the cross section. Assuming that the cross-section of the beam remains plane (Euler-Bernoulli theory) it is only necessary to determine the strain at one point and its distance to the neutral axis. In order to make the most of the materials, one should set the strain of the furthermost concrete fiber under compression as concrete's highest allowable compressive strain (3.5‰ according to ABNT (2014)). This can be seen by the point in Fig. 3b. Lastly, it was required to compute the maximum resistant bending moment. This part was done by solving equilibrium of bending moments, Eq. 2, after having determined the neutral axis.

$$\sum_{i=1}^{6} \int_{y_{mn}}^{y_{max}} F_i \, da = 0 \tag{1}$$

$$M_{max} - \sum_{i=1}^{6} \int_{y_{min}}^{y_{max}} F_i \left(y_i - y_{nl}\right) da = 0 \tag{2}$$

As shown in Fig. 3a, this beam might not necessarily be filled by concrete. Hence, if the pipe is hollow, the neutral axis should be at the symmetrical center. Consequently, the calculation procedure in these cases might be simplified to the equilibrium

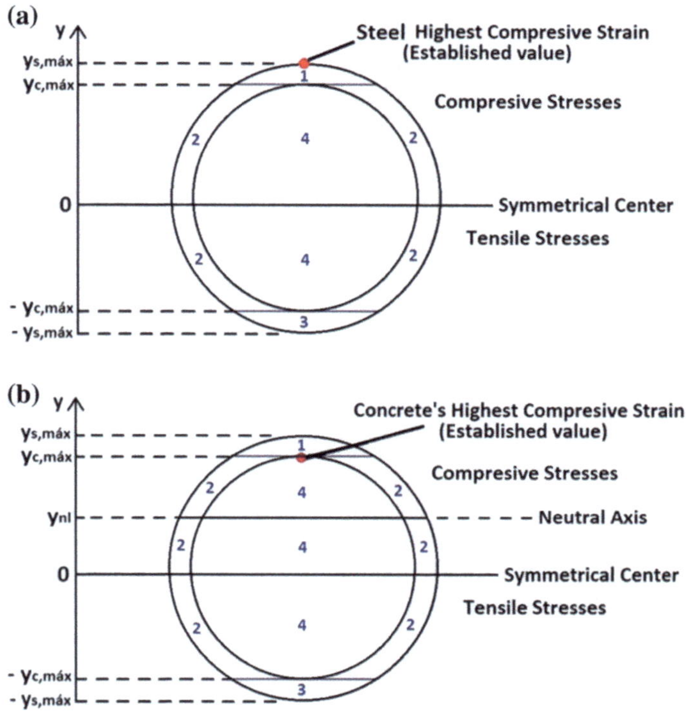

Fig. 3. Pipe's cross-section regions **a** hollow beam **b** composite (concrete and steel) beam

of bending moments, once the position of the neutral line is already determined (Fig. 3b).

As shown in Domingues et al. (2016), to solve this problem a couple of simplifying assumptions were taken into account. First of all, the structure was analyzed with the elements (pipe canopy, portico and foundation) working together without specifying the connections. Secondly, the steel and concrete/soil were also considered working together in the pipes (not considering the shear connectors and/or the roughness of the tube). Lastly, it is also important to highlight that the highest allowable compressive strain—following Brazilian Standards: ABNT (2008) and (2014)—is 3.5‰ for concrete and 10‰ for steel. This fact explains the limits imposed in this methodology.

3.1 Constitutive Frameworks

The constitutive frameworks for pipe canopies materials, steel and concrete, exposed in ABNT (2014), were adopted in this paper.

Concrete framework presents different stress-strain relationship when submitted by tensile or compressive stresses. For compressive stresses lower than 50% of the resistance, a linear stress-strain relationship can be assumed, adopting the secant modulus of elasticity as shown in Eq. 3.

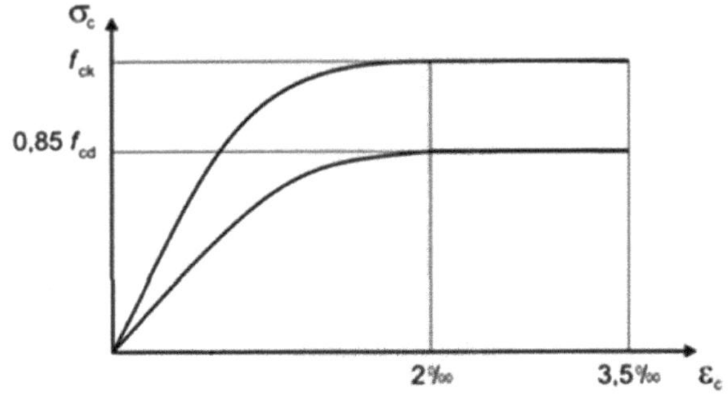

Fig. 4. Idealized compressive stress-strain diagram proposed by ABNT (2014)

$$Ecs = 0.85\,Eci \tag{3}$$

where Ecs is the secant modulus of elasticity and Eci is $5600\,\sqrt{f_{ck}}$. For analyzes using the Ultimate Limit State (ULS), the idealized stress-strain diagram is shown in Fig. 4.

Where σ_c is compressive stress imposed to concrete; ξ_c is concrete strain, f_{ck} is compressive strength of concrete and 0,85 f_{cd} is compressive design strength of concrete (following Eq. 4). In the present paper, a function f of a variable x shall be denoted as $\mathbf{f}[x]$.

$$\boldsymbol{\sigma}_\mathbf{c}[\varepsilon_c] = 0.85 f_{cd} \begin{cases} \left[1 - \left(1 - \frac{\varepsilon_c}{2‰}\right)^2\right] & 0 < \varepsilon_c \leq 2‰ \\ 1 & 2‰ < \varepsilon_c \leq 3.5‰ \\ 0 & 3.5‰ < \varepsilon_c \end{cases} \tag{4}$$

and with $f_{cd} = f_{ck}/\gamma_c$, where γ_c assumes the value of 1.4 in normal combinations of forces. In another context, when submitted by tensile stresses, ABNT (2014) proposes the bilinear diagram shown in Fig. 5.

Where σ_{ct} is tensile stress imposed to concrete, ξ_{ct} is concrete strain and f_{ctk} is tensile strength of concrete. Therefore, concrete's constitutive model can be described by Eq. 5.

$$\boldsymbol{\sigma}_\mathbf{c}[\varepsilon_c] = \begin{cases} 0 & \varepsilon_c \leq -0.15‰ \\ -0.9 f_{ctd} - 0.1 f_{ctd} \frac{\frac{-0.9 f_{ctd}}{E_{ci}} - \varepsilon_c}{\frac{-0.9 f_{ctd}}{E_{ci}} + 0.15‰} & -0.15‰ < \varepsilon_c \leq -0.9 \frac{f_{ctd}}{E_{ci}} \\ E_{ci}\,\varepsilon_c & -0.9 \frac{f_{ctd}}{E_{ci}} < \varepsilon_c \leq 0 \\ 0.85 f_{cd} \left[1 - \left(1 - \frac{\varepsilon_c}{2‰}\right)^2\right] & 0 < \varepsilon_c \leq 2‰ \\ 0.85 f_{cd} & 2‰ < \varepsilon_c \leq 3.5‰ \\ 0 & 3.5‰ < \varepsilon_c \end{cases} \tag{5}$$

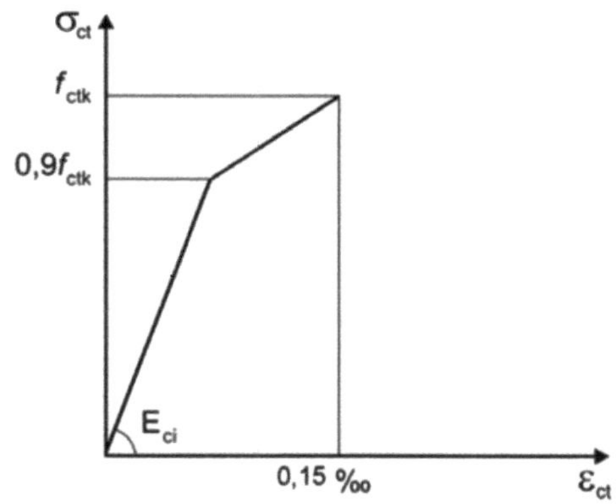

Fig. 5. Idealized tensile stress-strain bilinear diagram proposed by ABNT (2014)

Steel framework presents the same stress-strain relationship when submitted by tensile or compressive stresses. A simplified elastic-perfectly plastic diagram proposed by ABNT (2014) and shown in Fig. 6 was adopted in this paper.

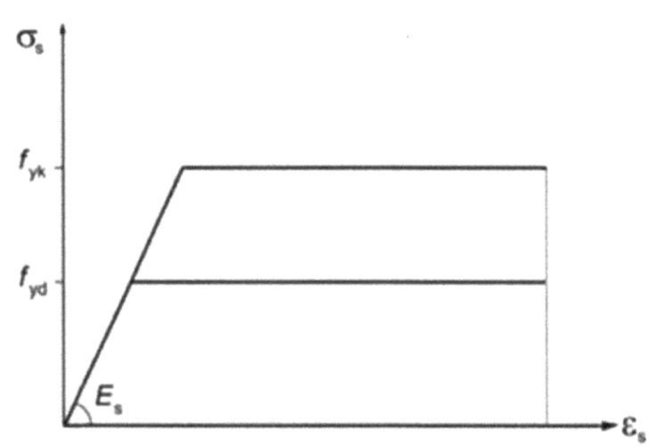

Fig. 6. Stress-Strain relationship proposed by ABNT (2014) for steels

Where is the yield resistance of steel and $f_{yd} = f_{yk}/\gamma_s$ is the design yield resistance of steel, in which γ_s assumes the value of 1.15 in normal combinations of forces. Steel constitutive model is defined by Eq. 6.

$$\boldsymbol{\sigma}_{\mathbf{s}}[\varepsilon_s] = \begin{cases} 0 & \varepsilon_s \leq -10\text{‰} \\ -f_{yd} & -10\text{‰} < \varepsilon_s \leq -\frac{f_{yd}}{E_s} \\ E_s\varepsilon_s & -\frac{f_{yd}}{E_s} < \varepsilon_s \leq \frac{f_{yd}}{E_s} \\ f_{yd} & \frac{f_{yd}}{E_s} < \varepsilon_s \leq 10\text{‰} \\ 0 & \varepsilon_s > 10\text{‰} \end{cases} \tag{6}$$

The equations used to calculate the equilibrium of stresses, for each region presented in Fig. 3, are shown in Eqs. 7 to 10.

$$F_1 = \int_{y_{c,max}}^{y_{s,max}} 2\sqrt{y_{s,max^2} - y^2}\boldsymbol{\sigma}_{\mathbf{s}} \left[\frac{(y - y_{nl})\varepsilon_{c,max}}{y_{c,max} - y_{nl}}\right] dy \tag{7}$$

$$F_2 = \int_{-y_{c,max}}^{y_{c,max}} 2\left(\sqrt{y_{s,max^2} - y^2} - \sqrt{y_{c,max}^2 - y^2}\right)\boldsymbol{\sigma}_{\mathbf{s}} \left[\frac{(y - y_{nl})\varepsilon_{c,max}}{y_{c,max} - y_{nl}}\right] dy \tag{8}$$

$$F_3 = \int_{-y_{s,max}}^{-y_{c,max}} 2\sqrt{y_{s,max} - y}\boldsymbol{\sigma}_{\mathbf{s}} \left[\frac{(y - y_{nl})\varepsilon_{c,max}}{y_{c,max} - y_{nl}}\right] dy \tag{9}$$

$$F_4 = \int_{-y_{c,max}}^{y_{c,max}} 2\left(\sqrt{y_{s,max^2} - y^2} - \sqrt{y_{c,max^2} - y^2}\right)\boldsymbol{\sigma}_{\mathbf{c}} \left[\frac{(y - y_{nl})\varepsilon_{c,max}}{y_{c,max} - y_{nl}}\right] dy \tag{10}$$

where each equation is related to one region, as follow: Eq. 7 with Region 1, Eq. 8 with Region 2, Eq. 9 with Region 3 and Eq. 10 with Region 4.

On the other hand, the equilibrium of moments, for each region presented in Fig. 3, are shown in Eqs. 11 to 14.

$$F_1' = \int_{y_{c,max}}^{y_{s,max}} (y - y_{nl})2\sqrt{y_{s,max^2} - y^2}\,\boldsymbol{\sigma}_{\mathbf{s}} \left[\frac{(y - y_{nl})\varepsilon_{c,max}}{y_{c,max} - y_{nl}}\right] dy \tag{11}$$

$$F_2' = \int_{-y_{c,max}}^{y_{c,max}} (y - y_{nl})2\left(\sqrt{y_{s,max^2} - y^2} - \sqrt{y_{c,max^2} - y^2}\right)\boldsymbol{\sigma}_{\mathbf{s}} \left[\frac{(y - y_{nl})\varepsilon_{c,max}}{y_{c,max} - y_{nl}}\right] dy \tag{12}$$

$$F_3' = \int_{-y_{s,max}}^{-y_{c,max}} (y - y_{nl})2\sqrt{y_{s,max} - y}\,\boldsymbol{\sigma}_{\mathbf{s}} \left[\frac{(y - y_{nl})\varepsilon_{c,max}}{y_{c,max} - y_{nl}}\right] dy \tag{13}$$

$$F_4' = \int_{-y_{c,max}}^{y_{c,max}} (y - y_{nl})2\left(\sqrt{y_{s,max^2} - y^2} - \sqrt{y_{c,max^2} - y^2}\right)\boldsymbol{\sigma}_{\mathbf{c}} \left[\frac{(y - y_{nl})\varepsilon_{c,max}}{y_{c,max} - y_{nl}}\right] dy \tag{14}$$

Similarly, each equation is related to one region as follow: Eq. 11 with Region 1, Eq. 12 with Region 2, Eq. 13 with Region 3 and Eq. 14 with Region 4.

4 Parametric Analysis

In this paper, the pipe canopy made of steel cylinders or steel tubes, the later ones filled by concrete or by soil were analyzed. As the solution depends on an iterative algorithm to determine the neutral axis, this procedure was done several times for different values of diameter and thickness and the results interpolated. To provide an acceptable accuracy, the range of variables studied is shown in Table 1. Besides, the values used for f_{yd} and f_{ck} are 300 MPa with a FoS of 1.15 and 30 MPa with a FoS of 1.4 respectively.

Table 1. Range of variables analyzed

Pipe canopy	Diameter variation (m)	Thickness variation (mm)
Cross section filled by concrete	0.1–2.0, steps of 0.1	5–25, steps of 5
Hollow cross section	0.1–2.0, steps of 0.1	5–25, steps of 5
Solid cross section	0.05–1.0, steps of 0.05	–

A pipe canopy filled by concrete may increase the structural resistance, but may also raise up the imposed loadings. Hence, according to Table 1 and shown in Fig. 7, 20 ordered pairs containing information about geometry and resistance were calculated for 5 different values of thickness. These results must be analyzed concomitantly with the site conditions, including the structural weight, in order to conclude whether the dimensions proposed for the pipes lead to allowable resistances or not. This comparison must also take into account safety factors suggested by the current local standards.

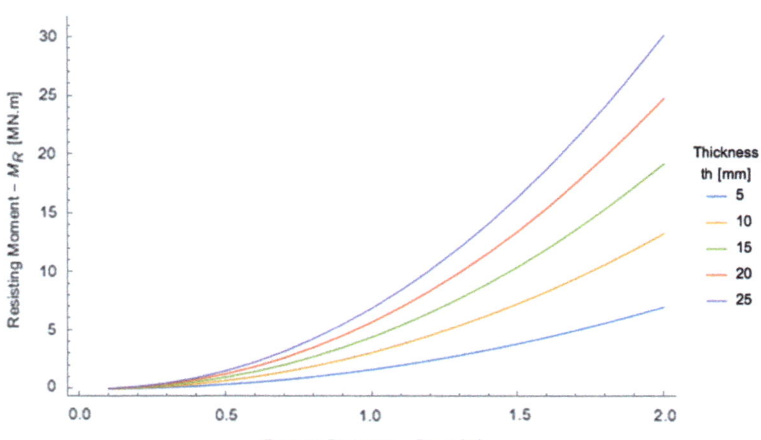

Fig. 7. Geometry and resistance for a composite pipe canopy

On the other hand, depending on the context, it might be favorable to save money and time spent in the process of removing the soil inside the pipes and filling it with concrete. In this case, since the soil inside the pipes does not contribute to the beam resistance, it was considered as hollow and the weight of the filling was added to the loadings imposed to the canopy. Similarly to the procedure applied to the beam filled by concrete, 20 ordered pairs containing information about geometry and resistance were calculated for 5 different values of thickness. These results, shown in Fig. 8, must be analyzed as explained above.

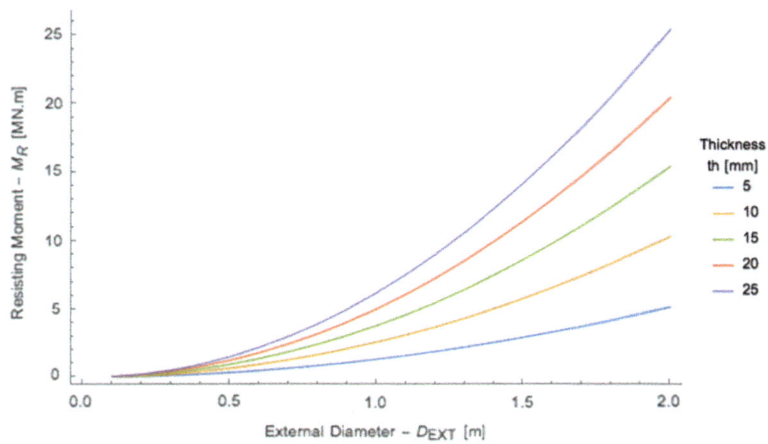

Fig. 8. Geometry and resistance for a pipe canopy filled by soil

Lastly, it can be difficult to drive steel cylinders underneath the soil mass without interfering on the superficial structures. Even so, their high resistance, observed even in small structures, may make their adoption feasible. Hence, also depending on the context, this solution may be suitable. In according to the methodology proposed, 20 ordered pairs, containing information about geometry and resistance, were calculated. These results, shown in Fig. 9, must also be analyzed as explained above.

5 Case Study

First of all, to use the data presented in this paper, it is required to know the loadings imposed on the structural system. Hence, in order to clarify the analysis procedure, the data exposed in Domingues et al. (2016) was used. The overburden was composed by the soil massif and traffic loading. To make feasible an analytical solution, the dynamic loadings from the traffic, proposed by ABNT (2013), were replaced by static loads.

According to Domingues et al. (2016), to calculate the traffic load, an avenue with 6 lanes and 25 m wide subjected to bus congestion was taken into account. The expected maximum load for these vehicles was 16.8 tons. In this context, the expected load is equivalent to one vehicle per lane, disregarding buses that are possibly located at the

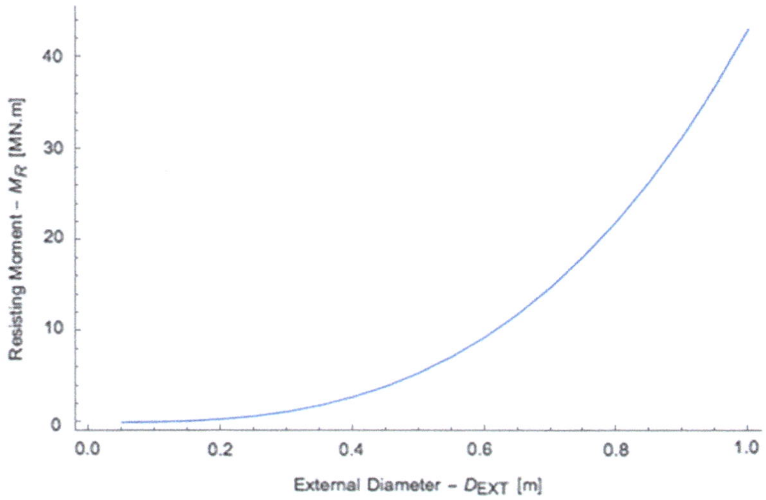

Fig. 9. Geometry and resistance for a massive pipe canopy

ends of the portico. This simplification converts live loads into dead ones seeking an extreme working condition for the support structure (pipe canopy, porticos and foundations). In this example, the loadings are summarized in Table 2.

Table 2. Loadings imposed on the structural system

Loading origin	Calculation	Load
Traffic load	$6\,\text{buses} \times 16.8\,\text{tons} \times \frac{9.8\,\text{kN}}{1\,\text{ton}} \times \frac{1}{(25\,\text{m} \times 11\,\text{m})}$	3.59 kPa
Pavement (0.60 m)	$0.60\,\text{m} \times 22\,\frac{\text{kN}}{\text{m}^3}$	13.20 kPa
Fill (1.40 m)	$1.40\,\text{m} \times 17\,\frac{\text{kN}}{\text{m}^3}$	23.80 kPa
Pipe canopies	Variable	–
$\sum$		40.59 kPa

The weight of the structural elements should be taken into account. In this context, the pipe canopies are variable, then the solution must be interactive.

As shown in Table 2, ground conditions were taking into account and thus a couple of geotechnical tests were performed to determine the loadings imposed. Moreover, superficial loadings and settlements limits are required to guide the solution. Following the proposal shown in Domingues et al. (2016), ground conditions are presented in Table 3.

In this case, the tunnel analyzed was excavated in a soft ground and the pipe canopies were driven in medium clay at two meters deep. These beams receive the loadings shown in Table 1, totaling 40.59 kPa plus their own weight. Thus, considering the distance between porticos of 25 m, the bending moment must be 3171 kN m added to the pipe canopies influence. In this context, using the tables presented in the

Table 3. Geotechnical profile

Material	Depth (m)	Average of SPT's	Extents of structural system and tunnel	Soil properties					
				Young's modulus (MPa)	Poisson	Specific weight (kN/m^3)	Cohesion (kPa)	Friction angle (°)	Dilatancy angle (°)
Pavement	0–0.6	–	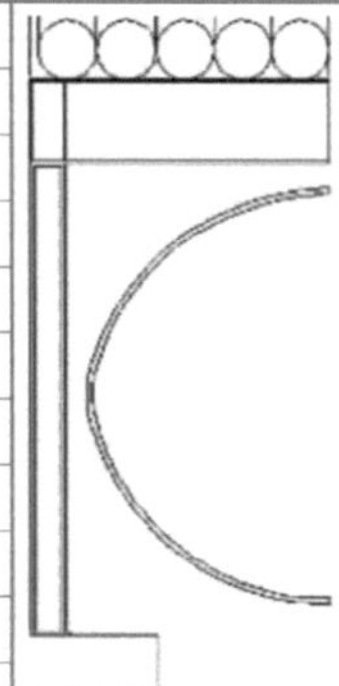	200	0.3	22	–	–	–
Fill	0–0.6	–		10		14	10	26	1
Fill	0.6–2	13		25		18	30	28	10
Medium clay	2–3	5		15		16	13	27	5
	3–4	3					25	27	
	4–5	4					36	26	
	5–6	6					36	26	
	6–7	6					27	27	
Stiff clay	7–8	10		15		16	30	27	5
	8–9	10							
	9–10	7							
	10–11	8		20		17.5	40	28	8
	11–12	8							
	12–13	8							
	13–14	11							
Very stiff silty clay	14–15	20		30		19	35	28	9
	15–16	20							
	16–17	12							
Very stiff sandy Silt	17–18	18		40		20	40	30	10
	18–19	23							
Hard sandy silt	19–20	30							
	20–21	36							
	21–22	43							

Domingues et al. (2016)

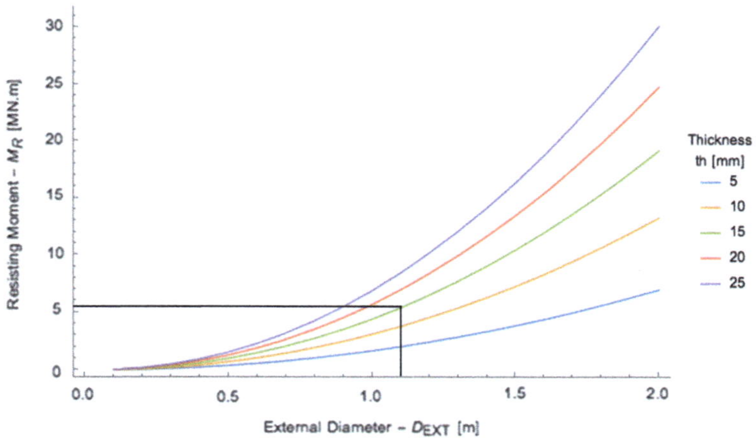

Fig. 10. Resisting moment of a beam in the form of steel tube filled with concrete

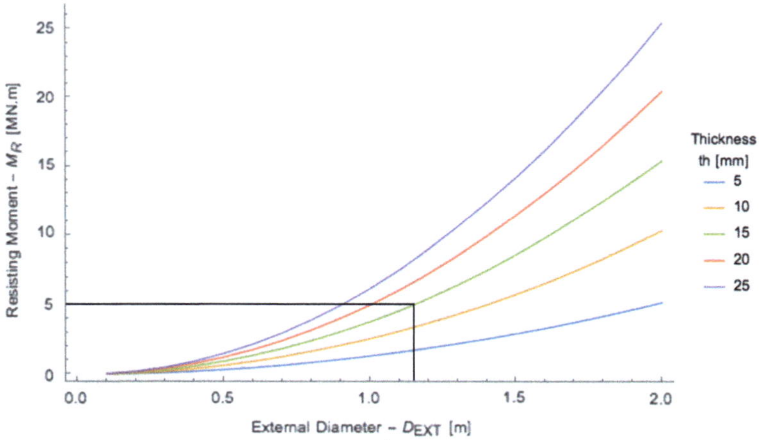

Fig. 11. Resisting moment of a beam in the form of steel tube

parametric analysis, a couple of solutions may be found. Figures 10, 11 and 12 present the results obtained for this case study.

As shown in Fig. 10, a composite pipe with 1.1 m in diameter and 15 mm of thickness, filled by concrete, may resists approximately 5250 kN m. Considering this geometry, the imposed loading is 5170.57 kN m, of which 3171 kN m were obtained from Table 2 and 1999.57 kN m represent the weight of a single pipe (steel + concrete). In another context, Fig. 11 presents the data obtained for the pipe filled by soil.

As shown in Fig. 11, a pipe with 1.15 m in diameter and 15 mm of thickness may resists 5000 kN m. Considering this geometry, the imposed loading is 4807.49 kN m, of which 3171 kN m were obtained from Table 2 and 1636.49 kN m represent the

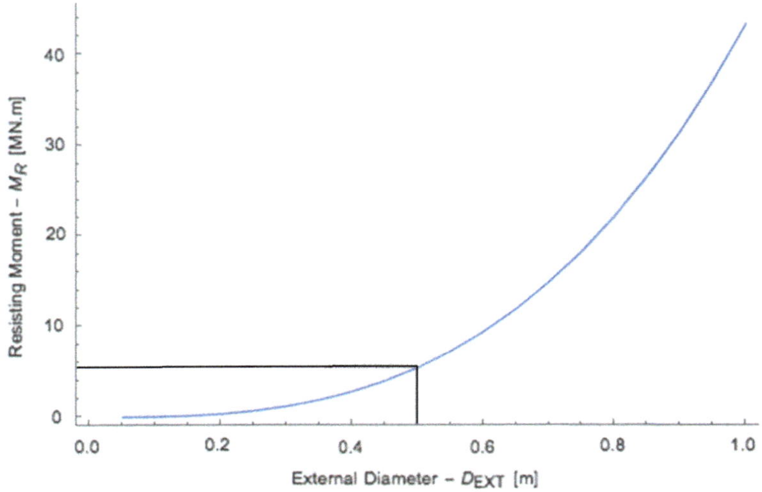

Fig. 12. Resisting moment of a beam in the form of steel cylinder

weight of a single pipe (steel + soil). On the other hand, Fig. 12 presents the data obtained for the steel cylinder beam.

As shown in Fig. 12, a steel cylinder with 0.5 m in diameter may resists 5200 kN m. Considering this geometry, the imposed loading is 4473.44 kN m, of which 3171 kN m were obtained from Table 2 and 1302.44 kN m represent the weight of a single cylinder.

It is emphasized that none of the above analyses consider the factors of safety for the imposed loadings, therefore they must be included according to the local standards. The cases presented are technically feasible for the case study. However, a factor of safety that is consistent with the local standards should be applied. Besides, economic aspect must be taken into account in order to elect the best of these possibilities.

6 Conclusions

The use of rigid pre-support system is usually a suitable technique for excavating a shallow tunnel. When it comes to pipe canopies, the common technique is to drive the steel tubes and fill them with concrete. Nevertheless, based on the analysis of the bending moments of pipe canopies, it was shown that it may be favorable to keep the soil inside the beam instead of replacing it by concrete.

According to the results presented in Domingues et al. (2016), using numerical analysis, there is a notable massif influence in the final settlements. Thus, the approach of this paper is limited to analytically evaluating the bending stability of the pre-support system, without extending the analysis to superficial settlements. However, it should be emphasized that the displacements and strains must be a function of the affected area and, therefore, it should be studied case-by-case. Hence, the graphs analysis is adequate for pre-design phase and, mainly, for decision-making regarding the constructive method. Thus, knowing the alternatives, it may be possible to evaluate which solution best suits the local costs and logistics factors.

Acknowledgements. The authors acknowledge the Brazilian Research Council (CNPq) and University of Brasília (UnB) for funding the first steps of this research.

References

ABNT: Design of Steel and Composite Structures for Buildings: NBR-8800/2010. Brazilian Association of Technical Standards, Rio de Janeiro, RJ, 237p (2008) (in Portuguese)

ABNT: Design and construction of foundations: NBR-6122/2010. Brazilian Association of Technical Standards, Rio de Janeiro, RJ, 91p (2010) (in Portuguese)

ABNT: Road and pedestrian live load on bridges viaducts, footbridges and other structures—NBR-7188/2013. Brazilian Association of Technical Standards Rio de Janeiro, RJ, 14p (2013) (in Portuguese)

ABNT: Design of Structural Concrete—Procedure—NBR 6118/2014. Brazilian Association of Technical Standards, Rio de Janeiro, RJ, 238p (2014) (in Portuguese)

Domingues, V.R., Albuquerque, B.C.P., Assis, A.P.: Structural conception of an undercrossing tunnel supported by pipe canopies. In: Society for Mining, Metallurgy and Exploration, The World Tunnel Congress 2016, San Francisco, California, United States

Domingues, V.R.: Construction techniques of undercrossing tunnels. University of Brasília, Master thesis, Brasília, Federal District, Brazil (2016)

Analysis of Offshore Rock Socketed Monopile Foundations Considering Stiffness Degradation

Nivya Basheer[1], Khalid Abdel-Rahman[2], Johannes Albiker[2], Tanusree Chakraborty[1(✉)], and Martin Achmus[2]

[1] Department of Civil Engineering, Indian Institute of Technology (IIT) Delhi, Hauz Khas, New Delhi 110 016, India
nivya.basheer@gmail.com, tanusree@civil.iitd.ac.in
[2] Institute for Geotechnical Engineering (IGtH), Leibniz University of Hannover, Appelstraße 9A, 30167 Hannover, Germany
{khalid, albiker, achmus}@igth.uni-hannover.de

Abstract. There are a number of offshore wind farms where the monopile is socketed into rock layers. Since it is socketed into rock, it may behave different from monopile embedded in soil. A numerical modelling of rock socketed monopile is done using finite element (FE) software Abaqus. A stiffness degradation method (SDM) is applied to FE model in order to predict the behaviour under cyclic loading conditions. Parametric studies are carried out by varying rock socketed depth (*d*), length of monopile below seabed (*L*), intensity of horizontal loading (*H*) and subsoil conditions to evaluate the long-term permanent deformation of offshore rock socketed monopile foundations. Suitable permanent deformation factors are proposed for offshore rock socketed monopiles for the first time in the literature. It is observed from the results that the deformation behaviour of the monopile changes from stiff to flexible with increase in rock socketing and in turn the pile head deflection going down. From the bending moment diagram, flexible and stiff behaviour of monopile can be identified and is an indicator of curvature of the deflection line of pile.

Keywords: Cyclic loading · Rock socketed monopiles · Stiffness degradation method · Numerical modelling · Lateral deformation

1 Introduction

Monopiles are one of the common foundation options for offshore wind turbines. Often the sea bed is made up of bed rock and rock socketing becomes necessary for installing the monopiles. In many places, the monopiles are being successfully socketed into rock for different wind farms, e.g., situated on the East coast of England in Northumberland, Blyth offshore wind project installed monopiles of diameter 3.5 m into bedrock of sandstone. At North Hoyle wind farm, installation of monopile (4.0 m diameter at seabed) consists of driving through upper layers of sand and clay, and drilling and

W.-C. Cheng et al. (eds.), *Tunneling in Soft Ground, Ground Conditioning and Modification Techniques*, Sustainable Civil Infrastructures,
https://doi.org/10.1007/978-3-319-95783-8_15

driving through rock layer of sandstone and mudstone. In Bockstigen wind farm in Sweden monopiles were socketed into rock (www.offshorecenter.dk, www.subacoustech.com, www.technology.stfc.ac.uk).

Germanische Lloyd (GL) rules and regulations give the design procedure for foundations of offshore wind energy converters in Germany (Achmus 2010). In this regulation the *p-y* method defined by API code is recommended in order to estimate the behaviour of piles under horizontal loading. But the use of *p-y* curves in estimating the behaviour of monopiles may be misleading because of the fact that these curves are formulated based on field testing of piles with number of cycles less than 200 and applicable for piles with diameters up to 2 m.

Many studies have been conducted in the past in order to understand the behaviour of laterally loaded and axially loaded monopiles (Achmus et al. 2008, 2009; Albiker and Achmus 2012; Achmus and Albiker 2014; Arshi and Stone 2011; Kellezi and Hansen 2003; Little and Briaud 1988; Schmoor and Achmus 2013) and some design guide lines are also available in the literature (Achmus et al. 2008; Schmoor and Achmus 2013; Thieken et al. 2014). But the rock-monopile interaction under cyclic loading is less discussed. Arshi and Stone (2011) conducted a series of small scale single gravity tests to investigate the performance of a monopile, combined monopile and bearing plate foundation where the pile is socketed into a weak rock. In the model studies, the weak rock layer is modelled using a weak sand and gypsum mix. The results of the study provide an insight into the effect of the various foundation elements (i.e. pile, plate and rock socket) and their contribution to the overall performance of the foundation system. Wang et al. (2007) discusses the behaviour of large-diameter rock-socketed CFST (concrete-filled steel tube) piles under lateral loads based on field tests and numerical analysis. The horizontal capacity and deformation of large-diameter rock-socketed piles are analyzed from the measured displacements and internal forces of piles. The interactive behaviour of pile-rock and the influence of backfilled sand on horizontal capacity are also discussed. Using the Finite Element Method (FEM) considering the properties of the pile-soil interface, the test results are simulated numerically. This result show that stress concentration effect in the region near the bottom of the steel tube should be considered in the design, because the socketed part of piles bears most of the lateral load.

From the literature review, it may be concluded that the offshore monopile foundations are mainly subjected to wave and wind loading and these loads are cyclic in nature. One of most important aspects of designing a monopile foundation is the deformation under cyclic horizontal loading and accumulation of permanent deformation with increasing number of cycles. In this paper, the deformation behaviour of rock socketed monopile foundation under cyclic loading is studied by numerical modelling of the pile-soil system using the FEM software Abaqus and by applying the stiffness degradation method (SDM) with increasing number of load cycles. Permanent deformation of the piles under cyclic loading is investigated for different rock socketing depths, different pile lengths and horizontal load magnitudes. A deformation accumulation parameter is proposed further.

2 Stiffness Degradation Method

This method was developed at the Institute for Geotechnical Engineering, Leibniz University of Hannover, Germany (Achmus et al. 2009). In a cyclic triaxial test, the degraded stiffness of soil after N cycles ($E_{s,N}$) can be expressed in terms of stiffness modulus after first cycle ($E_{s,1}$) using the equation,

$$\frac{E_{S,N}}{E_{s,1}} = N^{-b_1 X^{b_2}} \tag{1}$$

Here b_1 and b_2 are soil parameters and X is the cyclic stress ratio defined by

$$X = \frac{\sigma_{1,\text{cyclic}}}{\sigma_{1,\text{f}}} \tag{2}$$

where $\sigma_{1,\text{cyclic}}$ is the maximum principal stress in a cycle and $\sigma_{1,\text{f}}$ is the maximum principal stress at failure subjected to static loading. From cyclic triaxial test results documented in the literature, typical regression parameters b_1 and b_2 were found for dense sand to be $b_1 = 0.12$, $b_2 = 0.50$ and for medium dense sand $b_1 = 0.15$, $b_2 = 0.50$. Detailed numerical implementation of the SDM in the finite element code Abaqus (Abaqus documentation version 6.11-3) is discussed in Achmus et al. (2008).

3 Numerical Modelling

The model considered in the present work consists of a monopile of diameter 7.5 m and wall thickness of 9 cm. For simplicity the hollow cylindrical steel monopile (modulus of elasticity E = 210 GPa and Poisson's ratio ν = 0.2) is replaced by a solid cylindrical pile with same diameter such that bending stiffness of both piles remains the same. The monopile is installed into a layered soil with upper sand and lower rock layer as shown in Figs. 1 and 2. The sand is considered elasto-plastic with Mohr–Coulomb failure criterion. The stiffness modulus of the soil varies with depth according to the following equation,

$$E_s = \kappa \sigma_{at} \left(\frac{\sigma_m}{\sigma_{at}} \right)^{\lambda} \tag{3}$$

Here E_s is the oedometric stiffness modulus which varies with stress condition, σ_m is the current mean principal stress in the considered soil element and σ_{at}= 100 kN/m^2 is a reference (atmospheric) stress. The parameter κ determines the soil stiffness at the reference stress state and the parameter λ rules the stress dependency of the soil stiffness. The material parameters used to model different materials are listed in Table 1 (Das 2013; Jaeger et al. 2008; Achmus et al. 2009). At the contact of the pile soil interface, interface friction coefficient tan δ = 2/3 tan(ϕ) is considered. At pile-rock interface, tan δ = tan(ϕ) is taken. In the contact normal direction, hard contact is

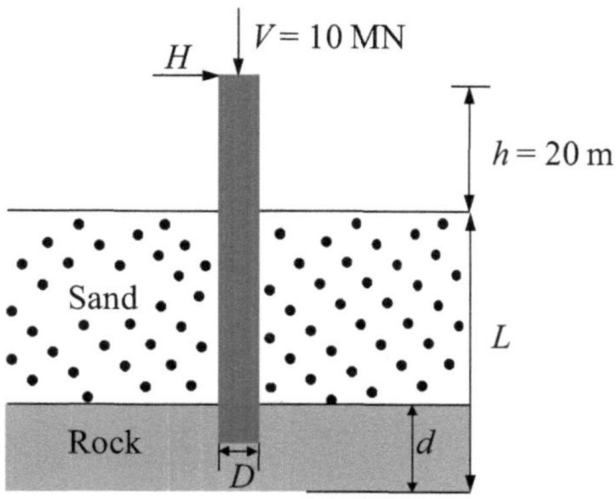

Fig. 1. Schematic sketch showing the subsoil conditions

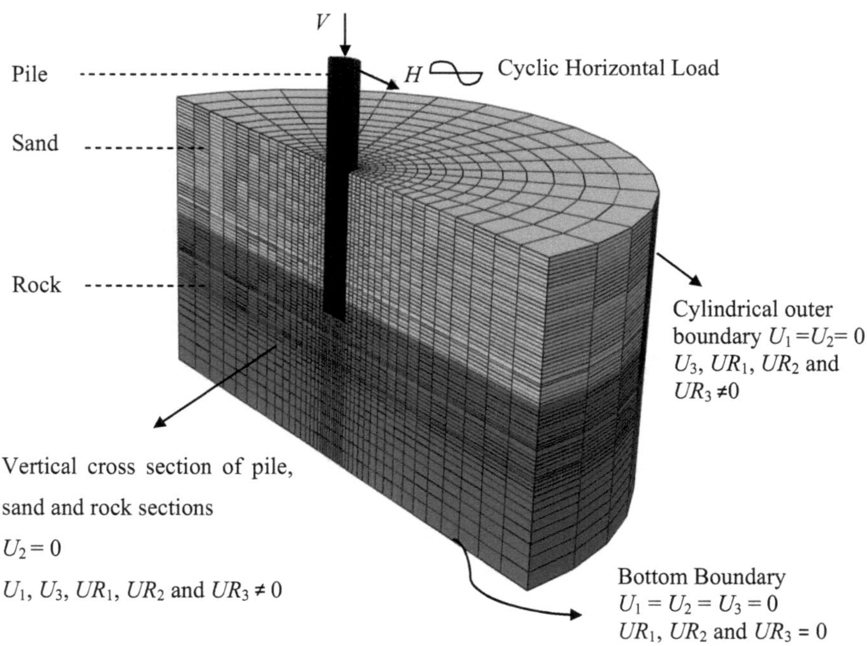

Fig. 2. Finite element model using Abaqus/CAE

considered at the pile-rock interface and soft contact is considered at the pile-soil interface with assumed interface stiffness taken same as the soil stiffness.

The vertical load (V) acting on the monopile is assumed as 10 MN which is the weight of the super structure. The variable horizontal load (H) is acting at a height (20 m) above the seabed level. Three different horizontal load magnitudes are assumed-10, 15 and 30 MN. Due to symmetry conditions, only the half of the pile-soil-model is considered. The diameter of the numerical model considered for analysis is sixteen

Table 1. Properties of materials

Soil type	Submerged unit weight (kN/m³)	Stiffness parameters		Friction angle	Dilatancy angle	Cohesion (kN/m²)	Poisson's ratio
	γ'_s	κ	λ	ϕ'	ψ	c	ν
Dense sand	11	700	0.55	37.5°	7.5	0.1	0.25
Hard rock	14	E = 70 GPa		–	–	–	0.2
Soft rock	12	E = 1.0 GPa		25°	0	50	0.33
Medium dense sand	11	550	0.6	35°	5°	0.1	0.25

times the pile diameter. The bottom boundary of the model is extended to 20 m below the base of the monopile. The steps involved in modelling using stiffness degradation method are shown in the Table 2.

Table 2. Steps taken to obtain the deformation behaviour of monopiles

Step	Input	Model and loading system	Output
1	Material, geometry of pile soil system	Finite element model	σ_{mean}
		Geostatic step (considering only the soil system)	
2	Material, geometry of pile soil system	Finite element model with SDM	Soil Stiffness after 10, 100, 1000, 10,000 cycles
	Varying stiffness modulus of soil with depth	Vertical loading (*V*) and horizontal loading (*H*)	
3	Material, geometry of pile soil system	Finite element model	Accumulated displacement after N cycle
	Degradation in secant modulus after *N* cycles	Vertical loading (*V*) and horizontal loading (*H*)	

4 Parametric Studies

A parametric study is conducted to estimate the effect of embedded length of the pile below seabed (*L*), rock socketed depth (*d*), horizontal load intensity (*H*) and the subsoil conditions on deformation behaviour of monopile foundation under long term cyclic loading. Table 3 lists different loading conditions and pile geometries considered in this study. The depths of rock socket is decided based on different literature (Choy et al. 2004; Srinivasamurthy and Pujar 2009; Liu et al. 2011). Three different subsoil conditions are simulated (i) two layered soil with upper layer dense sand and lower layer hard rock (case I), (ii) two layered soil with upper layer dense sand and lower layer soft rock (case II), three layered soil with upper layer medium dense sand, middle layer dense sand lower layer soft rock (case III). Hard rock is considered elastic ($E = 70$ GPa, $\nu = 0.2$) and soft rock elasto-plastic with Mohr-Coulomb criteria ($E = 1$ GPa, $\nu = 0.33$, $c = 50$ kN/m^2 and $\phi' = 25°$) (Table 1).

Table 3. Model conditions in parameter study

Loading condition	
V (MN)	10
H (MN)	10, 15, 30
Pile geometry	
D (m)	7.5
d (m)	0, 2, 4, 6, 8, 10
L (m)	30, 40
H (m)	20
t_p (m)	0.09
h/L	0.5, 0.67

5 Results and Discussions

5.1 Effect of Increase in Rock Socketing Depth

Figure 3 shows the lateral deflection of the pile with depth below seabed for different rock socketing depths (d)—2, 4 and 10 m and for different number of loading cycles (N)—1, 10, 100, 1000 and 10,000. From this Figure, it can be seen that with increased rock socketing of the pile, the deflection line of the pile reduces. Moreover, with increased rock socketing, the effect of loading cycle on the deflection response reduces. This is caused by the fact that the pile with zero socketing in the rock behaves like a stiff pile and with increased rock socketing, the behaviour gradually changes from stiff to flexible.

When a stiff pile is subjected to cyclic loading, the degradation of stiffness modulus of the surrounding soil after N number of cycles (E_N) occurs throughout the whole embedded length of the pile, while in case of a flexible pile the soil mobilization and degradation occur only in the upper layers, because in the rock socket, the pile is scarcely deformed, as it is noticed from the deflection lines shown in Fig. 4. This issue leads to a higher overall soil degradation and thus a higher cyclic deformation accumulation of a stiff pile compared to a flexible pile (Albiker and Achmus 2012). Based on the previous results, the deflection lines after different numbers of cycles for piles lie closer to each other with increased rock socketing, which means that deformation accumulation is decreasing as presented in Fig. 3.

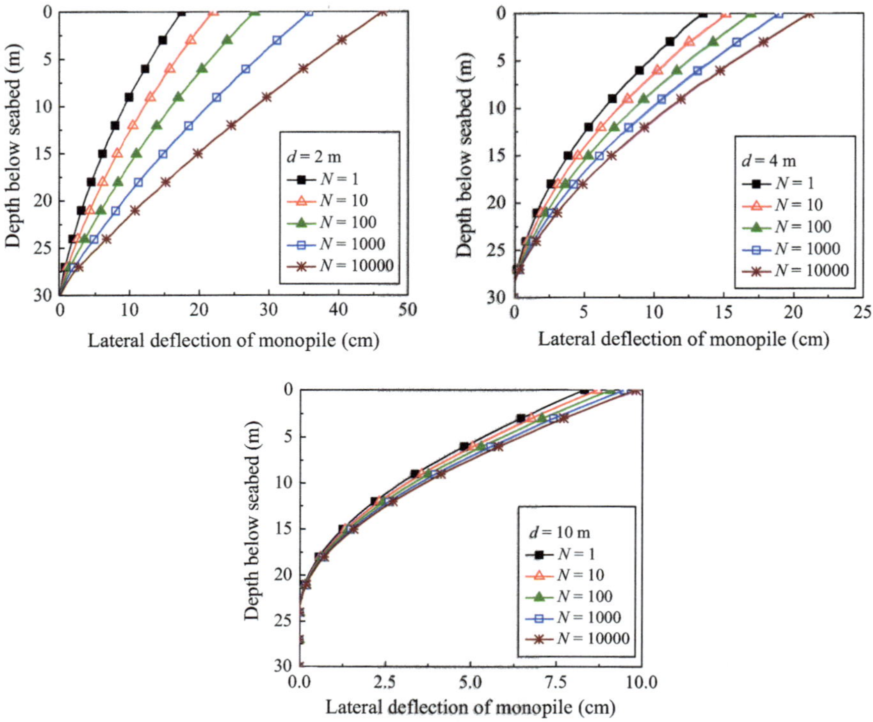

Fig. 3. Stiff to flexible behaviour with increase in rock socketing depth (monopile socketed into hard rock)

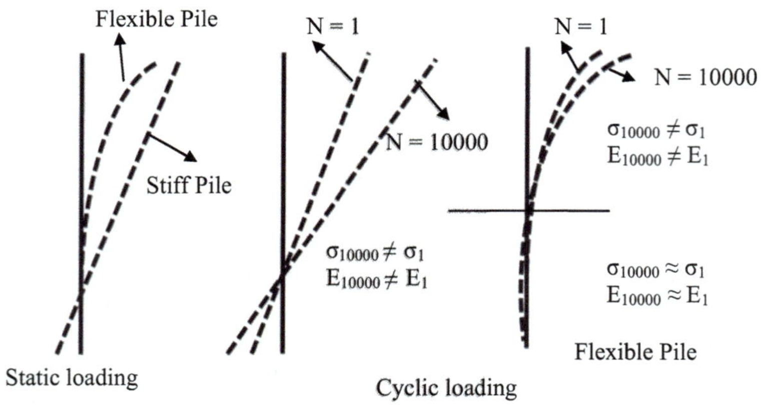

Fig. 4. Behaviour of piles when subjected to horizontal cyclic loading

5.2 Flexible Behaviour of a Monopile at Low Load Levels

Figure 5 presents the lateral deflection of monopiles at different magnitudes of lateral load. From the numerical analysis, it can be seen that at lower load levels (15 MN), the pile behaves more flexible which is in line with the observations made in Achmus and Albiker (2014). When the horizontal loading is relatively high (30 MN) with respect to the ultimate horizontal pile capacity, the soil is nearly in the failure state, which in turn leads to an apparently increased pile-soil system stiffness (stiffer behaviour) and therewith to a higher cyclic deformation accumulation as seen in Fig. 5.

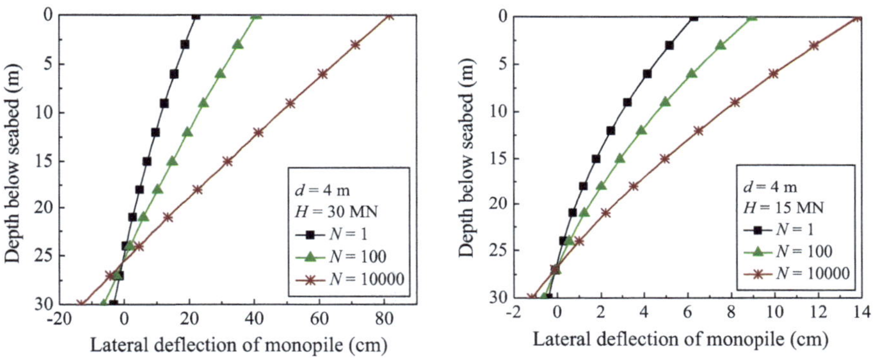

Fig. 5. Stiff and flexible behaviour of monopile socketed in soft rock at different load levels

5.3 Optimum Rock Socketed Depth

Figure 6 presents the normalized pile lateral deflection y_{sN}/y_{s1} where y_{sN} is pile deflection after N number of cycles and y_{s1} is the deflection after 1st cycle. From the Fig. 6, it can be seen that the accumulation of deformation at the pile surface decreases with increased rock socketing depth. In the case of a monopile of total length (L) of 30 m with different rock socketed lengths, it is found that above a certain depth of rock socketing, the accumulation of pile deformation is not changing or remains almost constant.

The rock socketing depth above which increase of socketing does not cause much change in accumulation of deflection of the monopile at the surface is taken as optimum depth of rock socketing. In case of a monopile of length 30 m, the optimum depth is found to be approximately 5 m, that means for a ratio between rock socketed length and total embedment length of $d/L = 1/6$, whereby this ratio might depend on the loading level.

5.4 Effect of Rock Properties on the Deformation Behaviour

Figure 7 shows lateral deflection of pile for different numbers of loading cycles and socketed in soft and hard rock. From the numerical results, it can be seen that a monopile socketed in hard rock behaves more flexible than that socketed in weak rock,

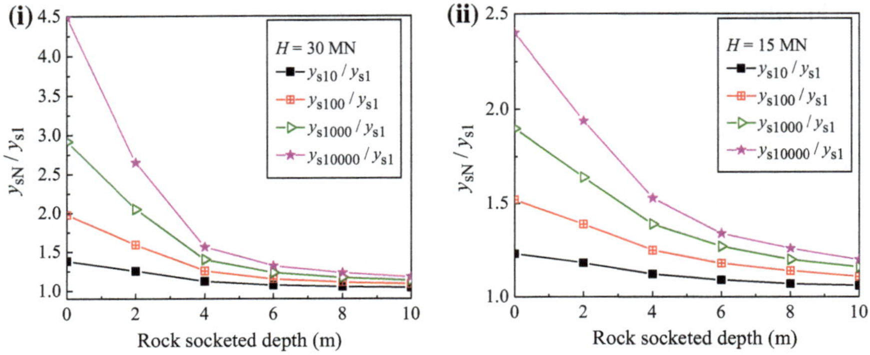

Fig. 6. Variation of accumulation rate of deformation with rock socketing depth for a monopile subjected to a cyclic lateral load (H) of (i) 30 MN (ii) 15 MN

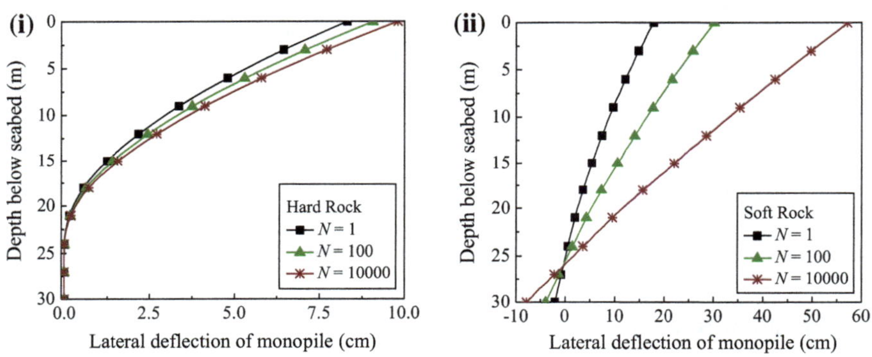

Fig. 7. Monopile of L = 30 m, d = 10 m socketed in (i) hard rock (ii) soft rock subjected to H = 30 MN

and the accumulation of head deformation for monopiles in weak rock is large compared to the other case (about 4 times larger for 4 m rock socketed piles and 6 times larger for 10 m rock socketed pile). In analogy to the statement given above when different load levels were analysed, also here it can be concluded that a higher stiffness of the rock leads to a reduced pile-soil system stiffness (flexible behaviour) and therewith to a reduced cyclic deformation accumulation. Figure 8 presents the normalized pile lateral deflection y_{sN}/y_{s1} for different rock socketing depths in soft and hard rock after increasing number of cycles. As the rock socketed depth increases, the rate of accumulation under cyclic loading reduces.

5.5 Evaluation of the Accumulation of Pile Head Deflection as a Function of Number of Cycles by Means of an Empirical Equation

There are several empirical equations to estimate the head deflection of a pile after *N* cycles such as the equation given by Little and Briaud (1988).

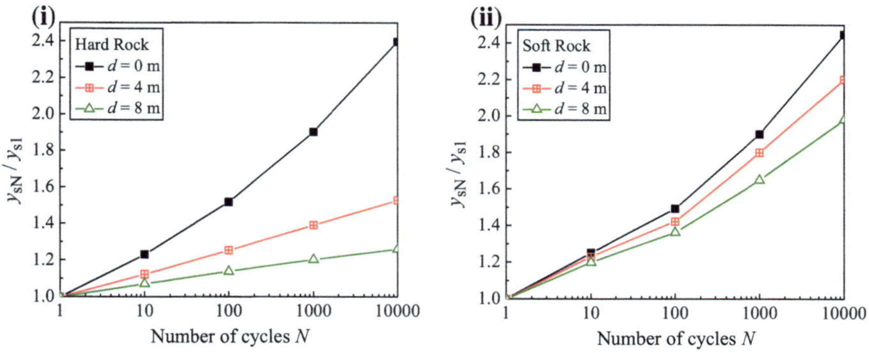

Fig. 8. Accumulation of deformation of monopile of L = 30 m socketed in (i) hard rock and (ii) soft rock subjected to H = 15 MN

$$y_N = y_1 N^m \tag{4}$$

Here y_N and y_1 are the pile head deflection after N number of cycles and 1st cycle respectively. The empirical parameter m in Eq. 4 is evaluated herein using the results (deflection at seabed level after N cycles) obtained from numerical analysis using a regression analysis. Figure 9 presents the variation in m for different rock socketed depths. From the Fig. 9, it can be seen that the value of m depends on length of monopile below seabed level (L), rock socketed depth (d), intensity of lateral loading (H) and subsoil conditions. The m increases with load intensity and decreases with increase of rock socketing and with increase of the pile length.

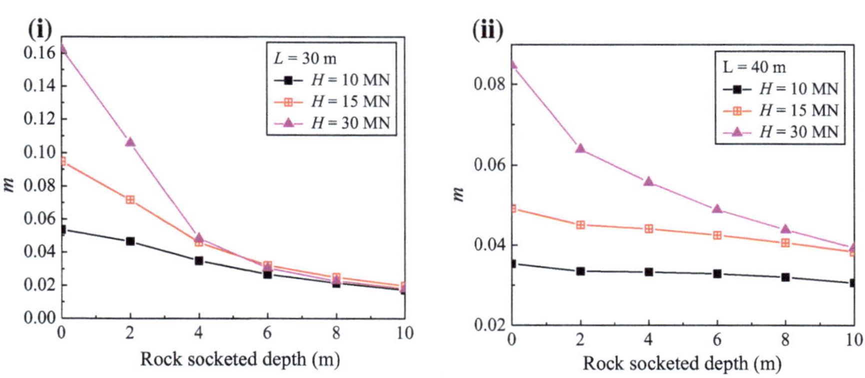

Fig. 9. Variation of degradation parameter m for monopile of (i) L = 30 m (ii) L = 40 m for case I

5.6 Bending Moment

Figures 10 and 11 represent the bending moment diagram for a monopile socketed in hard rock. The value of the maximum moment increases with increase in rock socketing. This is due to the change of pile behaviour from stiff to flexible with increased socketing into rock layer. And, due to the higher moment developing in the monopile, it has a tendency to fail in flexure even if the lateral deflection occurring is smaller compared to that of a monopile with very small rock socketed depth.

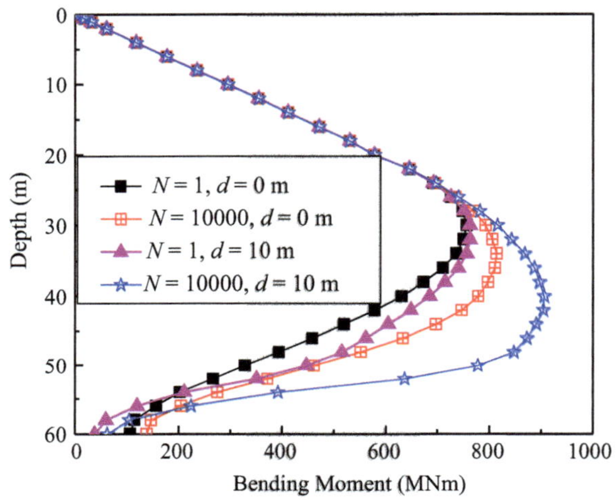

Fig. 10. Moment distribution for a monopile (L = 40 m) subjected to H = 30 MN and lever arm = 20 m

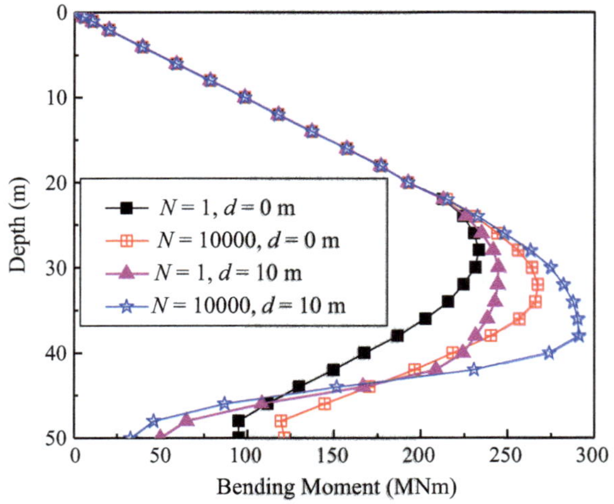

Fig. 11. Moment distribution for a monopile (L = 30 m) subjected to H = 10 MN and lever arm = 20 m

6 Conclusions

The behaviour of laterally loaded piles for varying geometry, soil conditions, lateral loading and rock socketing depths is analysed and studied numerically. From the results it can be concluded that the behaviour of the monopile changes from stiff to flexible with increase in rock socketed depth and that in turn the deformation accumulation of the pile at the surface is going down. The other observation is that accumulation rate of deflections follow the same trend. As the depth of the pile below seabed level increases, rock socketing depth has no significant effect on the accumulation of deformation at low load levels. For the evaluated cases, it is also concluded that a ratio between rock socketed length d and total pile embedment length L of approximately $d/L = 1/6$ might be seen as an orientation for sufficient cyclic stability. From the bending moment diagram, flexible and stiff behaviour of monopile can be identified and is an indicator of curvature of the deflection line of pile.

Acknowledgements. This work has been carried out with scholarship funded by DAAD for Ms. Navy Basheer for a period of six months. This work is a result of collaboration between Institute for Geotechnical Engineering, LUH, Germany and Department of Civil Engineering, IIT Delhi, India.

References

Abaqus/Standard User's Manual, Version 6.11-3: Dassault Systèmes Simulia Corporation, Providence, Rhode Island, USA, 2014

Achmus, M.: Design of axially and laterally loaded piles for the support of offshore wind energy converters. In: Indian Geotechnical Conference—2010, GEOtrendz, IGS Mumbai Chapter & IIT Bombay, 16–18 Dec 2010

Achmus, M., Kuo, Y.-S., Abdel-Rahman, K.: Behaviour of monopile foundations under cyclic lateral load. Comput. Geotech. **36**(5), 725–735 (2009)

Achmus, M., Abdel-Rahman, K., Kuo, Y.-S.: Design of monopile foundations for offshore wind energy plants. In: Proceedings of the 11th Baltic Geotechnical conference of Geotechnics in Maritime Engineering, 15–18 Sept 2008, Gdansk, Poland

Achmus, M., Albiker, J.: Prediction of accumulated deformations of cyclic laterally loaded piles in sand. In: Proceedings of the 8th European Conference on Numerical Methods in Geotechnical Engineering (NUMGE), Delft, The Netherlands, 2014

Albiker, J., Achmus, M.: Cyclic performance of horizontally loaded piles in layered subsoil. In: Proceedings of the 12th Baltic Sea Geotechnical Conference, Rostock, Germany, May 31st–June 2nd, 2012

Arshi, H.S., Stone, K.J.L.: An investigation of a rock socketed pile with an integral bearing plate founded over weak rock. In: Proceedings of the 15th European Conference of Soil Mechanics and Geotechnical Engineering, 12–15 Sept 2011, Athens, Greece, 2011

Choy, K.K., Pang, T.C.P., Li, W.W., Tse, S.H.V., Lam, S.C., Kung, W.C.F., Lau, C.W.J., Pappin, J.W., Ng, H.K., Lee, W.H., Wong, N.K.P., Lee, K.K.P., Cheng, M.L.: Code of Practice for Foundations. Technical Report. Buildings Department, Mongkok, Kowloon, Hong Kong (2004)

http://www.offshorecenter.dk/log/bibliotek/Blyth%20Wind%20Farm.pdf as seen 24th June 2017

http://www.subacoustech.com/wp-content/uploads/544R0503.pdf as seen 24th June 2017 http://www.technology.stfc.ac.uk/OWEN/workshop_3/ws3_final.pdf as seen 24th June 2017

Jaeger, J.C., Cook, N.G.W., Zimmerman, R.: Fundamentals of rock mechanics, 4th Edition, (2008) ISBN: 978-0-632-05759-7, 488 pages. April 2007, Wiley-Blackwell

Kellezi, L., Hansen, P.B.: Static and dynamic analysis of offshore monopile windmill foundation. In: Proceedings of the BGA International Conference on Foundations: Innovation, observation, design and practice, 2–5 Sept 2003, Dundee, UK

Little, R.L., Briaud, J.L.: Full Scale Cyclic Lateral Load Tests on Six Single Piles in Sand. Miscellaneous paper GL-88-27. Texas: Geotechnical Division, Texas A&M University (1988)

Liu, X., Zhang, D., Su, Y.: A method for computing embedded depth in rock of the laterally loaded rock-socketed piles. In: Proceedings of the IEEE Conference, 2011

Schmoor, K.A., Achmus, M.: On the validation of reliability and partial safety factors for axially loaded piles in dense sand. In: Proceedings of the 4th International Symposium on Geotechnical Safety and Risk (4th ISGSR), 4–6 December 2013, Hong Kong, 2013

Srinivasamurthy, B.R., Pujar, K.L.: Socketing of bored piles in rock. In: Proceedings of the Indian Geotechnical Conference (IGC) 2009, Guntur, India, 2009

Thieken, K., Achmus, M., Schmoor, K.: On the ultimate limit state design proof for laterally loaded piles. Geotechnik **37**(1), 19–31 (2014)

Wang, J.-H., Chen, J.-J., Li, Y.-L., Fan, W.: The behavior of large diameter rock-socketed piles under lateral loads. In: Proceedings of the 17th International Offshore and Polar Engineering Conference, 1–6 July 2007, Lisbon, Portugal

Seismic Re-qualification of Caisson Supported Dhansiri River Bridge

Begum Emte Ajom[1(✉)] and Arup Bhattacharjee[2]

[1] Design of Civil Engineering Structures, J.E.C, Jorhat, Assam, India
begumemte@gmail.com
[2] Jorhat Engineering College, Jorhat, Assam, India

Abstract. There are multitudinous major river bridges in India which were built prior to the development of seismic codes and it is very difficult to predict the performance of those bridges during earthquake. Since bridges are the lifeline structures, it is essential to requalify these structures in light of the new and better understanding of seismic resistant design philosophies. This paper aims to carry out a requalification study of an important river bridge supported on caisson foundations (or Well Foundations). Field investigation and laboratory tests on soil samples from the bridge site are carried out and the data achieved are used as input values for the soil model. Two different types of earthquakes, each having different dynamic properties are considered for the study. The effective stress site response analysis is carried out and the liquefaction potential of the bridge site is evaluated. Analysis revealed that a large number of the soil layers are liquefied under the applied earthquake motions. Considering the liquefied soil, seismic analysis of the bridge is carried out. The seismic analysis gives the damage levels in terms of bending moments and displacements of the well foundation. The moment of resistance of the well section is calculated and the maximum bending moments under the considered earthquakes are checked with the moment capacity of the well. It is found that the well is safe under both the earthquakes. Hence, no strengthening or retrofitting strategies of the bridge structure are required for this study.

1 Introduction

The extent of earthquake damage to the structures depends both on the details of the structure and the quality of the ground on which the structure is built. It is well established that geotechnical aspects, i.e., soil can be a major contributor to damages to the structures. It is, therefore, important to consider the seismic safety of various geotechnical structures.

Well foundations (or caisson foundations) have their origin in India and are most often used in the construction of important bridges and other structures that require foundation beneath the water bodies. Many of these bridges were constructed before the development of Indian Seismic Code (IS 1893: 2002). Therefore, re-examining these structures designed and built earlier has become essential. Krishna et al. (2014) suggested the methodology for seismic requalification studies of the structures. Sarkar et al. (2014) carried out a requalification study of pile foundations considering the soil

W.-C. Cheng et al. (eds.), *Tunneling in Soft Ground, Ground Conditioning and Modification Techniques*, Sustainable Civil Infrastructures,
https://doi.org/10.1007/978-3-319-95783-8_16

liquefaction. Bhattacharjee et al. (2014) conducted the seismic requalification of Showa Bridge which failed during the Kobe earthquake. Damala et al. (2017) conducted the seismic requalification of Saraighat Bridge over river Brahmaputra supported by well foundation.

Dhansiri River Bridge is one of the major river bridges in a high seismic zone situated in North Eastern part of India in the vicinity of the Himalayan faults and was built and opened for the traffic in the year 1964. It is important to note that the seismic liquefaction effects were rigorously brought to the attention of engineers after the 1964 Niigata earthquake and 1964 Alaska earthquake and also the first Indian Seismic code was developed in the year 1964. Therefore, it is essential to carry out seismic requalification of the said bridge considering the soil liquefaction of the bridge site. This study is mainly focused on the analysis of the sub structure considering the sub soil conditions of the structure.

2 Overview of the Selected Bridge Site

2.1 Bridge Details

An RCC bridge named 'Sankar Barua Setu' over Dhansiri River, generally termed as the Dhansiri River Bridge on NH-37 in Golaghat district of Assam (India) is selected for this study. The location map of the bridge is shown in Fig. 1. The bridge has five spans of 40.79 m each with a total length of 203.95 m. There are six numbers of piers and abutments, supported on well foundations. The internal and external diameters of the well foundation are 4.57–7.32 m respectively. The depth of each well foundation is 24 m. The structural drawing and hydraulic data of the bridge are collected from, Jorhat NH division of PWD, Assam and is shown in Fig. 2. The inertia load on each well foundation due to the load of the superstructure is calculated to be 11,397 kN. As the concrete and steel data are not available for the considered bridge, the properties are assumed as M30 grade concrete, Fe415 grade steel and a minimum amount of vertical reinforcement is considered as 0.2% of gross cross-sectional area of the well staining (IRC 78: 2000).

Fig. 1. Location of the Dhansiri river bridge. *Source* Google maps

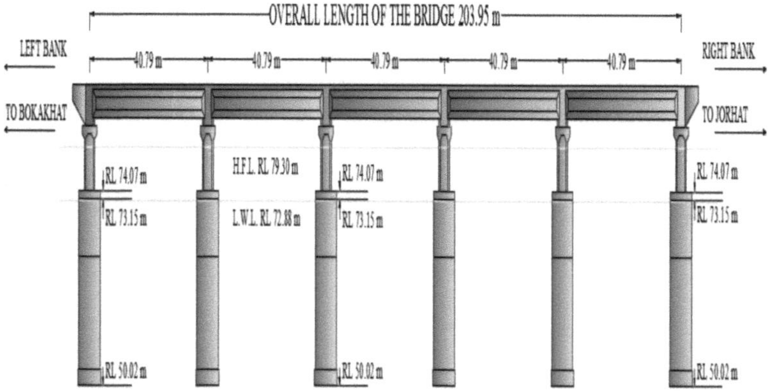

Fig. 2. Schematic view of the Dhansiri river bridge. *Source* PWD NH division, Jorhat

2.2 Local Soil Conditions

Standard Penetration Test (SPT) is conducted near the vicinity of the bridge to collect the geotechnical data. Standard penetration resistances (N values) are determined and soil samples are collected over 1 m interval up to a depth of 24 m. Data obtained from field and laboratory tests (e.g. unit weights, relative densities, shear wave velocities, etc.) are used for calculating the input parameters of the soil model.

The soil stratigraphy of the bridge site consists of loose to medium density sand layers over 24 m of depth with a clay layer at a depth of 5 m from the surface. The shear wave velocities for all layers are estimated (Maheswari et al. 2008) from the observed standard penetration resistance. The bore log of the geotechnical investigation showing different soil properties and respective N values is shown in Table 1.

Table 1. Typical soil borehole and laboratory tests data for the bridge site

Depth (m)	Soil type	Relative density, D_r (%)	SPT-N values	Saturated soil mass density, ρ_{sat} (kg/m^3)	Shear wave velocity, V_s (m/s)	Friction angle, ϕ (°)	Cohesion, c (kPa)
1	Medium sand	53	8	1640	178.8	33.5	–
2	Medium sand	42	6	1500	164.0	32	–
3	Medium sand	64	13	1810	206.9	35	–
4	Loose sand	23	2	1440	117.8	31	–
5	Clay	–	3	2690	133.1	0	37
6	Medium sand	52	7	1640	171.8	33.5	–

(*continued*)

Table 1. (*continued*)

Depth (m)	Soil type	Relative density, D_r (%)	SPT-N values	Saturated soil mass density, ρ_{sat} (kg/m^3)	Shear wave velocity, V_s (m/s)	Friction angle, ϕ (°)	Cohesion, c (kPa)
7	Medium sand	46	10	1570	191.2	33.5	–
8	Medium sand	53	8	1590	178.8	33.5	–
9	Medium sand	49	15	1630	216.1	33.5	–
10	Loose sand	28	6	1320	164.0	31	–
11	Medium sand	36	6	1460	164.0	32	–
12	Medium sand	45	10	1610	191.3	32	–
13	Medium sand	37	8	1520	178.8	32	–
14	Medium sand	41	9	1550	185.3	32	–
15	Medium sand	41	10	1540	191.3	32	–
16	Medium sand	46	18	1580	228.3	33.5	–
17	Medium sand	46	16	1600	220.3	33.5	–
18	Loose sand	35	10	1460	191.3	31	–
19	Medium sand	40	10	1520	191.3	32	–
20	Medium sand	41	14	1600	211.7	32	–
21	Medium sand	60	25	1910	252.0	35	–
22	Medium sand	57	24	1610	248.9	35	–
23	Medium sand	52	25	1590	252.0	33.5	–
24	Medium sand	56	29	1820	263.5	35	–

3 Input Motions

The entire North Eastern region of India, including Assam, is regarded as one of the most seismically active regions of the world and according to IS 1893 (part I): (2002), this region is classified as one of the highest seismic zone in the country with an expected PGA of 0.36 g. Therefore two different types of earthquakes are considered for this study. The Loma Prieta Earthquake (1989) motion recorded at station GilroyNo.1 EW with a peak ground acceleration (PGA) value of 0.47 g is selected as it has a higher PGA than the expected PGA of the region. Also, a regional earthquake is considered pertaining to PGA 0.36 g, scaled from the seismic ground motion recorded during the Sikkim Earthquake (2011) at the IIT Guwahati campus. Acceleration time histories of the considered earthquakes and their corresponding Fast Fourier Transformation (FFT) are presented in Fig. 3.

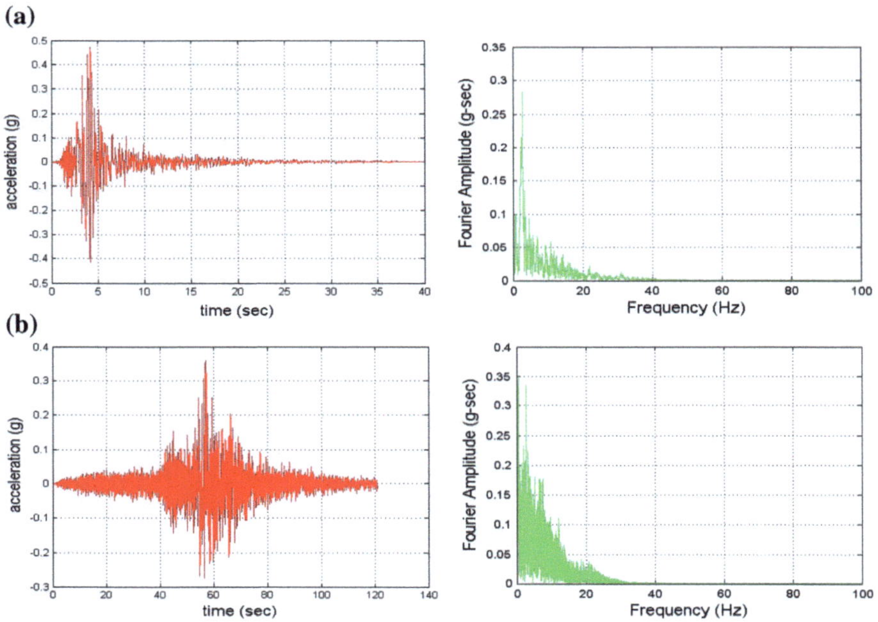

Fig. 3. Acceleration time histories and their corresponding FFTs of different earthquakes **a** Loma Prieta earthquake **b** Sikkim earthquake

4 Numerical Modeling of the Soil Column and Soil-Foundation Interaction

4.1 Site Response Analysis

Using the data obtained from the field investigation and laboratory tests, site response and liquefaction analyses are carried out in Open System for Earthquake Engineering Simulations (OpenSees) finite-element analysis software framework. A multi-layered single column of soil is modeled in 2D with periodic boundary conditions to emulate a 1D analysis and is subjected to an earthquake excitation at the base (Fig. 4). The soil column is modeled with 24 layers, each of 1 m thick. Based on convergence analysis, 48 elements are considered across a 24 m depth, one element of which is 2 m wide and 0.5 m deep. Nine node quadrilateral elements are used to model the soil. These elements are able to track displacements of the solid, and pore-water pressure in the fluid. Out of nine nodes, four corner nodes (shown in red in Fig. 4) have three degrees of freedom (DoF), two translational and one pore pressure, and the interior nodes (shown in blue in Fig. 4) have only two translational DoF. The effective stress approach is adopted to investigate the liquefaction potential of the bridge site. The analysis is carried out in undrained condition to facilitate the development of pore pressure. The nodes at the base of the soil column are fixed against vertical translation to represent the base as fixed. To perform a 1D free-field analysis, periodic boundary conditions are assumed. In OpenSees, periodic boundary conditions are achieved by setting a multi-point constraint between nodes at the same depth from the surface. This multi-point constraint controls the constrained nodes to have the same horizontal and vertical displacements at either end of the soil column. The soil is modeled using the Pressure Dependent Multi Yield02 (PDMY02) material for sand and Pressure

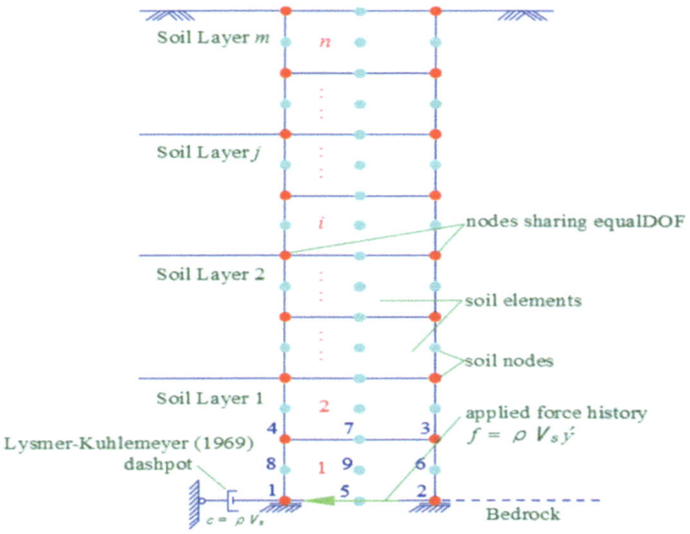

Fig. 4. Schematic representation of the site response model

Table 2. Parameter values for the soil material models

Depth (m)	Shear modulus, G (kPa)	Bulk modulus, B (kPa)	Phase transformation angle, ϕ_{PT} (°)	Non-negative constants				Void ratio, e
				contact1	contact3	dilate1	dilate3	
1	52,429.88	233,000	25.5	0.045	0.15	0.06	0.15	0.7
2	40,344	220,000	26	0.067	0.23	0.06	0.27	0.77
3	77,481.77	240,000	26	0.028	0.05	0.1	0.05	0.65
4	19,982.65	160,000	31	0.087	0.18	0	0	0.85
5	47,654.99	300,000	–	–	–	–	–	–
6	48,404.99	233,000	25.5	0.045	0.15	0.06	0.15	0.7
7	57,395.18	233,000	25.5	0.045	0.15	0.06	0.15	0.7
8	50,831.41	233,000	25.5	0.045	0.15	0.06	0.15	0.7
9	76,119.71	233,000	25.5	0.045	0.15	0.06	0.15	0.7
10	35,502.72	160,000	31	0.087	0.18	0	0.18	0.85
11	39,268.16	220,000	26	0.067	0.23	0.06	0.23	0.77
12	58,919.06	220,000	26	0.067	0.23	0.06	0.23	0.77
13	48,593.55	220,000	26	0.067	0.23	0.06	0.23	0.77
14	53,220.94	220,000	26	0.067	0.23	0.06	0.23	0.77
15	56,357.36	220,000	26	0.067	0.23	0.06	0.23	0.77
16	82,351.01	233,000	25.5	0.045	0.15	0.06	0.15	0.7
17	77,651.34	233,000	25.5	0.045	0.15	0.06	0.15	0.7
18	53,429.71	160,000	31	0.087	0.18	0	0.18	0.85
19	55,625.45	220,000	26	0.067	0.23	0.06	0.23	0.77
20	71,707.02	220,000	26	0.067	0.23	0.06	0.23	0.77
21	121,292.6	240,000	26	0.028	0.05	0.1	0.05	0.65
22	99,741.45	240,000	26	0.028	0.05	0.1	0.05	0.65
23	100,971.4	233,000	25.5	0.045	0.15	0.06	0.15	0.7
24	126,366.7	240,000	26	0.028	0.05	0.1	0.05	0.65

Independent Multi Yield (PIMY) material for clay. The most significant material parameters for soil model are reported in Table 2. Validation of the seismic response of the soil mesh modeled in OpenSees is completed by comparing the PGA profile predicted using OpenSees with that predicted using DEEPSOIL. Loma Prieta earthquake (1989) acceleration-time series (Fig. 3a) is used as the input motion. The comparison of PGA profile using both the codes is shown in Fig. 5. The OpenSees analysis correlates well with the results obtained using the nonlinear 1D analysis in DEEPSOIL.

4.2 Liquefaction Potential

The pore pressure ratios (computed as the ratio of excess pore pressure to initial effective vertical stress) obtained from the four corner nodes of each element with depth of the soil layers is shown in Fig. 6. The soil layers from 6 to 20 m and 6 to 23 m are liquefied when subjected to Loma Prieta and Sikkim earthquake respectively.

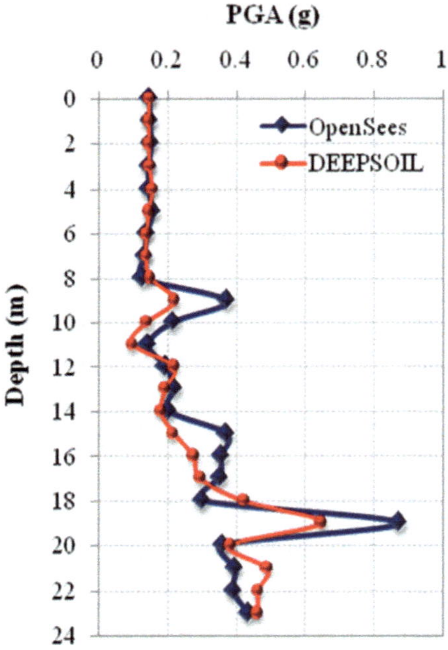

Fig. 5. PGA profile of the soil column as compared to DEEPSOIL results

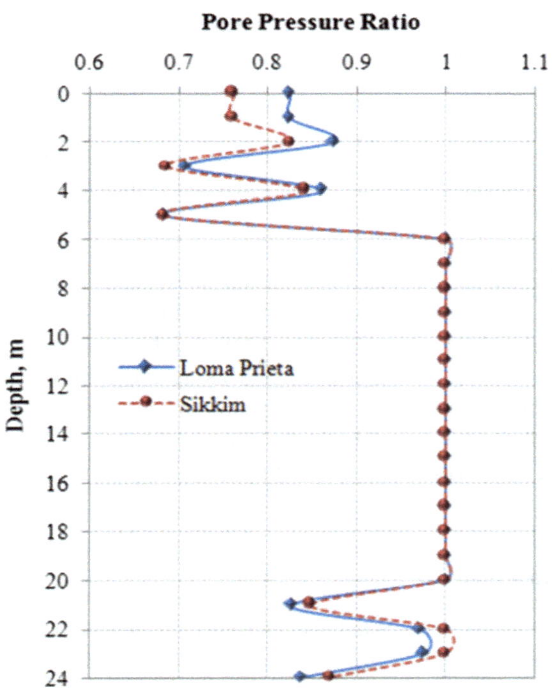

Fig. 6. Variation of pore water pressure ratio with depth for the two different earthquakes

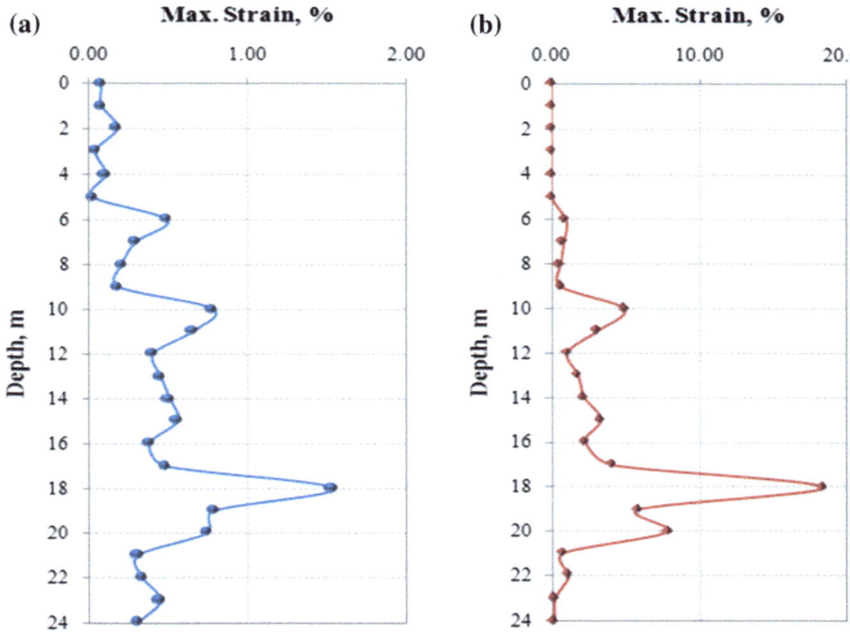

Fig. 7. Variation of maximum strain percentage with depth under **a** Loma Prieta earthquake **b** Sikkim earthquake

Variation of maximum strain with depth for different earthquakes is shown in Fig. 7a, b. From both the figures, it can be seen that the strain percentage within the liquefied zone is more compared to the other layers. This is due to liquefaction of the soil.

The time histories of pore pressure ratio for the liquefied layers under Loma Prieta earthquake are shown in Fig. 8. The pore pressure ratios of the liquefied layers approaches 1.0 at 5 s and it continues up to about 20 to 25 s of the earthquake indicating that the soil medium is completely liquefied during this time.

4.3 Dynamic Soil-Foundation Interaction

A two-dimensional finite element model is developed in OpenSees for seismic analysis of the soil-foundation system. The well foundation is modeled using the nonlinear beam-column element (Neuenhofer and Filippou 1998). The depth of the well is 24 m having 96 elements. The well is connected to the soil column in such a way that the tip of the well rests 2 m above the assumed bed rock level. The fiber section is used to model the cross-section of the well as it considers the inner and the outer diameter of the well. The uniaxial materials are used for modeling the concrete and the reinforcing steel and these materials are designated as Concrete02 (Yassin 1994) and Steel02 (Filippou et al. 1983), respectively in OpenSees. The input material properties of well foundation are summarized in Table 3.

The soil column modeled for site response analysis is connected to the well foundation with zero-length nonlinear springs which represents the flexibility of the

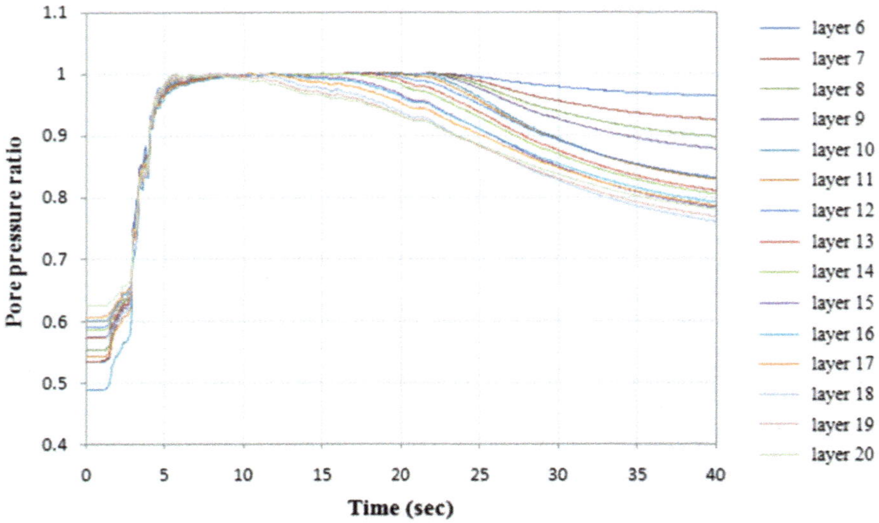

Fig. 8. Pore pressure ratios versus time under Loma Prieta earthquake

Table 3. Parameter values for modeling the well foundation

Concrete compressive strength	Elastic modulus of concrete	Ultimate stress in concrete	Ultimate strain in concrete	Tensile strength of concrete	Steel yield stress	Elastic modulus of steel
30,000 kPa	25×10^6 kPa	10,800 kPa	0.0035	3834 kPa	415,000 kPa	2×10^8 kPa

soil-foundation interface. Three types of soil-interface springs are used to model the soil-foundation interface viz. lateral resistance (*p-y*), skin friction (*t-z*), and end bearing resistance (*q-z*). Each spring is defined by an ultimate resistance (p_{ult}, t_{ult}, and q_{ult}) and the displacement at which 50% of the ultimate resistance is mobilized (i.e. y_{50} for *p-y* and z_{50} for the *t-z* springs) (Boulanger et al. 1999). The parameters defining the soil springs are chosen in accordance with recommendations from API (1993). The values of p_{ult}, t_{ult}, q_{ult} and the corresponding y_{50} and z_{50} are summarized in Table 4.

The gapping effects are modeled by incorporating residual resistance or drag force along the sides of the well. The drag coefficient, C_d, is 0.3 for the drag resistance within a fully mobilized gap (Romney et al. 2014).

From the site response analysis, it is found that a greater portion of the soil column is liquefied. One approach for approximating the effects of liquefaction on *p-y* behavior for design purposes is to apply a reduction factor, the p-multiplier (m_p) to the *p-y* resistance (Ashford et al. 2011). In this study, the *p-y* springs of the liquefied soil are modeled by reducing the strength and stiffness of the springs using the p-multiplier suggested by Brandenberg et al. (2005).

Figure 9 illustrates the schematic of the finite element model developed in OpenSees with the various well, springs and soil components. The nonlinear finite element analysis of the soil-foundation system is divided into three stages to simulate in situ soil conditions both pre- and post-construction and to incorporate the effects of staged

Table 4. Summary table of the calculated *p-y*, *t-z* and *q-z* soil-interface spring value

Depth (m)	p_{ult} (kN)	y_{50} (m)	t_{ult}/q_{ult} (kN)	z_{50} (m)
0	0	0.00351	241.3404	0.002295
1	46.2346	0.00418	241.3404	0.002295
2	115.323	0.01389	192.7275	0.002753
3	247.705	0.00542	316.0410	0.002187
4	241.611	0.00672	86.19300	0.003364
5	228.938	0.00509	258.5790	0.004583
6	491.177	0.00910	229.8480	0.002286
7	751.223	0.01193	201.1170	0.002497
8	949.747	0.02860	241.3404	0.002300
9	1008.44	0.01620	201.1170	0.002244
10	983.870	0.01422	114.9240	0.003193
11	1275.03	0.01675	143.6550	0.002625
12	1639.42	0.01975	160.8936	0.002047
13	1789.00	0.01989	155.1474	0.002705
14	1977.58	0.01570	189.6246	0.002783
15	2795.82	0.02072	189.6246	0.002772
16	3228.32	0.04864	201.1170	0.002427
17	3640.12	0.03096	201.1170	0.002475
18	3695.69	0.02968	143.6550	0.002784
19	5445.58	0.02589	195.3708	0.004583
20	3893.54	0.01759	189.6246	0.002767
21	7158.26	0.03791	287.3100	0.002241
22	4930.47	0.02025	28,142.99	0.034000

construction of the structural components. In the first stage, soil gravity load is applied to simulate pre-construction sub-surface conditions. A nonlinear transient analysis is performed with large time steps simulating a static analysis, in which the Krylov-Newton algorithm (Scott and Fenves 2010) is used to solve the equilibrium equations. In the second stage, well nodes and fixities are defined followed by the material and element properties. Loads and masses are assigned to the nodes. Before application of the structure gravity loads, the well and the soil column are connected by the nonlinear springs. A transient analysis is performed for the application of the gravity loading. In the application of both soil and bridge structure gravity loads, large numerical damping is introduced to simulate static loading. In third and final stage, the nonlinear time history analysis is carried out by applying the earthquake motion to the base of the soil column. Numerical convergence is achieved for all the stages.

5 Results and Discussion

The bending moments and relative displacements of well foundation subjected to two different earthquakes Loma Prieta earthquake and Sikkim earthquake are determined.

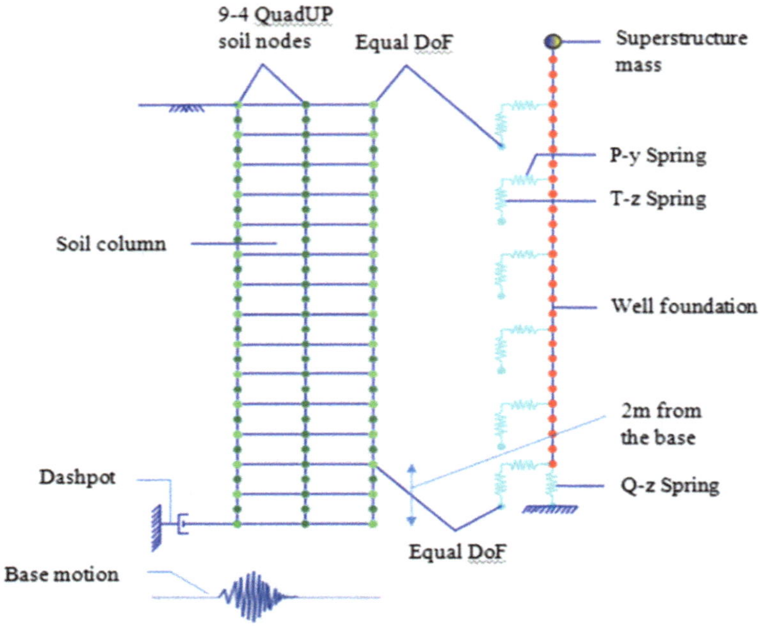

Fig. 9. Schematic representing the finite element model for soil-foundation system

5.1 Bending Moment on Well Foundation

Figure 10 shows the variation of bending moments along the depth of the well for both the earthquakes. The maximum bending moments which occur near the midpoint of the well foundation are 435 and 1300 kN-m under Loma Prieta and Sikkim earthquake respectively. The well foundation, when subjected to Sikkim earthquake undergoes more bending moment compared to the Loma Prieta earthquake. Though the PGA (0.47 g) of the Loma Prieta earthquake is more, the bending moment is higher in case of Sikkim earthquake due to higher amplitude of Sikkim earthquake over Loma Prieta earthquake (Fig. 3a). The moment capacity of well foundation is determined as per Rai et al. (2006). The ultimate moment of resistance of the well section is calculated to be 225 MN-m. The moment produced in the well foundation is well within the ultimate moment of resistance of the well and thus the failure of the well foundation due to bending moment can be ruled out.

5.2 Horizontal Displacement of the Well Foundation

The performance criteria of the bridge will be governed by the pier head displacement to arrest the possible deck unseating failures. Therefore, the displacement of the top of the well is checked which would lead to the displacement of the pier. Figure 11 shows the displacement of the top of the well with respect to the base of the well foundation for both the earthquakes at the end of the respective ground motions. The displacement of the top of the well is very small for both the earthquakes. The maximum relative displacements at top of well foundation are 0.08 and 0.36 mm for Loma Prieta and

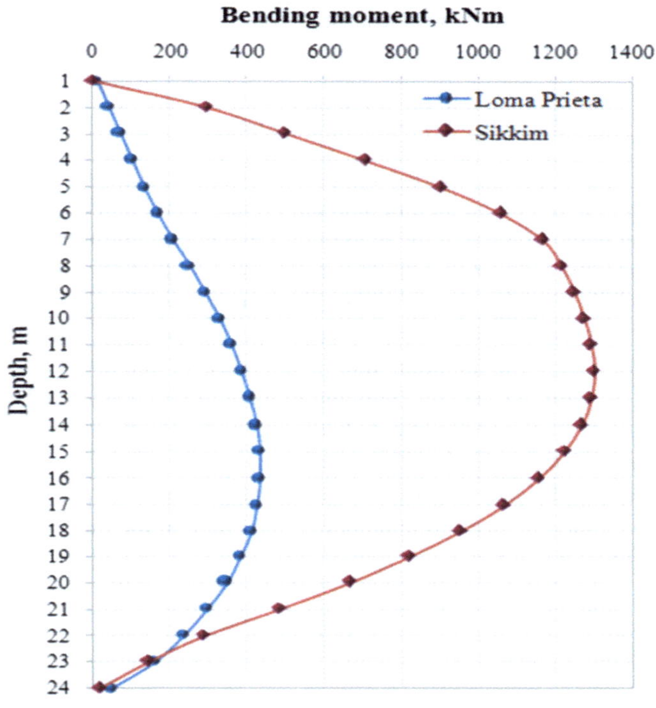

Fig. 10. Variation of bending moment along the depth of the well for the two different earthquakes

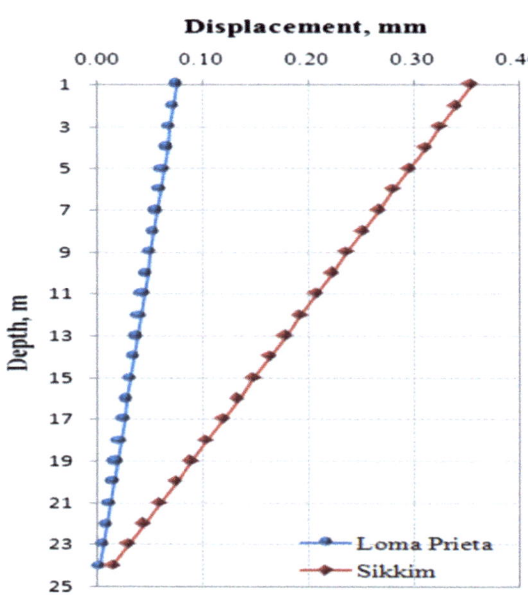

Fig. 11. Depth versus relative displacement of the well for the two different earthquakes

Sikkim earthquake respectively which is negligible. Therefore, there is no chance of falling off the deck of the bridge.

5.3 Effect of Liquefaction on Well Foundation

Figure 12 shows the variation of bending moment along the depth of well foundation at various time interval viz. at the time of application of ground motion, at the time when bending moment is the maximum, at the start of liquefaction, during liquefaction and at the end of liquefaction. From the figure, it is observed that the maximum bending moment is 435 kN-m, which occurs before the initiation of liquefaction (at 3.435 s). During liquefaction, the bending moment reduces to 20 kN-m, unlike pile foundations where the bending moment increases during liquefaction due to low stiffness of the foundation and the surrounding soil (Puri and Prakash 2008; Sarkar et al. 2014). Based on this observation, it may be inferred that there is no effect of liquefaction on the bending moment of the well foundation because of the high rigidity of the well foundation which needs relatively low stiffness contribution from the surrounding soil. It is also observed that the bending moment is maximum at the time when the PGA is attained (Fig. 3a).

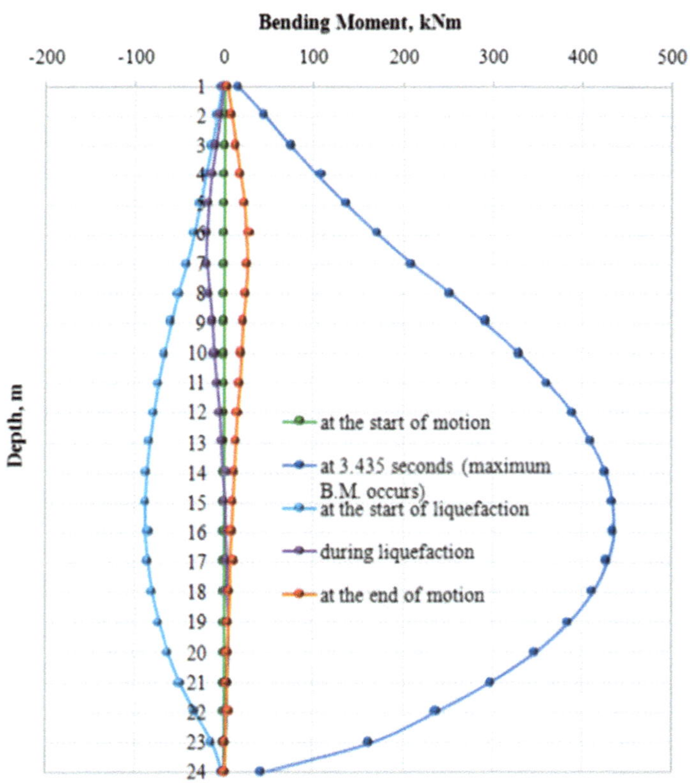

Fig. 12. Variation of bending moment along the depth of the well at different time steps for Loma Prieta earthquake

6 Conclusion

Seismic requalification is vitally important to verify the design and construction methods adopted earlier to meet today's or future requirements in all respects. In this paper, a requalification study is carried out for a caisson supported river bridge. Two types of earthquakes are considered with varying PGAs and dynamic properties and the effects of these earthquakes on the well foundation are observed considering soil liquefaction. The failure mechanism of the bridge is obtained in terms of bending moments and relative displacements of well foundation. It is found that the bending moment of the well is directly dependent on the amplitude of the input motions and not on the peak acceleration of the input motions. The bending moments of the well foundation are much less than the moment capacity of the well. Therefore, failure of the bridge due to bending of the well foundation has been ruled out. The horizontal displacement of the top of the well is also found to be very less. Therefore, there is no chance of falling off the deck of the bridge. Also, there is no effect of liquefaction on the bending moment of the well foundation because of the high rigidity of the well foundation which needs relatively low stiffness contribution from the surrounding soil. The overall conclusion of the study is that the bridge is safe and hence no seismic strengthening or retrofitting is required. The methodology presented in this paper can be used to carry out similar studies.

Acknowledgements. The first author would like to thank the supervisor of this work Dr. Arup Bhattacharjee, J.E.C., Assam for his inspiring guidance and constant encouragement in completing the work. The authors acknowledge Jorhat NH division of PWD, Assam for providing the structural drawing and hydraulic data of the bridge.

References

American Petroleum Institute: Recommended Practice for Planning, Designing and Construction Fixed Offshore Platforms. American Petroleum Institute, Washington, D.C. (1993)

Ashford, S.A., Boulanger, R.W., Brandenberg, S.J.: Recommended design practice for pile foundations in laterally spreading ground. In: PEER Report 2011/04, Pacific Earthquake Engineering Research Center, College of Engineering, University of California, Berkeley (2011)

Boulanger, R.W., Curras, C.J., Kutter, B.L., Wilson, D.W., Abghari, A.: Seismic soil-pile-structure interaction experiments and analyses. J. Geotech. Geoenviron. Eng. **125** (9), 750–759 (1999)

Brandenberg, S.J., Boulanger, R.W., Bruce, L., Kutter, B.L., Dongdong, C.D.: Behavior of pile foundations in laterally spreading ground during centrifuge tests. J. Geotech. Geoenviron. Eng. (2005). https://doi.org/10.1061/(ASCE)1090-0241(2005)131:11(1378)

Brandenberg, S.J., Zhao, M., Boulanger, R.W., Wilson, D.W.: p-y plasticity model for nonlinear dynamic analysis of piles in liquefiable soil. J. Geotech. Geoenviron. Eng. (2013). https://doi.org/10.1061/(ASCE)GT.1943-5606.0000847

Choobbasti, A.J., Zahmatkesh, A.: Computation of degradation factors of p-y curves in liquefiable soils for analysis of piles using three-dimensional finite-element model. Soil Dyn. Earthq. Eng. (2016). https://doi.org/10.1016/j.soildyn.2016.07.017

Chopra, A.K.: Dynamics of Structures: Theory and Applications to Earthquake Engineering, 4th edn. Prentice Hall, Englewood Cliffs, New Jersey (2012)

Cubrinovski, M., Ishihara, K.: Empirical correlation between SPT N-value and relative density for sandy soils. Jpn. Geotech. Soc. **39**(5), 61–71 (1999)

Dayal, U., Jain, S.K.: Liquefaction analysis of a bridge site in Assam (India). In: International Conferences on Recent Advances in Geotechnical Earthquake Engineering and Soil Dynamics, p. 40 (2001)

Dammala, P.K., Bhattacharya, S., Krishnaa, A.M., Kumara, S.S., Dasguptaa, K.: Scenario based seismic re-qualification of caisson supported major bridges—a case study of Saraighat Bridge. Soil Dyn. Earthq. Eng. (2017). https://doi.org/10.1016/j.soildyn.2017.06.005

Fard, M.Y., Babazadeh, M., Yousefzadeh, P.: Soil liquefaction analysis based on geotechnical exploration and in situ test data in the Tabriz Metro Line 2. In: International Conference on Case Histories in Geotechnical Engineering, p. 31 (2013)

Filippou, F.C., Bertero, V.V., Popov, E.P.: Effects of Bond Deterioration on Hysteretic Behavior of Reinforced Concrete Joints. Earthquake Engineering Research Center, Berkeley, California (1983)

Gazetas, G.: Foundation vibrations. In: Fang, H.Y. (ed.) Foundation Engineering Handbook, 2nd edn. Chapman and Hall, New York (1991). Chapter 15

Gazetas, G., Mylonakis, G.: Seismic soil-structure interaction: New evidence and emerging issues. In: Proceeding, Geotechnical Earthquake Engineering and Soil Dynamics, ASCE, Reston, VA, pp. 1119–1174 (1998)

Idriss, I.M., Boulanger, R.W.: Semi-empirical procedures for evaluating liquefaction potential during earthquakes. In: Eleventh ICSDEE & 3rd ICEGE, pp. 32–56 (2004)

IS 1893: (Part-I)-2002 Indian standard criteria for earthquake resistant design of structures. Bureau of Indian Standards, New Delhi (2002)

IS 1893: Criteria for earthquake resistant design of structures. Bureau of Indian Standards, Govt. of India, New Delhi (2002)

IRC 78: Standard specifications and code of practice for road bridges, section VII – foundations and substructure. The Indian Road Congress, New Delhi (2000)

IRS: Manual on the design and construction of well and pile foundations. In: Research designs and standards organization, Lucknow—226011 (1985)

Janalizadeh, A., Zahmatkesh, A.: Lateral response of pile foundations in liquefiable soils. J. Rock Mech. Geotech. Eng. **7**, 532–539 (2015)

Joyner, W.B., Chen, A.T.F.: Calculation of nonlinear ground response in earthquakes. Bull. Seismol. Soc. Am. **65**(5), 1315–1336 (1975)

Kramer, S.L.: Geotechnical Earthquake Engineering. Prentice Hall Inc., Upper Saddle River, New Jersey (1996)

Krishna, A.M., Bhattacharya, S., Choudhury, D.: Seismic requalification of geotechnical structures. Indian Geotech. J. **44**(2), 113–118 (2014). https://doi.org/10.1007/s40098-014-0115-5

Kumar, S.S., Krishna, A.M.: Seismic ground response analysis of some typical sites of Guwahati City. Int. J. Geotech. Earthq. Eng. **4**(1), 83–101 (2013)

Lysmer, J., Kuhlemeyer, A.M.: Finite dynamic model for infinite media. J. Eng. Mech. Division ASCE **95**, 859–877 (1969)

Maheswari, R.U., Boominathan, A., Dodagoudar, G.R.: Development of empirical correlation between shear wave velocity and standard penetration resistance in soils of Chennai city. In: The Fourteenth World Conference on Earthquake Engineering, Beijing, China (2008)

Mazzoni, S., McKenna, F., Scott, M.H., Fenves, G.L., et al.: OpenSees Command Language Manual. PEER University of California, Berkeley (2006)

Neuenhofer, A., Filippou, F.C.: Geometrically nonlinear flexibility-based frame finite element. J. Struc. Eng. **124**, 704 (1998). **125**(1), 116–116 (1999)

Ravichandran, N., Krishnapillai, S.H., Bhuiyan, A.H., Eleanor, L., Huggins, E.L.: Simplified finite-element model for site response analysis of unsaturated soil profiles. Int. J. Geomech. (2015). https://doi.org/10.1061/(ASCE)GM.1943-5622.0000489

Rai, D.C., Kumar, K., Kaushik, H.B.: Ultimate flexural strength of reinforced concrete circular hollow sections. Indian Concr. J. **12**, 39–45 (2006)

Puri, V.K., Prakash, S.: Pile design in liquefying soil. In: The Fourteenth World Conference on Earthquake Engineering, Beijing, China, 12–17 Oct 2008

Romney, K.T., Barbosa, A.R., Mason, H.B.: Developing a soil bridge interaction model for studying the effects of long-duration earthquake motions. In: Proceedings of the 10th National Conference in Earthquake Engineering, Earthquake Engineering Research Institute, Anchorage, AK (2014)

Saran, S.: Solid Dynamics and Machine Foundations. Galgotia Publications pvt. Ltd., New Delhi (1999)

Sarkar, R., Bhattacharya, S., Maheshwari, B.K.: Seismic requalification of pile foundations in liquefiable soils. Indian Geotech. J. (2014). https://doi.org/10.1007/s40098-014-0112-8

Sarkar, R., Maheshwari, B.K.: Influence of soil nonlinearity and liquefaction on dynamic response of pile groups. The Fourteenth World Conference on Earthquake Engineering, Beijing, China, 12–17 Oct (2008)

Scott, M.H., Fenves, G.L.: Krylov subspace accelerated Newton algorithm: application to dynamic progressive collapse simulation of frames. J. Struct. Eng. (2010). https://doi.org/10.1061/(ASCE)ST.1943-541X.0000143

Su, D., Yan, W.M.: A multidirectional p–y model for lateral sand–pile interactions. The Japanese Geotechnical Society. Soils Found. **53**(2), 199–214 (2013)

Yassin, M.H.M.: Nonlinear analysis of prestressed concrete structures under monotonic and cycling loads. Ph.D. Thesis, University of California, Berkeley, California (1994)

Zhang, S.Y., Conte, J.P., Yang, Z., Elgamal, A., Bielak, J., Acero, G.: Two-dimensional nonlinear earthquake response analysis of a bridge-foundation-ground. Earthquake Spectra **24** (2), 343–386 (2008). https://doi.org/10.1193/1.2923925

Numerical Investigation on Muck Pressures During EPB Shield Tunneling with Varying Discharge Ratio Based on Coupled PFC3D/FLAC3D Method

Tongming Qu, Shuying Wang(✉), Jinyang Fu, Qinxin Hu, and Junsheng Yang

School of Civil Engineering, Central South University, Changsha, Hunan, China
sywang@csu.edu.cn

Abstract. In this study, an efficient numerical method coupled PFC3D and FLAC3D was presented to simulate the tunneling behaviors of EPB shield. Groups of tunneling cases with different muck-discharging ratios were conducted in PFC3D and FLAC3D with data exchanging each other in every step. The results show the muck pressures in the back of spokes are invariably less than those behind the opening of cutterhead. Moreover, pressure differences exist permanently not just between left side and right side in chamber, but in different longitudinal positions within the domain affected by tunneling, and the values of these pressure differences are subjected to muck-discharging ratio in chamber. These findings extend knowledge regarding the tunneling mechanisms of EPB shield. Besides, the proposed coupling method can be popularized with all studies about the interaction between engineering devices and soil as well.

Keywords: Earth pressure balanced (EPB) shield · Coupling FLAC3D/PFC3D Discharge ratio · Discrete element method (DEM) · Muck pressure

1 Introduction

It is well-known that the speed of muck entering and exiting the chamber in shield determines responses in ground. Different ground responses correspond to different distribution and transmission features of muck pressure in chamber. Being restricted to the difficulties of conventional research approaches such as laboratory tests and common numerical tools in capturing data and describing tunneling behaviors, few existing literatures pay attention to the muck pressures associated with various discharging ratios during tunneling. However, this point is critical to know the working mechanisms and improve the control level of shield. This paper investigates response characteristics on muck pressures during tunneling of EPB shield using FLAC3D/PFC3D coupled numerical modeling. Particularly, the pressure differences within the influence domain by tunneling of shield are studied in detail.

W.-C. Cheng et al. (eds.), *Tunneling in Soft Ground, Ground Conditioning and Modification Techniques*, Sustainable Civil Infrastructures,
https://doi.org/10.1007/978-3-319-95783-8_17

2 Implementation of Coupling FLAC3D/PFC3D Numerical Scheme

2.1 Calculation Principles

The simulation of coupling FLAC3D/PFC3D is achieved by exchanging the velocity and force between finite zones and discrete balls in each calculation step. As the Fig. 1 shows, the data exchange between the FLAC3D zones and the PFC3D walls is achieved through the Socket I/O, with the PFC3D and FLAC3D programs being a server and client, respectively.

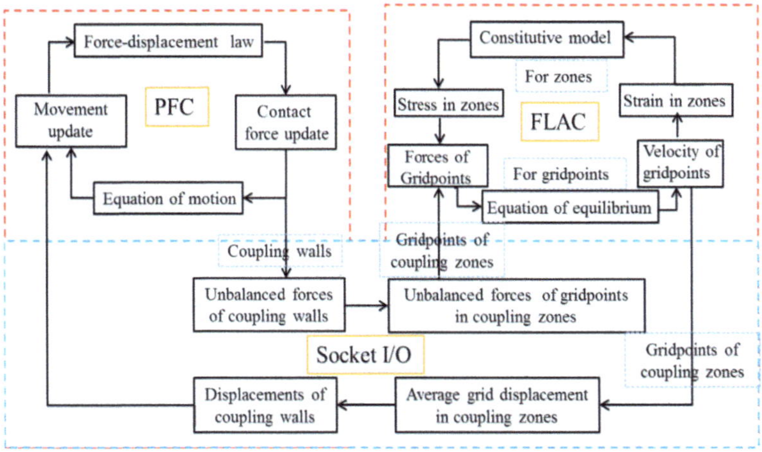

Fig. 1. Coupling procedures between the FLAC3D zones and the PFC3D zone

The displacements of gridpoints at the coupling sides of the FLAC3D zones are developed based on the equilibrium equations. On the one hand, the element strain is developed based on the node displacement, and then the stress is calculated according to its constitutive relations. As a result, the stress in zones and forces of gridpoints are renewed. On the other hand, the displacements at the coupling faces in FLAC3D are assigned to the walls of the PFC3D through the Socket I/O and the positions of both walls and particles are then updated. Based on the force-displacement law between adjoining particles, the contact forces are updated and unbalanced forces of coupling walls are assigned to the gridpoints at the coupling sides of the FLAC3D zones. Essentially, the complete coupling process is achieved by providing dynamic boundary conditions (force and displacement, respectively) for FLAC3D and PFC3D models each other.

2.2 Calibration of Micro-parameters

The Xiangjiang sand in Changsha, China, was simulated as the ground passed by the EPB shield. The micro-parameters were calibrated by numerical direct shear modelling and previous experienced formula. In reality, the sand includes particles of different sizes, for example, the very fine ones. It is nearly impossible to simulate fine particles using the DEM in a large model at present. However, it is both reasonable and practical to use particles with the same diameters to investigate the macroscopic shear behaviour of the soil, as long as the models containing uniformly sized particles show the same shear behaviours as those they include particles of various sizes (Coetzee 2017). In this study, particles with the radius of 8 mm were therefore adopted for PFC model, considering the practicability of the modelling. Finally, the micro-parameters of the sand, such as the friction coefficient of the particles, were obtained by careful calibration. Both the micro and macro parameters of numerical sand are listed in Table 1.

Table 1. Physical and mechanical parameters of the Xiangjiang sand and its microscopic parameters in the numerical modeling

Xiangjiang sand	Macroscopic parameters	Values	Simulated sand	Microscopic parameters	Values
	Density (kg m^{-3})	2630		Particle radius (mm)	8
	Cohesion (kPa)	0		Friction coefficient	3.3
	Friction angle (°)	37.5		Normal stiffness (N m^{-1})	1.15e6
	Elastic modulus (MPa)	40		Tangential stiffness (N m^{-1})	4.32e5
	Poisson ratio	0.2			

2.3 Numerical Model of EPB Shield

The engineering scale such as the excavation of EPB shield is immense, and it is nearly impossible to simulate tunnelling by a real shield at a 1:1 scale even using the state-of-the-art discrete element algorithm. This paper does not aim to offer quantitative applications for specific engineering but focus a qualitative investigation of ground responses by shield tunnelling. Here, a scaled model of an EPB shield with a diameter of 0.88 m was modelled (Fig. 2). The coupled numerical model included initial ground with dimensions of 6120 mm × 4200 mm × 2640 mm. And the DEM model was centrally located with a size of 1080 mm × 720 mm × 1140 mm in this model.

2.4 Tunnelling Parameters

In this study, three kinds of discharging rates in chamber including more than 1, equal to 1, and less than 1 were conducted numerically to analyse the ground behaviours associated with tunnelling.

Due to the limited capacities of computer resources, the excavation speed was taken as 30 mm/s, which was much faster than the commonly adopted advancing speed in field but does not affect the target results if the muck amount in the chamber keeps steady. Correspondingly, groups of enlarged muck discharge speeds were chosen considering the discharge ratios in chamber. It should be noted that only the volume of

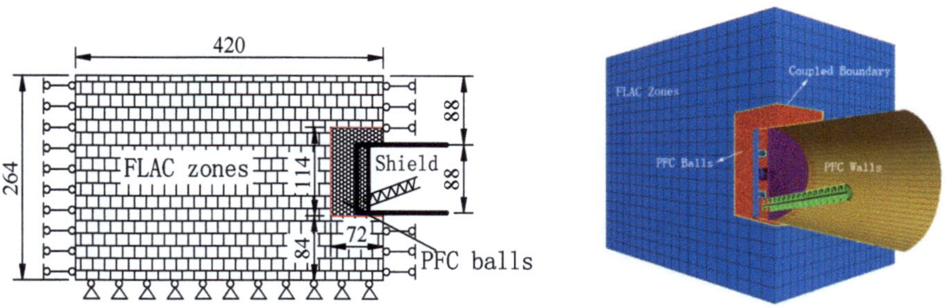

(a) Longitudinal cross-section (Unit: cm) **(b) Three-dimensional model (one-half)**

Fig. 2. FLAC3D/PFC3D coupled numerical mode

particles is considered and the porosity volume among particles is ignored. Here, actual discharge ratios of 1.1, 1.02, 1.0, 0.98, 0.96, respectively were selected and the penetration of 10 mm/rad was used, so the rotation velocity of cutterhead was set to be 3 rad/s considering the designated advancing speed. Precisely, the tunnelling parameters are listed in Table 2.

Table 2. Tunneling parameters of the EPB shield for the FLAC3D/PFC3D coupled numerical modeling

Tunneling cases	Discharging ratio	Advancing speed (mm/s)	Rotation rate of the cutterhead (r/s)	Exiting speed (mm/s)	Distance of one tunneling cycle (mm)
1	0.96	30	3	680.56	150
2	0.98	30	3	796.23	150
3	1.00	30	3	887.6	150
4	1.02	30	3	981.4	150
5	1.10	30	3	1639.2	150

3 Numerical Results

3.1 Muck Pressure in Cross-Section Ahead of Chamber Bulkhead

All the monitoring values of muck pressures in field are localized, due to the fact that only limited stress sensors are available on the bulkhead. It is thus clear that a global description of muck pressure in cross-section ahead of chamber bulkhead would contribute us to assess the monitoring values in field more objectively. In this section, the contour maps of muck pressure ahead of chamber bulkhead before and after tunnelling a cycle were investigated under groups of discharge ratios (Fig. 3).

Figure 3a shows the longitudinal stress distribution of the muck particles immediately in front of the chamber bulkhead prior to tunnelling. It can be seen that the longitudinal stress of the muck generally shows an upward trend with progressive

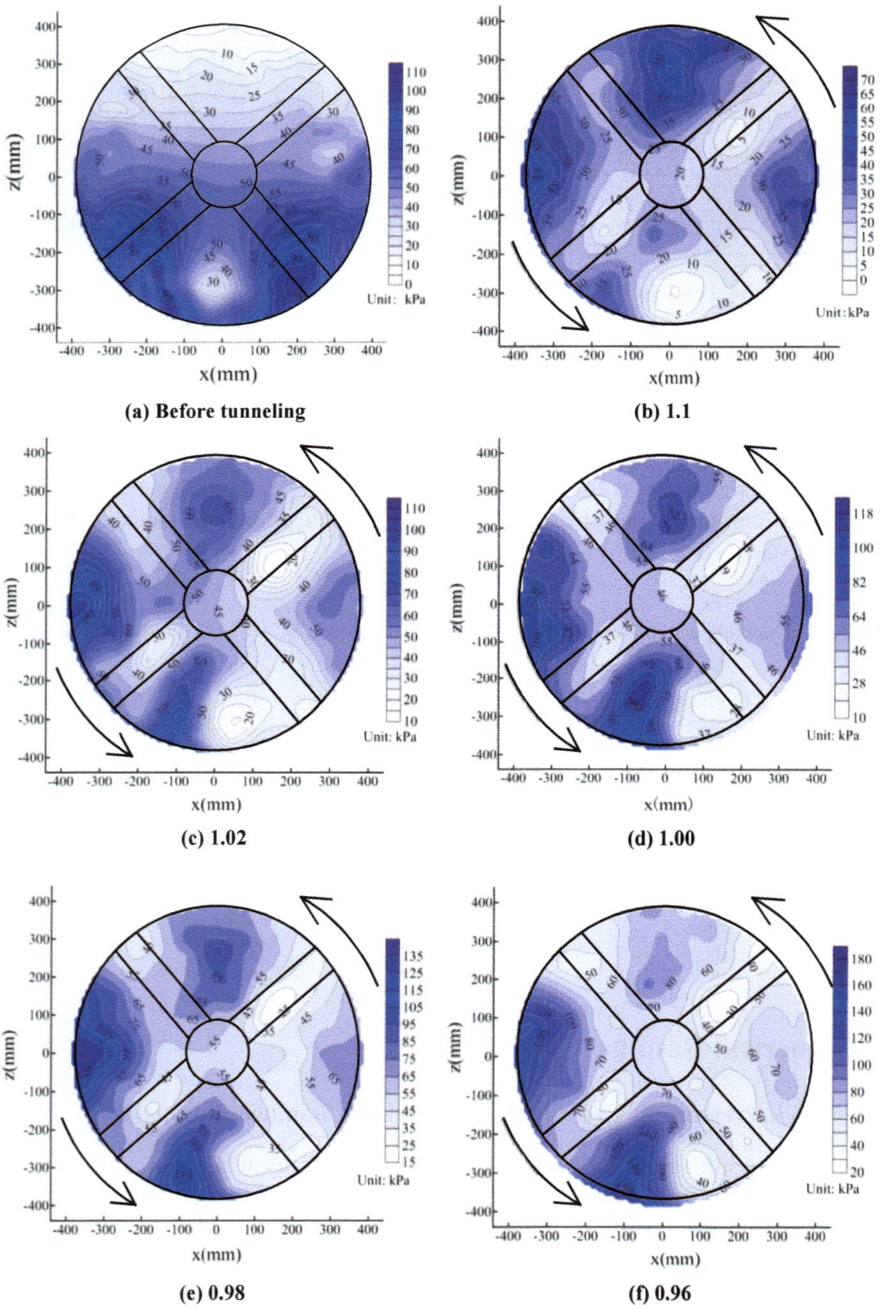

Fig. 3. Distributions of the longitudinal stresses of the muck immediately in front of the chamber bulkhead for different discharge ratios

depth. Due to the presence of the screw conveyor at the bottom of the chamber, the longitudinal stress drops locally around the screw conveyor inlet, which agrees with the findings well presented by Bezuijen et al. (2005) based on their field monitoring results.

Overall, the contour patterns of groups of tunnelling cases remain the same with each other from Fig. 3b–f. Three common points could be found: (1) the muck pressures located in the back of spokes are less than their counterparts behind the opening of cutterhead. The reason is that these spokes bear the soil pressure from natural ground during tunnelling, and the muck pressures behind the spokes are therefore unloaded due to an increasing distance. In contrast, the muck pressures behind the opening of cutterhead have to solely resist the stresses at the tunnel face. As a result, the muck pressures behind spokes are less than those behind the opening of cutterhead. (2) The muck pressures at the exiting area near the screw conveyor remain lower values, because muck near exiting area is continuously discharged and the muck remains a loose flowing state compared with those in other areas during tunnelling. (3) Leftside and rightside pressures at the same altitude were noticeably different during excavation. Specifically, the muck pressures on the left are higher than their right counterparts while cutterhead rotates counter-clockwisely facing the tunnelling direction. The reason is that the stress and density of muck increase with an increase in depth as a result of the presence of gravity, and the spokes and mix arms rotate from the uppermost position to the bottommost one with greater compression (much higher pressure). In contrast, when spokes and mix arms rotate from the bottommost place to the uppermost position, muck stress may be loosened up (lower pressure). Therefore, the leftside muck pressures exceed the rightside ones because spokes always squeeze the leftside muck but loosen the rightside muck in the case that cutterhead keeps rotating counter-clockwisely. Similar findings were also found by Mosavat and Mooney (2015).

Besides, there are clear differences among cases with various discharge ratios. The first point is that the muck pressures are generally lower in the case of a larger discharge ratio. Furthermore, the bigger the discharge ratio is, the smaller pressure differences between left and right one are. This is due to the fact that the rotation of cutterhead has a more evident disturbance to a denser muck. In the case of loose muck in chamber, by contrast, the rotation of cutterhead affects the muck slightly. Therefore the discharge ratio in shield chamber negatively affects pressure differences between the left and right side in chamber.

3.2 Average Muck Pressure Along the Longitudinal Direction

To study the changes in pressures of both muck and ground during excavation, average muck pressures at some cross sections (Fig. 4) along the tunnelling direction were selected. The average pressures before shield tunnelling and after tunnelling a cycle in groups of discharge ratios were investigated. Figure 5 shows features of average muck pressure along the longitudinal direction.

By comparing these figures, it could be found that the muck pressures near the bulkhead are larger than those round the cutterhead in the shield chamber. And the pressures ahead of the cutterhead are larger than those in chamber after tunnelling a cycle. There is a distinct pressure difference at different sections after tunnelling although few differences were found before advancing. A pressure difference before

(a) Five selected monitoring sections **(b) Arrangement of the measurement spheres**

Fig. 4. Arrangement of the monitoring positions

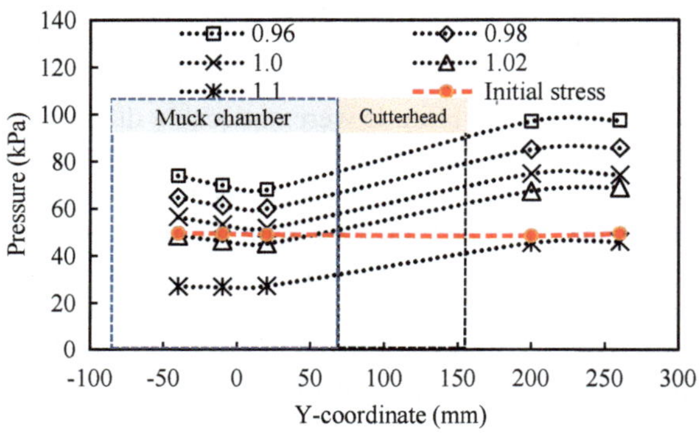

Fig. 5. Variation of the average longitudinal stress within the whole tunnelling cross section against longitudinal location

and behind the cutterhead takes place owing to two reasons: First of all, the soils near the cutterhead are extruded by rotating spokes during the tunnelling of shield. Specifically, this kind of extrusion always exists once the EPB shield advances. Secondly, due to the rotating of cutterhead, soils which are extruded originally by spokes get into opening areas of cutterhead periodically, which gives rise to a sudden pressure difference, pushing the soil to come into muck chamber. After coming into muck chamber, the soil consumes a large amount of potential energy due to the movement. The pressures in front of the cutterhead are thus larger than those approximately behind cutterhead during the tunnelling of EPB shield.

Besides, the EPB shield advances with bulkhead of chamber pushing the muck in chamber forward during excavation and a larger pressure thus comes out near the bulkhead. As a result, the muck pressures near the bulkhead are larger than those just behind the cutterhead and this pressure difference in chamber always exists as well.

On the other hand, the discharge ratio can have a clear impact on the muck pressure. The values of muck pressure decline with an increase in discharge ratio. Especially, when the discharge ratio is 1.02 or 1.1, the average muck pressures are less than those before tunnelling. The reason is that the muck amount in chamber decreases gradually if the discharge ratio is larger than 1, and a decreasing muck amount means a lower pressure. In contrast, when the discharge ratio is less than 1, the muck amount into chamber increase gradually and thus a larger pressure is expected.

In general, the pressure immediately behind the cutterhead is the lowest while the pressures near the bulkhead and in front of the cutterhead remain a higher value during tunnelling of EPB shield. Additionally, the pressures at the tunnel face originate from the muck pressure in chamber and constantly extruded by spokes during excavation. Thus the pressures at the tunnel face increase with an increase of muck pressures in chamber.

3.3 Pressure Difference Between Some Specific Cross Section

As previously mentioned, the pressure difference between cross section 4 and cross section 3, and the one between cross section 1 and section 3 always exist during excavation of EPB shield. The relations between these pressure differences and discharge ratios in chamber were drawn in Fig. 6. It is found that both pressure difference before and after the cutterhead (section 4 and 3) and difference between the section near the bulkhead and immediately behind the cutterhead (section 1 and 3), decline gradually with an increasing discharge ratio. Particularly, the pressure difference evaporates when the discharge ratio is 1.1. For the pressure difference between the section near the bulkhead and the one immediately behind the cutterhead, the bulkhead extrudes the muck in chamber slightly when there is only limited muck amount in chamber, which resulted from a constantly large discharge ratio during excavation. With regards to the pressure difference between before and after the cutterhead, this value remains nearly steady under kinds of discharge ratios in chamber, which is due to the fact that this difference originates the extrusion of spokes during the tunnelling and it is determined by both the configuration of cutterhead and the value of penetration, and has nothing to do with the discharge ratio. As for the difference between the section immediately in front of the cutterhead and the one near the bulkhead in chamber, when the discharge ratio is less than 1.0, this difference increases with a drop in discharge ratio. In contrast, when the discharge ratio exceeds 1.0, this difference keeps almost steady as the discharge ratio goes up.

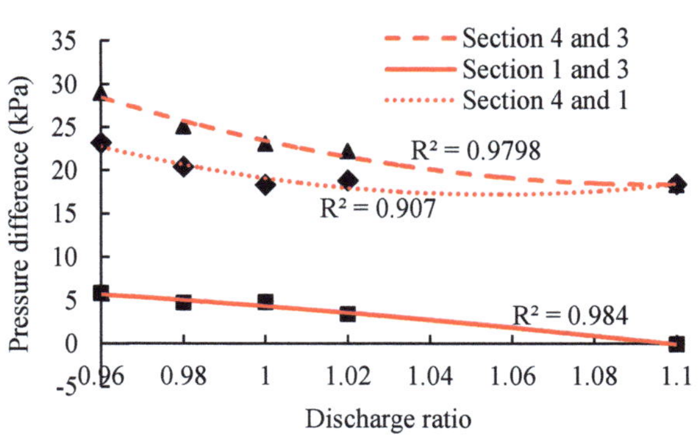

Fig. 6. Pressure differences among some specific section against the discharge ratio

3.4 Changes of Muck Pressure in Certain Monitoring Point with Shield Advancing

To research influences on muck pressure by tunnelling parameters. Monitoring point M1 was chosen near the bulkhead (Fig. 4). Practical formulas of muck pressure was carried out to compare with these numerical outputs. According to the size of model, the buried depth of monitoring point M1 is 1101 mm, and the coefficient of lateral earth pressure at rest (K_0) is 0.538. Thus upper and lower limit of control pressure can thus be determined (Koizumi 1997) as P_{up} = 76.87 kPa, P_{lo} = 11.89 kPa.

Adding these upper and lower limit pressures to numerical outputs in Fig. 7, contrasts were conducted and some conclusions can be obtained: firstly, the cycle time of muck pressure in chamber remains a similar value even though in the cases of varied discharge ratios, which means the tunnelling parameters have a few impact on the cycle time of muck pressure. In addition, the linear fitting of all these monitoring data could describe the tendency in average value of muck pressure in chamber. Precisely, when the discharge ratio is larger than 1, the slope of fitting pressure curve is negative, indicating that muck pressure in chamber decreases gradually during tunnelling. For the case whose discharge ratio is larger than 1, the slope of fitting pressure curve keeps positive, meaning that the muck pressure increases as the shied advances. As for the case that discharge ratio is 1, the pressure curve remains horizontal overall. This phenomenon shows the discharge ratio negatively affects the variation tendency of muck pressure. Finally, after multiple tunnelling cycles, it is predicted that the muck pressure whose discharge ratio is more than 1 or less than 1 will exceed these calculated safe scopes.

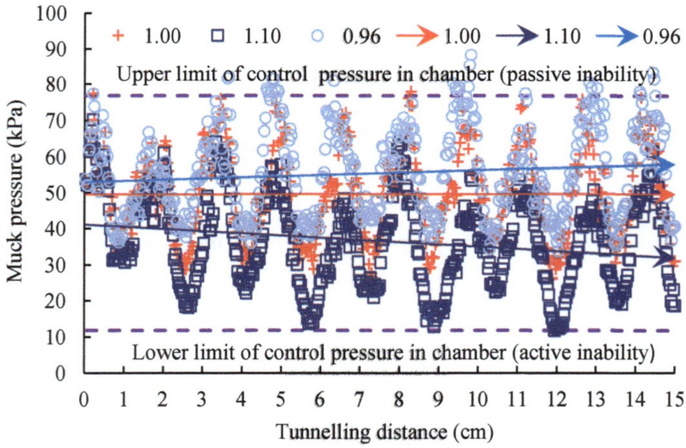

Fig. 7. Changes in muck pressure under groups of discharge ratios during tunnelling

4 Conclusion

The muck pressures located in the back of spokes are less than those behind the opening of cutterhead, and the muck pressures at the exiting area near the screw conveyor remain lower values during excavation of EPB shield. Leftside and rightside pressures at the same altitude in chamber were noticeably different during excavation. Specifically, the muck pressures on the left are higher than their right counterparts while cutterhead rotates counter-clockwisely facing the tunnelling direction. Furthermore, despite of various discharge ratios, the pressure immediately behind the cutterhead is the lowest while the pressures near the bulkhead and those in front of the cutterhead remain a higher value during tunnelling of EPB shield. It is generally found that both the pressure difference before and after the cutterhead and the difference between the section near the bulkhead and immediately behind the cutterhead, decline gradually with an increasing discharge ratio. Finally, the discharge ratio determines the change tendency of muck pressure, and the instability at the tunnel face is the accumulated results of multiple tunnelling cycles.

In general, this study presents a detailed picture about the muck behaviours in EPB shield chamber and facilitates the engineers making a more sensible decision during the construction of shield tunnelling. Besides, this method could be generally applicable for the optimal design and construction technologies of any engineering devices interacting with soil. On the other hand, although the simulated results match well with the prevalent calculation formula and physical common sense, considering the possible loss of precision due to exchanging data in the interface between the PFC and FLAC, further verification based on engineering data in field should be performed in the future.

Acknowledgements. The financial supports from National Key R&D Program of China (No. 2017YFB1201204) and the National Natural Science Foundation of China (No. 51778637) are acknowledged and appreciated.

References

Bezuijen, A., et al.: Pressure gradients at the tunnel face of an Earth Pressure Balance shield. The World Tunnel Congress, Istanbul, The Republic of Turkey (2005)

Coetzee, C.J.: Review: calibration of the discrete element method. Powder Technol. **310**, 104–142 (2017)

Koizumi, A.: The investigation, design, and construction of shield tunnel, Tokyo (1997)

Physical and Mechanical Characterization of Himalayan Dolomite

Sunita Mishra and Tanusree Chakraborty(✉)

Department of Civil Engineering, Indian Institute of Technology (IIT) Delhi, Hauz Khas, New Delhi 110 016, India
sunita.mishra256@gmail.com,
tanusree.chakraborty@gmail.com

Abstract. In the present work, dynamic stress-strain response of Himalayan dolomite is tested under high loading rates using 38 mm split Hopkinson pressure bar (SHPB) device. The physical and static mechanical properties of dolomite e.g. density, specific gravity, static compressive strength and elastic modulus values are also determined. Petrological studies of dolomite are carried out through X-ray diffraction (XRD) test and scanning electron microscope (SEM) test. In the SHPB tests, it is observed from the stress-strain response that the dynamic peak stress increases with increasing strain rate whereas the elastic modulus does not show any clear trend with an increase in strain rate. Dynamic force equilibrium at the incident and transmission bar ends of the rock samples is attained in all tests till the failure of the rock samples. Dynamic increase factor (*DIF*) for the rock is determined at a particular strain rate by comparing the dynamic to static peak compressive stress.

1 Introduction

Design, development, and building of civil infrastructure in the mountainous regions involve many complexities in terms of diverse geological and geomorphological features of the region—the Chenab river bridge in the Himalayas, the Gotthard Base tunnel in the Alps are to name a few. The young mountain ranges of the Himalayas and the Alps contain joint planes, shear seams, active fold, and fault zones. Moreover high in situ stresses and high level of seismicity in these regions pose severe challenges to the construction of infrastructure. In addition to this, unanticipated loads caused by natural hazards, e.g. landslide, earthquake and manmade hazards, e.g. blast and projectile penetration add to the difficulties already existing therein. It may be noted that the loads caused by hazardous events like earthquake and blast are highly transient in nature generating high strain rates in rock and strain rate caused by blast may reach up to 10^4/s (Ngo et al. 2007; Dusenberry 2010) which in turn affects both the stiffness and the strength properties of the rocks. Thus, in order to ensure the sustainable design of civil infrastructure in the mountains, it becomes necessary to characterize the rocks under static and dynamic loading conditions.

In the present work, only the dynamic compression response of rocks have been discussed and reported. Dynamic compression tests of rocks at low to high strain rates have been performed by several researchers through split Hopkinson pressure bar

W.-C. Cheng et al. (eds.), *Tunneling in Soft Ground, Ground Conditioning and Modification Techniques*, Sustainable Civil Infrastructures,
https://doi.org/10.1007/978-3-319-95783-8_18

(SHPB) and dynamic triaxial tests (Perkins and Green 1970; Christensen et al. 1972; Lindholm et al. 1974; Lundberg 1976; Blanton 1981; Klepaczko 1990; Olsson 1991; Grady 1995; Zhao et al. 1999; Frew et al. 2000; Fukui et al. 2004; Li et al. 2005; Qi et al. 2009; Fuenkajorn and Kenkhunthod 2010; Ming et al. 2014). Perkins and Green (1970) performed dynamic uniaxial compression tests using SHPB on three rocks at strain rates from 10^{-4} to 10^{4}/s and reported SHPB test data for rocks for the first time in the literature. They also conducted uniaxial compressive strength test of porphyritic tonalite at strain rates up to 10^{3}/s and at varying temperatures. They observed that rocks exhibit increased stiffness and higher stress with increasing strain rate and decreasing temperature. Christensen et al. (1972) performed SHPB test on rocks for nugget sandstone at strain rates from 10^{2} to 10^{3}/s at confining pressure of 206.84 MPa. All rocks showed an increase in strength with the increasing loading rate. The dynamic stress-strain curves exhibited similar trends as observed in quasi-static testing of the same materials. Lundberg (1976) studied energy absorption in two different rocks, i.e. Bohus granite and Solenhofen limestone. It was observed that the energy absorbed by the rocks increases markedly when the applied load reaches the critical value of 1.8 and 1.3 times the static compressive strength for Bohus granite and Solenhofen limestone, respectively. Blanton (1981) performed dynamic triaxial compression test at strain rates from 10^{-2} to 10/s for charcoal granodiorite up to a confining pressure of 0.45 GPa, for Berea sandstone and Indiana limestone up to a confining pressure of 0.25 GPa. It was observed from the results that the differential stress at failure was relatively constant up to a strain rate of 1/s, however, increased when the strain rate was higher than 1/s. Zhao et al. (1999) conducted a dynamic uniaxial compressive test on Bukit Timah granite in Singapore at four different loading rates (10^{0}, 10^{1}, 10^{3} and 10^{5} MPa/s). It was concluded from the tests that for each log scale increase in loading rate the compressive strength of the rock increases by 15%. They also observed that there were small changes in elastic modulus and Poisson's ratio values with increase in loading rate. Frew et al. (2000) conducted uniaxial compressive SHPB on limestone by using a copper disk at the impact end of the incident bar as pulse shaper which results in dynamic stress equilibrium of samples and maintains constant strain rates over the test duration. Fukui et al. (2004) reported the loading rate dependency for the shear strength of a rock Sanjome andesite with discontinuities. They observed that if loading rate is increased by an order of magnitude, the cohesion of rock increases by 6.1% and this rate of increase in cohesion remains approximately same for uniaxial compressive strength, uniaxial tensile strength, indirect tensile strength and fracture toughness. Li et al. (2005) reported dynamic stress-strain response of Bukit Timah granite loaded at a medium strain rate of 20–60/s using SHPB. It was observed from the results that the dynamic fracture strength of the granite is directly proportional to the cube root of strain rate whereas the elastic modulus remains unchanged with increasing strain rate. At higher strain rate, the rocks show a higher amount of energy absorption and the particle size of the fragments at the end of the test becomes smaller. Statistical analysis of strain rate effects on rock strength was carried out by Grady (1995) and Qi et al. (2009) over a wide range of strain rates on dolomite, limestone, and basalt. The strain rate of 1/s was mentioned as the initial threshold of dynamic strain rate. It may be summarized from the literature review that dynamic compressive strength test of rocks using SHPB have been carried out on different rock types, e.g. granite, Barre granite, basalt, volcanic tuff, Kawazu tuff, red sandstone, Indiana limestone, porphyritic tonalite, oil shale, granodiorite, coal, kidney stone, Tennessee marble and Akyoshi marble up to 2000/s strain rate (Lu et al. 2010) and strain rate has significant effect on the mechanical behavior of rocks.

2 Test Plan

The objective of the present work is to characterize Himalayan dolomite under strain rate dependent loading at different levels of strain rates. The dolomite rock blocks collected herein are of unweathered nature. The rock blocks are collected from a hydropower project site in Bilaspur, India under National Thermal Power Corporation (NTPC) Ltd., India. The rock samples have been tested for both physical and mechanical properties. The physical properties, e.g. dry density, saturated density and specific gravity, the petrological studies, e.g. X-ray diffraction test (XRD) and scanning electron microscope test (SEM) and the static mechanical properties, e.g. uniaxial compressive strength, and static elastic modulus of the rock are determined by the Indian Institute of Technology (IIT) Delhi laboratories. Static uniaxial compressive strength tests on dry and saturated rock samples have been carried out using the automatic uniaxial compression device for rock samples with an aspect ratio (*L*/*D*) = 2:1. All static tests have been carried out following the specifications given in (ASTM D3148-02 2002; ISRM 1985; ISRM 1979).

The strain rate dependent tests are carried out using SHPB in the Rel Inc. laboratory, Michigan, USA. The stress-strain response of the rocks under dynamic loading, force equilibrium at incident and transmission bar ends of the rock sample, peak stress and dynamic elastic modulus are studied. The dynamic tests are performed using a 38-mm diameter SHPB for all three rocks at different strain rate levels. In the present study, the low strain rate range is defined for 10–100/s, medium strain rate is defined for 100–250/s, and high strain rate is defined for 250–500/s. For this test, the rock samples are prepared with a diameter of 38 mm and aspect ratio of 0.5:1. The results obtained from the static tests are presented in Table 1 and those from the dynamic tests are presented in Table 2 for dolomite. The striker bar is propelled using a compressed air gas gun. The strain rate in the dynamic tests is controlled by varying the striker bar length and the striking velocity. The sample number (S.N.), sample length (L_s), length of the striker bar used (L_{st}) and the striking velocity applied (V_{st}) for a particular strain rate are also reported in the above-mentioned tables. The dynamic increase factor (*DIF*), i.e. the ratio of dynamic to static peak stress is calculated for each test at different strain rates.

Table 1. Physical and static properties

Rock	ρ_d (kg/m^3)	ρ_{sat} (kg/m^3)	G	σ_c (MPa)	E_t (GPa)
Dolomite	2731.82	2723.20	2.70	38.59	3.75

ρ_d = Dry density, ρ_{sat} = Saturated density, G = Specific gravity, σ_c = Uniaxial compressive strength, E_t = Static elastic modulus

Table 2. Dynamic properties: Dolomite

S.R.R.	S.N.	L_s (mm)	L_{st} (mm)	V_{st} (m/s)	$\dot{\varepsilon}$ (/s)	σ_{dc}(MPa)	ε	E_d(GPa)	*DIF*
Low	D15	19.02	304.8	20.54	49	395.88	0.006	80.79	10.25
	D3	18.90		20.55	69	398.76	0.008	63.29	10.33
	D14	19.05		20.32	76	313.15	0.006	65.51	10.27
	D12	18.31		20.31	76	396.70	0.008	77.78	8.11
Medium	D8	19.09	139.7	29.50	125	342.17	0.008	47.39	8.86
	D10	19.08		27.71	132	403.28	0.008	45.51	10.45
	D9	18.58		24.70	132	326.36	0.007	46.75	8.45
	D5	18.29		29.50	144	433.33	0.009	47.51	11.22
High	D7	19.23	139.7	30.47	174	434.88	0.014	31.46	11.26
	D1	18.38		35.19	176	403.38	0.008	57.46	10.45
	D11	18.15		38.10	201	421.93	0.014	30.97	10.93

S.R.R. = Strain rate range, S.N. = Sample number, L_s = Sample number, L_{st} = Striker bar length, V_{st} = Velocity of the striker bar, $\dot{\varepsilon}$ = Strain rate achieved, σ_{dc} = Dynamic unaxial compressive strength, ε = Strain at peak stress, E_d = Dynamic elastic modulus, *DIF* = Dynamic Increase Factor

3 Test Setup for SHPB Device

Figure 1 shows the schematic diagram of the compression SHPB test setup in the Rel Inc. laboratory. The setup comprises of an incident bar, a transmission bar and striker bars of different sizes. The bars are made up of C300 maraging steel. The incident bar length is 2.59 m and diameter is 38.1 mm. The transmission bar length is 2.43 m and diameter is 38.1 mm. The dimensions of the incident and transmission bars allow one-dimensional loading of the sample. In the present work, two different lengths of striker bars are used, e.g. 152.4 mm and 304.8 mm; the diameter of striker bar is 38.1 mm. Compressed air is used to launch the striker bar on the incident bar. The striker bar is propelled by a compressed air gas gun at varying pressure magnitudes which generate stress waves inside the striker bar. The striker bar hits the impact end of the incident bar and remains in contact till the stress wave travels from one end of the striker bar to the other end. The stress wave upon reaching the other end of the striker bar gets reflected back. As a result, the contact between the striker bar and the incident bar is lost. The time duration taken by the stress wave to travel from one end of the striker bar to the other is the time duration for loading of the sample. The time duration is given by

$$\Delta t = \frac{2L_s}{c_{bar}} \quad (1)$$

where L_s is the length of the striker bar and c_{bar} is one-dimensional longitudinal stress wave velocity in the bar. Thus, using longer striker bar increases the loading time and the rock sample gets time to respond. As a result lower strain rate develops when longer

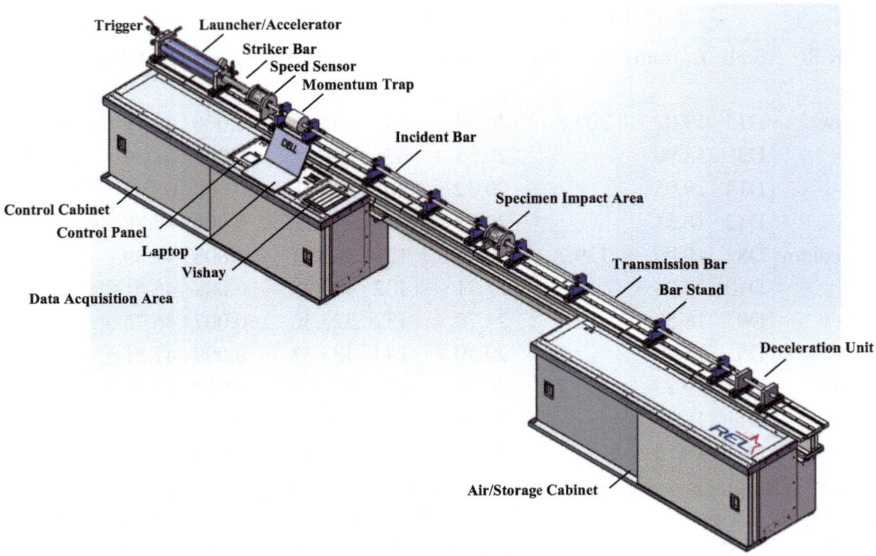

Fig. 1. 3D schematic diagram of split Hopkinson bar apparatus

striker bar is used. It may be observed from the Table 2 that for low strain rate level, the length of the striker bar used is 304.8 mm whereas a striker bar of length 139.7 mm is used for medium and higher strain rates. For smaller striker bars, the time taken by the stress wave to propagate from one end to the other end of the bar is less and hence the loading time duration is also less for the same amount of the load intensity which results in higher strain rate. The strains in the incident and transmission bars are measured using two strain gauges, one mounted on the incident bar and the other mounted on the transmission bar. The strain gauges are 6.35 mm in length, OMEGA 120Ω in resistance and dynamic gauge with a tolerance of ±0.35%. Strain gauges are attached to the incident and the transmission bars with M-bond adhesive. In order to read the strain signal, Vishay 2310B signal conditioner and amplifier with ¼ Wheatstone bridge has been used with a Picoscope5242 having sampling rate of 1 in 8 ns. Momentum trap, though present in the setup, has not been used in the present experiments.

The impact of striker bar on the incident bar causes a longitudinal elastic compressive stress wave which propagates through the incident bar. The strain pulse generated within the incident bar is designated as incident strain pulse $\varepsilon_i(t)$. The strain pulse generated in the incident bar is recorded by the strain gauge mounted on the incident bar. Upon reaching the bar-specimen interface, a part of the pulse, designated as reflected strain pulse $\varepsilon_r(t)$ is reflected back in the incident bar and the remaining part of the compressive pulse passes through the specimen. Upon reaching the transmission bar end of the specimen, the pulse propagates through the transmission bar and it is then termed as transmitted strain pulse $\varepsilon_t(t)$. The histories of strain $\varepsilon(t)$, strain rate $\dot{\varepsilon}(t)$ and stress $\sigma(t)$ within the sample in dynamic compression test are given by

$$\varepsilon(t) = \frac{C}{L}\int_{0}^{t} (\varepsilon_i - \varepsilon_r - \varepsilon_t) \tag{2}$$

$$\dot{\varepsilon}(t) = \frac{C}{L}(\varepsilon_i - \varepsilon_r - \varepsilon_t) \tag{3}$$

$$\sigma(t) = \frac{A}{2A_0}E(\varepsilon_i + \varepsilon_r + \varepsilon_t) \tag{4}$$

where L is the length of the sample, C is the one-dimensional longitudinal stress wave velocity in the bar, E is the elastic modulus of the bar material, A is the cross-sectional area of the bar and A_0 is the initial area of the sample. Assuming that stress equilibrium and uniform deformation of the sample prevails during dynamic loading, i.e. $\varepsilon_i + \varepsilon_r = \varepsilon_t$, the strain, strain rate, and stress are given by

$$\varepsilon(t) = -\frac{C}{L}\int_{0}^{t} \varepsilon_r dt \tag{5}$$

$$\dot{\varepsilon}(t) = -\frac{C}{L}\varepsilon_r \tag{6}$$

$$\sigma(t) = \frac{A}{A_0}E\varepsilon_t \tag{7}$$

The data processing method for the SHPB test is based on two assumptions - one is force equilibrium on both sides of the specimen and the other is the one-dimensional uniform deformation of the sample. The force equilibrium may be achieved by maintaining good contact between the bars and sample, reducing the friction between bars and supports, keeping the bars coaxial and using pulse shapers (Wang and Meyer 2010; Abotula et al. 2010). However, for rock samples, maintaining force equilibrium becomes challenging due to anisotropic nature of rock and propagation of crack inside the rock. Moreover, due to the brittle behavior of rocks, it becomes important that the equilibrium is achieved before brittle failure of the sample (Frew et al. 2000). Researchers have discussed the technique of pulse shaping in order to achieve force equilibrium in SHPB test for brittle materials like a rock (Frew et al. 2002). Xia and Yao (2014) have shown that wide variety of incident pulses can be produced by varying the geometry of the pulse shaper which can be used for different materials under investigation. In order to ascertain the force equilibrium in the tests reported herein, mild steel and copper pulse shapers have been used in all tests. The pulse shapers used herein are circular disks of 12.7 mm diameter and different thicknesses. Moreover, the incident and transmission bar ends of the specimen are prepared through rigorous grinding up to a smoothness of order 4.1 μm. Moreover, 38 mm diameter SHPB setup has been used to capture the effect of anisotropy on the stress-strain response of the rock samples. The sample thickness has been maintained half of its diameter for achieving stress uniformity throughout the length of the sample. The data processing is done to obtain stress-strain plots, strain, and strain rate time histories, force time histories at the interfaces of incident bar-sample and transmission bar-sample. From these plots, peak stress, average strain rate and strain at peak stress are studied.

4 Results and Discussion

4.1 Physical Properties

The dry density value of dolomite is determined for five samples and average values are determined to be 2731.82 kg/m^3. The saturated density value of dolomite is also measured for five samples and average values are determined to be 2723.20 kg/m^3. The saturated density value is observed to be smaller than the dry density value. The specific gravity value of dolomite is estimated to be 2.70. The density and specific gravity values are compared with the available data from the literature and observed to be well in agreement (Kumar 2007).

4.2 Results of Petrological Study

Petrological studies are performed by using XRD and SEM techniques to identify the mineralogical contents of the rocks. The SEM and XRD tests are done in the central research lab facility of Indian Institute of Technology (IIT) Delhi. The SEM Central Facility at IIT Delhi has a ZEISS EVO Series Scanning Electron Microscope EVO 50 which has magnification capacity of 5× to 1,000,000× and field of view of 6 mm at the analytical working distance. The SEM image of dolomite is shown in Fig. 2a. The SEM images show the grain texture of dolomite. The mineral content of dolomite is determined through X-ray diffraction tests. Figure 2b shows the results of X-ray diffraction graph of dolomite. The Himalayan dolomite has a dull yellowish white color and becomes greyish in color upon weathering. The mineral content of dolomite is ascertained through XRD test to be carbonate (80%) with biotite and mica. The XRD graphs of dolomite are compared with the available data from the literature and observed to be well in agreement (Kumar 2007).

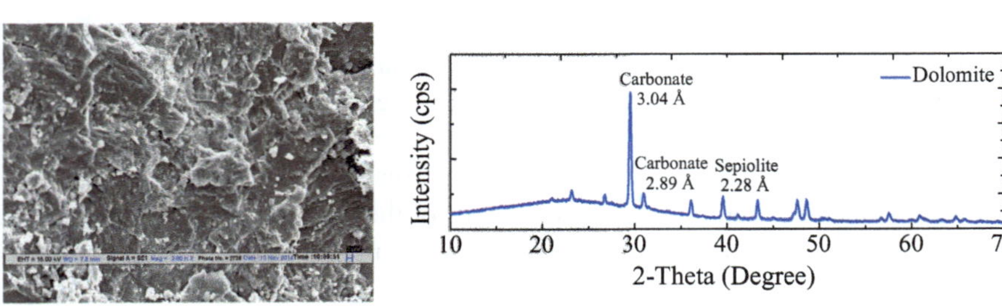

Fig. 2. Scanning electron microscope images and X-ray diffraction diagram of dolomite

4.3 Static Uniaxial Compressive Strength

The static uniaxial compressive strength and elastic modulus of the rock are determined under dry conditions following ASTM and ISRM standards (ASTM D3148-02 2002; ISRM 1979; ISRM 1985) and presented in Table 1. The tests for each rock type are repeated three times and the average strength values are noted. The static uniaxial compressive strength of dolomite is found to be 38.59 MPa. The elastic modulus from the stress-strain graph at 50% of peak stress value is calculated to be 3.75 GPa (ASTM D3148-02 2002; Kumar 2007).

4.4 Dynamic Stress-Strain and Force Equilibrium for Dolomite

Figure 3a, b, c show the stress-strain plots for dolomite at three different strain rate ranges e.g. high, medium and low respectively. Figure 3a shows the stress-strain plots for dolomite at different high strain rate values. For high strain rate range, the peak stress values obtained are 434.888 MPa at 174/s, 403.388 MPa at 176/s and 421.930 MPa at 201/s. Figure 3b shows the stress-strain plots for dolomite at different medium strain rate values giving different peak stress. For medium strain rate range, the peak stress values obtained are 342.170 MPa at 125/s, 403.280 MPa at 132/s, 326.366 MPa at 132/s and 433.338 MPa at 144/s. Figure 3c shows the stress-strain plots for dolomite at different high strain rate values giving different peak stress. For low strain rate range, the peak stress values obtained are 395.882 MPa at 49/s, 398.760 MPa at 69/s, 313.150 MPa at 76/s and 396.7 MPa at 76/s.

Figure 3d_1, d_2, e_1, e_2, f_1, f_2 present the force equilibrium obtained in all tests for all the strain rate ranges at the incident and the transmission bar ends of rock specimens. It may be observed that at high strain rate levels, the force equilibrium achieved is not satisfactory due to weathered nature of dolomite. However, at medium and low strain rate ranges, force equilibrium is obtained nicely. Figure 3g, h, i present the strain rate with respect to time for high, medium and low strain rate levels, respectively. The region of strain rate-time curve which is used to determine the strain rate is highlighted in the Figures. Figure 3j, k show the peak stress and elastic modulus obtained from the stress-strain curves for dolomite. The peak stress, elastic modulus and strain at peak stress values are also presented in Table 2 at varying strain rate levels. It can be seen that the peak stress increases with increase in the strain rate. The peak stress increases by 0.04% from low (49–76/s) to medium (125–144/s) level of strain rate and by 11.6% from the medium (125–144/s) to a high level of strain rate (174–201/s). It can be observed from Fig. 3k that the elastic modulus does not show any clear trend under a change in dynamic strain rates.

4.5 Dynamic Increase Factor and Proposed Correlation Equation

Dynamic increase factor (*DIF*) for dolomite has been determined by comparing the dynamic peak stress with the static peak stress. The *DIF* values of the rock with respect to varying strain rate are reported in Table 2. The *DIF* values are also plotted in Fig. 3f for dolomite. From Fig. 3f, it may be seen that the dynamic strength of dolomite is 8.11–11.26 times that of the static strength for strain rates varying from 49 to 201/s.

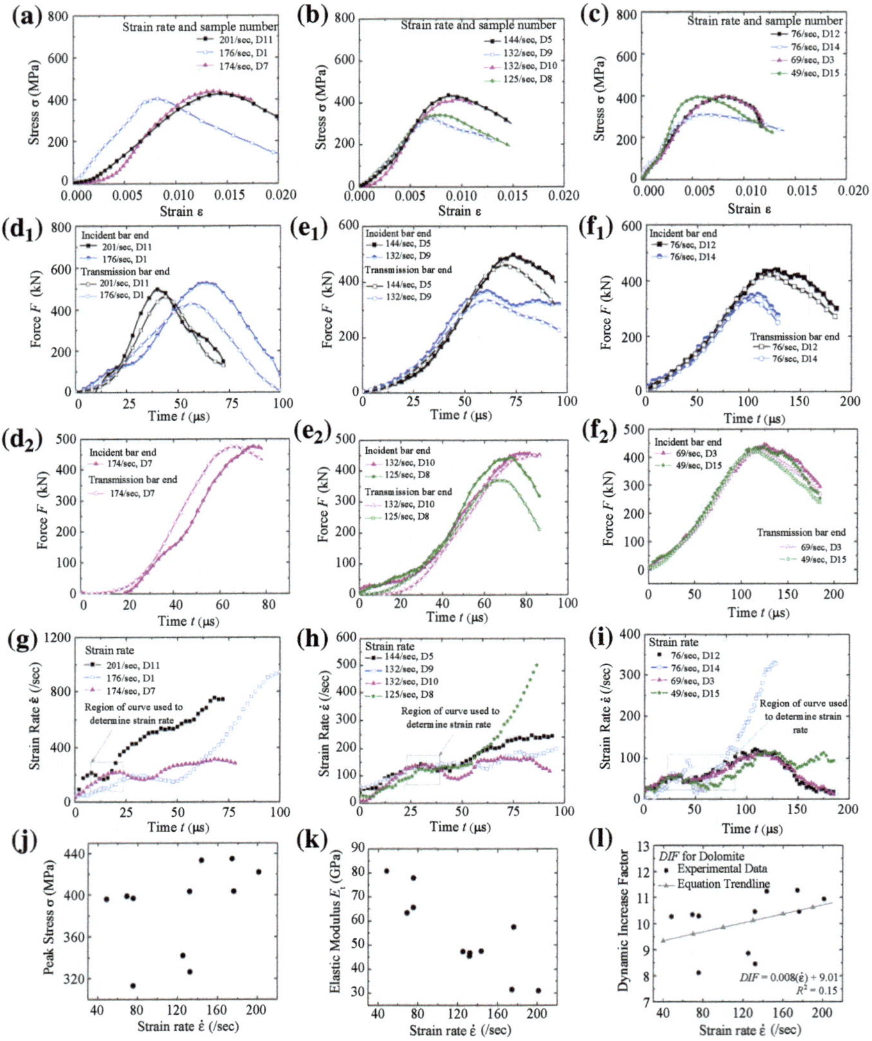

Fig. 3. Dolomite stress-strain curves for **a** high strain rate range, **b** medium strain rate range, **c** low strain rate range, force equilibrium curves for $\mathbf{d_1}$, $\mathbf{d_2}$ high strain rate range $\mathbf{e_1}$, $\mathbf{e_2}$ medium strain rate range $\mathbf{f_1}$, $\mathbf{f_2}$ low strain rate range, strain rate-time curves for, **g** high strain rate range, **h** medium strain rate range, **i** low strain rate range, data points for, **j** peak stress-strain rate, **k** elastic modulus-strain rate, **l** dynamic increase factor-strain rate

5 Conclusions

Characterization of Himalayan dolomite has been performed in the present work for a strain rate range varying from 49 to 201/s through uniaxial compressive 38 mm diameter SHPB test. The dynamic stress-strain response, peak stress, elastic modulus and force equilibrium at the incident and transmission bar ends of the rock specimen are studied. The physical properties and static stress-strain behavior of the rock are also investigated. The peak stress increases by 0.04% from low (49–76/s) to medium (125–144/s) strain rate and by 11.6% from the medium (125–144/s) to high (174–201/s) strain rate. The elastic modulus does not show any clear trend with an increase in strain rate. The dynamic compressive stress 421.93 MPa at 201/s is nearly eleven times of the static compressive stress. The dynamic increase factor varies from 8.11 to 11.26 times of the static compressive stress.

References

ASTM D3148-02.: Standard test method for elastic moduli of intact rock core specimens in uniaxial compression. American Society for Testing and Materials (2002)

ASTM D4543-08.: Standard practices for preparing rock core as cylindrical test specimens and verifying conformance to dimensional and shape tolerances. American Society for Testing and Materials (2008)

Abotula, S., Shukla, A., Chona, R.: Dynamic constitutive behavior and fracture initiation toughness of Hastelloy X under thermo-mechanical loads. Proceedings of the Society of Experimental Mechanics, Implast Conference. Wright-Patterson Air Force Base RBSM (2010)

Blanton, T.L.: Effect of strain rates from 10^{-2} to 10 sec^{-1} in triaxial compression tests on three rocks. Int. J. Rock Mech. Min. Sci. Geomech. Abs. **18**(1), 47–62 (1981)

Christensen, R.J., Swanson, S.R., Brown, W.S.: Split Hopkinson bar tests on rock under confining pressure. Exp. Mech. **12**, 508–541 (1972)

Dusenberry, D.O.: Handbook for Blast Resistant Design of Buildings, 1st edn, p 512. John Wiley and Sons, Hoboken, NJ (2010)

Frew, D.J., Forrestal, M.J., Chen, W.: A split Hopkinson pressure bar technique to determine compressive stress-strain data for rock materials. Exp. Mech. **41**(1), 40–46 (2000)

Frew, D.J., Forrestal, M.J., Chen, W.: Pulse shaping techniques for testing brittle materials with a split Hopkinson pressure bar. Exp. Mech. **42**(1), 93–106 (2002)

Fuenkajorn, K., Kenkhunthod, N.: Influence of loading rate on deformability and compressive strength of three thai sandstones. Geotech. Geol. Eng. **28**, 707–715 (2010)

Fukui, K., Okubo, S., Ogawa, A.: Some aspects of loading-rate dependency of sanjome andesite strengths. Int. J. Rock Mech. Min. Sci. **41**(7), 1215–1219 (2004)

Grady, D.E.: Shock wave properties of brittle solids. In: Schmidt, S. (ed.) Shock Compression of Condensed Matters, pp. 9–20. AIP Press, New York (1995)

ISRM: International society of rock mechanics commission on testing methods, suggested method for determining point load strength. Int. J. Rock Mech. Min. Sci. Geomech. Abs. **22**, 51–60 (1985)

ISRM: International society of rock mechanics commission on testing methods, suggested methods for determining the uniaxial compressive strength and deformability of rock materials. Int. J. Rock Mech. Min. Sci. Geomech. Abs. **16**(2), 138–140 (1979)

Klepaczko, J.R.: Behavior of rock-like materials at high strain rates in compression. Int. J. Plast **6**, 415–432 (1990)

Kumar, R.: Testing and constitutive modelling of the strain-softening behaviour of some rocks. Ph.D. Thesis, Indian Institute of Technology (IIT) Delhi (2007)

Lindholm, U.S., Yeakley, L.M., Nagy, A.: The dynamic strength and fracture properties of dresser basalt. Int. J. Rock Mech. Min. Sci. Geomech. Abs. **11**, 181–191 (1974)

Li, X.B., Lok, T.S., Zhao, J.: Dynamic characteristics of granite subjected to intermediate loading rate. Rock Mech. Rock Eng. **38**(1), 21–39 (2005)
Lu, Y.B., Li, Q.M., Ma, G.W.: Numerical investigation of the dynamic compressive strength of rocks based on split Hopkinson pressure bar tests. Int. J. Rock Mech. Min. Sci. **47**(5), 829–838 (2010)
Lundberg, B.: A split Hopkinson bar study of energy absorption in dynamic. Int. J. Rock Mech. Min. Sci. Geomech. Abs. **13**, 187–197 (1976)
Ming, L., Xian-biao, M., Li-li, C., Rong-rong, M., Guang-hui, Z.: High strain rate effects on mechanical behavior of coal-serial sandstone. Electron. J. Geotech. Eng. **19**, 6035–6046 (2014)
Ngo, T., Mendis, P., Gupta, A., Ramsay, J.: Blast loading and blast effects on structures–an overview. Electron. J. Struct. Eng. Spec. Issue: Load. Struct. **7**, 76–91 (2007)
Olsson, W.A.: The compressive strength of tuff as a function of strain rate from 10^{-6} to 10^{3}/sec. Int. J. Rock Mech. Min. Sci. Geomech. Abs. **28**(1), 115–118 (1991)
Perkins, R.D., Green, S.J.: Macroscopic description low and medium strain rates. Int. J. Rock Mech. Min. Sci. **7**, 527–535 (1970)
Qi, C.Z., Wang, M.Y., Qihu, Q.: Strain-rate effects on the strength and fragmentation size of rocks. Int. J. Impact Eng. **36**, 1355–1364 (2009)
Wang, Z.G., Meyer, L.W.: On the plastic wave propagation along the specimen length in SHPB test. Exp. Mech. **50**, 1061–1074 (2010)
Xia, K., Yao, W.: Dynamic rock tests using split Hopkinson (Kolsky) bar system–a review. J. Rock Mech. Geotech. Eng. **7**, 27–59 (2014)
Zhao, J., Li, H.B., Wu, M.B., Li, T.J.: Dynamic uniaxial compression tests on a granite. Int. J. Rock Mech. Min. Sci. **36**, 273–277 (1999)

Mechanical Assessment of Crushed Rocks Derived from Tunnelling Operations

Diego Maria Barbieri[1(✉)], Inge Hoff[1], and Mai Britt Engeness Mørk[2]

[1] Department of Civil and Environmental Engineering, NTNU, Trondheim, Norway
{diego.barbieri,inge.hoff}@ntnu.no

[2] Department of Geosciences and Petroleum, NTNU, Trondheim, Norway
mai.britt.mork@ntnu.no

Abstract. The Norwegian Public Roads Administration is currently running the "Ferry-free coastal route E39" project, which reduces the travel time along the Norwegian coast from Trondheim to Kristiansand. The plan includes the creation of several long tunnels, which will generate a surplus of blasted rocks; these could be used in the road unbound layers close to the place of production. The research presented here has three goals. The first aim is to map the geology encountered along the E39 road alignment. The second aim is to check whether the rocks fulfil the existing code requirements for road unbound layers, defined in terms of Los Angeles and micro-Deval limit values. The third aim is to investigate the crushing and the variation in grain size of the unbound materials during both construction phase and service life phase. The construction stage is achieved by a full scale testing to assess rock soundness after rolling, the service life stage is simulated by the repetition of a specific load in the triaxial cell apparatus. The current tunnelling operations located south of Bergen are producing blasted rocks, they adequately represent the geology spread along the entire E39 alignment. Three types of crushed rocks are selected and tested. The major part of the rocks excavated are suitable for direct use in pavement unbound layers. The most significant modification in grain size distribution curve takes place during the compaction phase for all the materials.

1 Introduction and Background

Norwegian Public Roads Administration (NPRA) is currently running the "Ferry-free coastal route E39" project, which improves the viability along the Norwegian coast for a total length of about 1100 km from Trondheim to Kristiansand (NPRA 2017; Dunham 2016). The project includes the building of several bridges and tunnels, while aiming for creating a sustainable infrastructure. The extended tunnelling systems will generate a surplus of blasted rocks. They could be used as viable substitutes for natural aggregates in the road unbound layers close to the place of production. This is beneficial from both economic and environmental point of views (Aatheesan et al. 2008; Arulrajah et al. 2013; Núñez et al. 2008; Onyango and Busch 2007; Otto et al. 2015). The usage of demolition materials in pavement applications is a sustainable option to minimise the waste while reducing the demand for scarce quarried materials and

W.-C. Cheng et al. (eds.), *Tunneling in Soft Ground, Ground Conditioning and Modification Techniques*, Sustainable Civil Infrastructures,
https://doi.org/10.1007/978-3-319-95783-8_19

lowering carbon footprints (Fladvad et al. 2017). The transport distance should be within 20–30 km to represent a competitive solution (Neeb 1992).

The need for recycling and reusing materials is a topic of global interest and concern. In Norway the production of natural aggregate has been declining since 1997, while the production of crushed rocks has been increasing since 2003. In 2015 almost 70 million tons of crushed rocks were produced in Norway. The average yearly aggregate consumption per capita is 11 tons; approximately half part of this figure is used for road construction (Erichsen and Aasly 2016).

The existing requirements for road unbound layers are connected to relatively simple tests: the Norwegian pavement design manual N200 (NPRA 2014) sets limits in terms of sieving curve (CEN 2012), Los Angeles (CEN 2010) and micro-Deval (CEN 2011) values. By respecting the specified thresholds, the road should perform adequately without encountering premature damage (Barbieri et al. 2017).

The research presented here has three goals. In the first place, the geology encountered along the E39 road alignment is mapped, this task is essential to identify the bedrock distribution where the tunnels are excavated. In the second place, the study investigates whether the blasted rocks fulfil the existing requirements for road unbound layers established by the design manual. In the third place, the crushing and variation in grain size of unbound materials is investigated during both construction phase and service life phase. The grain size distribution curve is closely connected to the material performance in terms of resilient modulus and deformation (Hicks and Monismith 1971; Lekarp et al. 2000a, b; Li and Selig 1994), therefore the variation in grain size largely affects the material behaviour. Furthermore, this investigation could bring to a better understanding between the mentioned standard tests and the behaviour of the crushed rocks in situ.

The building site "Svegatjørn-Rådal" is located south of Bergen and the current tunnelling operations are producing blasted rocks; which adequately represent the geology spread along the entire E39 alignment. Three types of crushed rocks (designated as material M1, M2, M3) are selected and tested.

2 Methodology

2.1 Geology and Materials

Knowledge about the geology encountered by the tunnelling operations is needed to map the origin and distribution of the materials (NGU 2017; Ramberg et al. 2013). Figure 1 displays the geology of the southern part of Norway and the alignment of E39 highway.

The highway alignment comes across different types of bedrocks. The major part of the rocks is igneous and supracrustal of Precambrian ages (1700–900 $\times$ 10^6 years); they mainly comprise granite, granodiorite and granitic to dioritic gneiss. There are also areas with Caledonian rocks, these locations are anyway at maximum 20–30 km far from the most widespread geology aforementioned. Igneous and metamorphic rocks occur close to Bergen, where various brownish and red colours stand for gabbro,

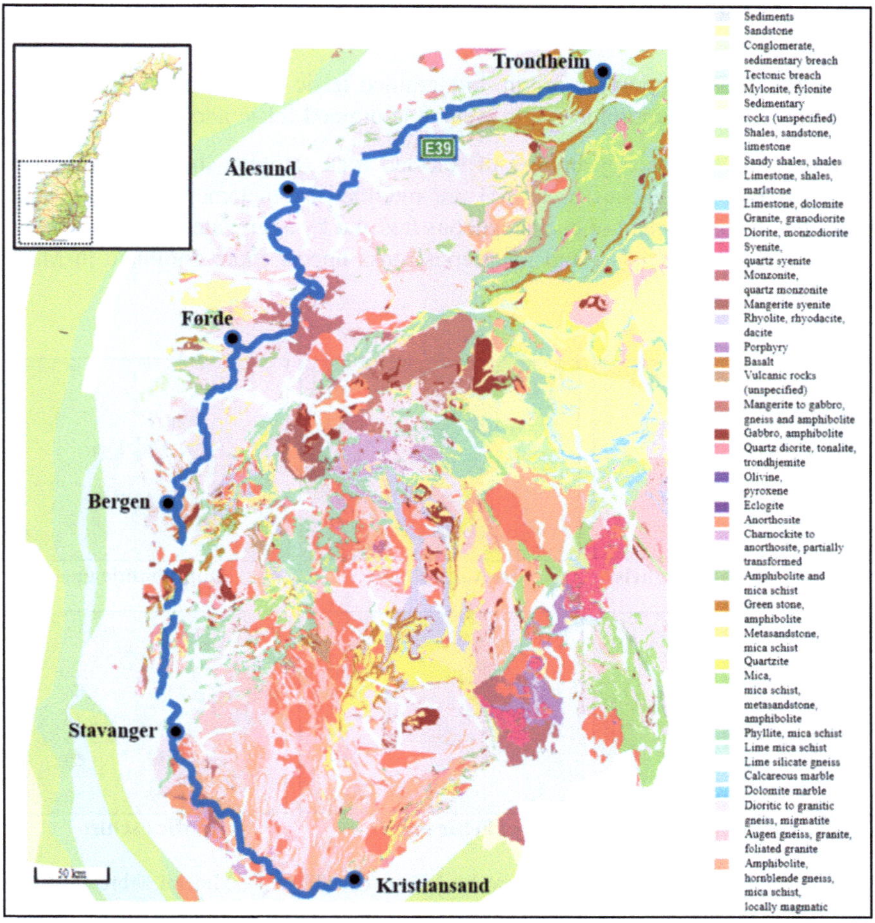

Fig. 1. Bedrock geology of southern Norway and alignment of E39 highway

diorite, anorthosite, granite and augen gneiss. Low grade metamorphic rocks such as phyllite occur in the Boknafjord area close to Stavanger.

NPRA is responsible for the current operations in the construction site "Sveg-atjørn-Rådal": tunnel connections will improve the traffic condition between Bergen and Os. Tunnels are excavated with means of explosive slurry injected by drilling jumbos. The blasted rocks are in size up to one meter. The material is then crushed by mobile impact crushers, sieved and stored in large areas. The blasted rocks derived from this building site adequately represent the variety in the geology spread along the entire highway alignment.

The materials collected come from Lyshorn tunnel, designed to connect the locality of Endelausmarka (Os municipality) to Rådal with a length of 9.3 km. Three types of rocks are selected and tested.

- Material M1. Mafic igneous origin, partly modified by metamorphism (amphibolite);
- Material M2. Metamorphic origin, fine-grained felsic and micaceous rocks;
- Material M3. Metamorphic origin, very fine-grained felsic and micaceous rocks.

Thin-section microscopy images of selected rock samples show mineralogy and grain sizes (Fig. 2). Igneous rocks M1 are modified by metamorphism, e.g. amphibolization and replacement of coarse igneous feldspar by aggregates of fine epidote and feldspar. Finer-grained felsic and micaceous rocks appear more dominant in M2 and especially in M3.

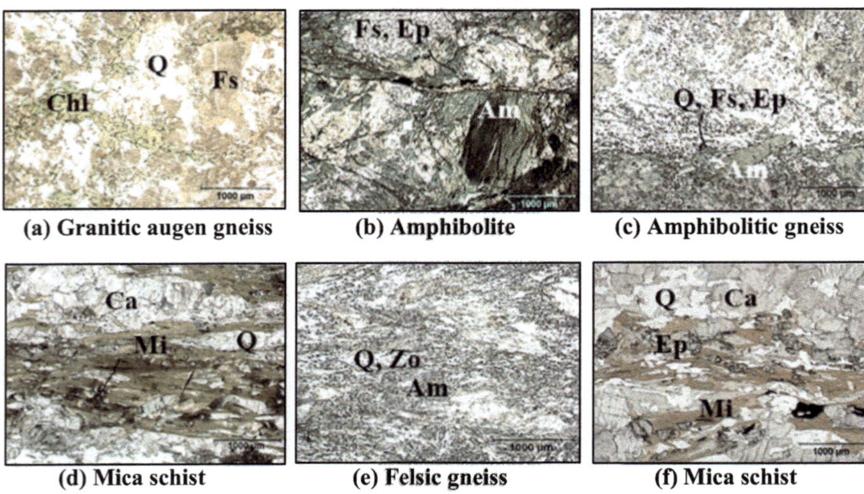

Fig. 2. Mineralogy and grain sizes M1 (**a**, **b**), M2 (**c**, **d**), M3 (**e**, **f**). Mineral abbreviations Am amphibole, Ca calcite, Chl chlorite, Ep epidote, Fs feldspar, Q quartz, Mi mica/biotite, Zo zoisite (optical micrographs, transmitted plane-polarized light, 1 mm scale bar)

2.2 Standard Tests for Road Unbound Layers Materials

The pavement design manual N200 (NPRA 2014) allows for the use of crushed rocks. It is possible to use this resource in the road base layer as paved crushed rocks and in the road subbase layer as unsorted crushed rocks if Los-Angeles standard test (LA value) and micro-Deval standard test (MDE value) are fulfilled. The LA limit values are respectively 30 and 35 for base layer and subbase layer, the MDE limit value is 15 for both of them.

Further requirements in terms of upper and lower grain size distribution curve are demanded for the base layer (Fig. 3). The distribution curve of the subbase layer must be within 20/120 mm. The major part of the E39 highway corresponds to traffic class F, which is associated to a minimum of 10 million repetitions of 10-ton equivalent standard axle load.

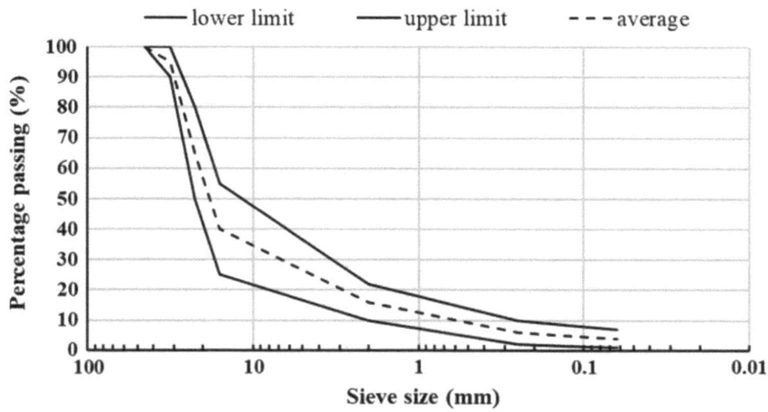

Fig. 3. Grain size distribution curve for base layer

Materials M1, M2, M3 are tested according to the Los Angeles standard test (CEN 2010) and the micro-Deval standard test (CEN 2011).

2.3 Modelling Construction Phase and Service Life Phase

The crushed rocks behaviour is studied according to two phases: construction phase and service life phase. These steps are respectively represented by the compaction action of a drum roller and the repetition of a 10-ton standard axle load. The variation in grain size due to crushing largely affects the most important engineering properties of granular materials in terms of resilient modulus and deformation.

The crushing of a subbase layer is investigated for this purpose. Materials M1, M2, M3 have gradation 20/120 mm.

Bomag BW 213 DH is the single steel drum roller taken into consideration for the construction phase. Its weight is 12.3 t, the drum axle load is 7.6 t, the wheel axle load is 5.2 t, the static linear load is 3.55 kg/mm and the working width is 2.1 m. The compactor is interpreted with a one-degree-of-freedom lumped parameter scheme (Fig. 4). m_d and m_f are the drum and frame masses, F_C is the centrifugal force, F_s is the contact force between drum and soil and g is the acceleration of gravity, frame inertia is neglected (Mooney and Dietmar 2007; Yoo and Selig 1980). For the compactor considered, the parameters are $m_d = 7560$ kg, $m_f = 1475$ kg, $F_C = 284$ kN. The drum acceleration a_d varies, a cautious value of 3 g is considered (Gorman and Mooney 2003). The contact force F_s is then assessed per equilibrium of forces and it is equal to 150 kN.

COMSOL software is used to simulate the two mentioned stages to assess the stress distribution; during the construction phase rocks are expected to experience a state of stress which is larger than the one expected during the service life phase (Kwon et al. 2008).

In the former stage, the subbase is subjected to the compaction load. The layer is 300 mm thick, 3 m wide, 0.5 m long and the drum is 2.1 m wide; the drum contact area is rectangular. The widths are halved thanks to the symmetry of the problem (Fig. 5a).

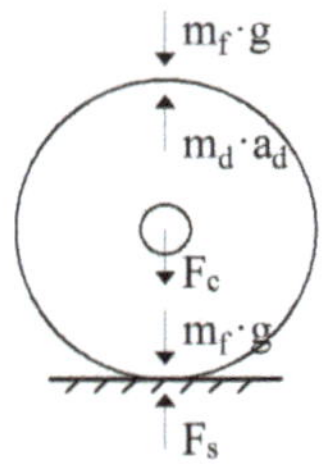

Fig. 4. Drum roller one-degree-of-freedom scheme (after Yoo and Selig 1980)

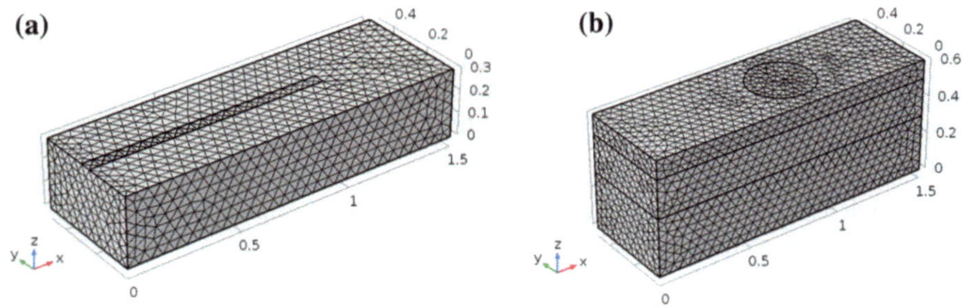

Fig. 5. Modelling of the subbase (**a**) and subbase, base, top layers (**b**)

In the second stage, the subbase and the other layers of the road are subjected to the vehicle loads, these are considered as repetitions of the 10-ton standard axle load. This configuration is modelled by adding a base layer and a top layer, respectively 100 and 200 mm thick, on the previous subbase layer. The distance between the axes of the wheels is 1.7 m and the tyre contact area is circular. The model parameters for the top layer are Young's modulus E = 3 GPa, Poisson's ratio $\upsilon = 0.4$ and density $\rho = 2300$ kg/m^3. The problem is symmetric so the widths are halved (Fig. 5b).

Both subbase and base are unbound layers and their mechanical behaviour can be properly described by a non-linear law defining the resilient modulus M_R. The k–θ model is a non-linear, stress dependent power function (Hicks and Monismith 1971). The model is given as follows in its dimensionless form:

$$M_R = k_1 \sigma_a \left(\frac{\theta}{\sigma_a} \right)^{k_2}, \quad (1)$$

where θ is the bulk stress sum of the principal stresses, σ_a is a reference pressure (100 kPa) and k_1 and k_2 are model parameters assessed from the regression of the test results.

Repeated Triaxial Load Test (RTLT) can define the resilient modulus parameter. The RTLT equipment available can be used to test materials up to 30 mm in size: all the samples are prepared according to the sieving curve reported as dashed line in Fig. 3, which corresponds to a base layer. In order to prepare a specimen, the desired amount of water is added to the material and rests for 24 h to let the moisture distribute uniformly. Four equal layers are compacted with the vibratory hammer. Each material set comprises two specimens, they have a diameter of 150 mm and height of 240 mm, the amount of water is 1%. A Kango 950X vibratory hammer (total weight 35 kg, frequency 25–60 Hz, amplitude 5 mm) is used to compact the specimen layers for 30 s. Figure 6 displays the average bulk density for each material set (CEN 2003). RTLTs are performed according to the multi-stage loading procedure in five different sequences with different confining stresses (CEN 2004). Water is the confining medium and applies a uniform constant pressure in all directions (triaxial or confining stress), a hydraulic jack exerts an additional vertical dynamic stress (deviatoric stress), which stepwise increases at different levels of confining stress. Two and three Linear Variable Differential Transducers (LVDTs) measure the axial and radial deformations, respectively. Figure 7 displays the resilient modulus curves and the values of k_1, k_2 parameters for the three materials under investigation.

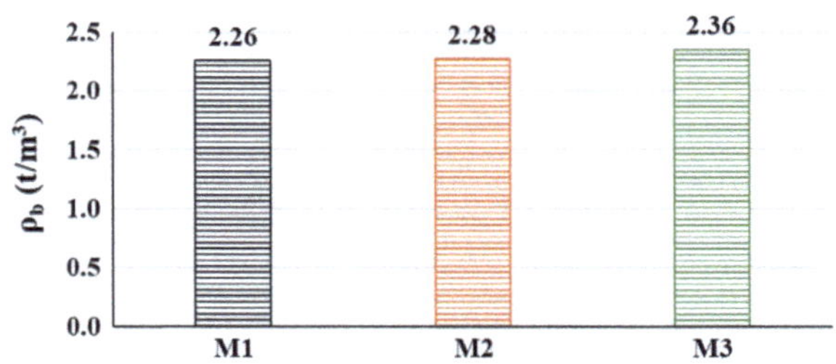

Fig. 6. Bulk density of the specimens tested with RTLT

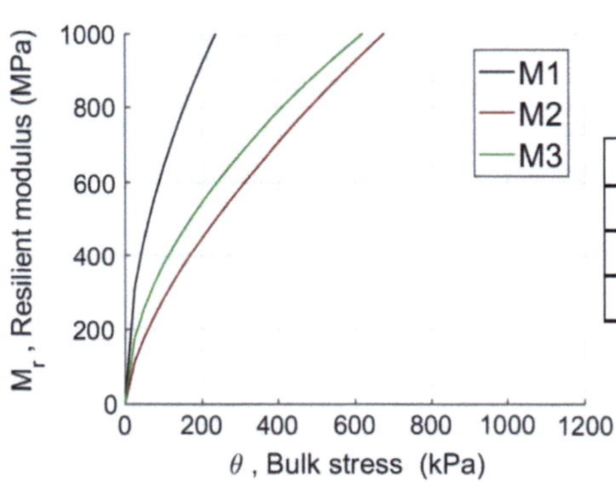

Material	k_1	k_2
M1	6378	0.52
M2	2816	0.66
M3	3737	0.54

Fig. 7. Resilient modulus k–θ model: curves and model parameters

2.4 Road Construction Phase

Field operations investigate the crushing of the materials M1, M2, M3: the sieving curves referring to before and after drum roller action are compared. The crushing of a subbase layer is studied: materials M1, M2, M3 have dimension 20/120 mm. They are placed in three areas close to each other: each area is approximatively 7 m long, 5 m wide and 300 mm high, it is made of about 15 t of rocks. The compactor exerts its action just on one side of the rocks (Fig. 8). It is assumed that, before the compacting action takes place, the grain size distribution associated with the not compacted stripe (marked in blue) is the same of the compacted stripe (marked in red). Four portions are identified in each area: two do not undergo compaction and two undergo compaction. Each portion is 1 m wide, 1 m width, 300 mm high, weighs about 200 kg and numbered from 1 to 12.

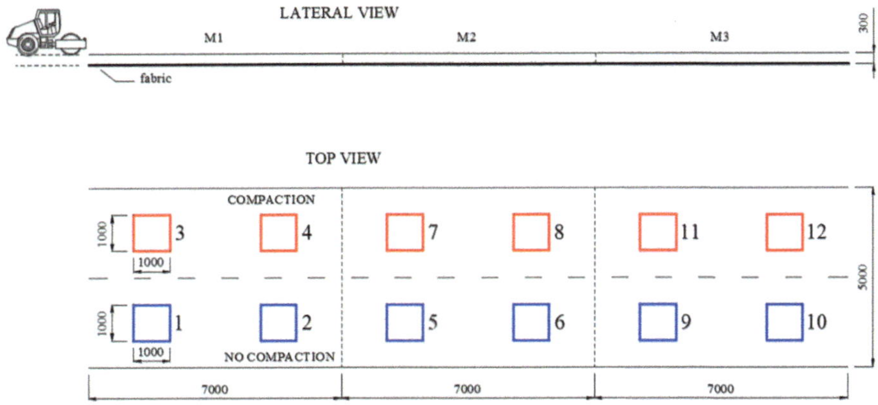

Fig. 8. Layout of the field test (dimensions in mm)

Figure 9 depicts the main stages of the test fulfilment: firstly, the existing surface is compacted with a single steel drum roller and operators lay out a polypropylene not woven geotextile (Fig. 9a), which prevents losing the small fractions produced during the test. The materials are tipped from a truck and distributed uniformly with the help of a gravedigger. A single steel drum roller then compacts one side of the placed crushed rocks (Fig. 9b). Bomag BW 213 DH accomplishes four passages as specified in the manual code (NPRA 2014).

Figure 10 displays the bulk densities of the layer before and after the roller compaction. The construction phase is completed and the twelve portions are highlighted with coloured spray (Fig. 9c). The rocks are collected from these portions by hand to keep the gradation as unaltered as possible and moved into plastic bags (Fig. 9d); finally, the material is sieved. Researchers have used various parameters or measures to represent the amount of particle breakage that takes place during loading (Gupta 2016); either as the variation of a particular grain diameter (Lade et al. 1996) or as the shift of the whole grain size distribution curve (Hardin 1985). This research refers to coefficient of uniformity C_u, coefficient of curvature C_c and particle breakage factor B_{10} (Lade et al. 1996). These parameters are respectively defined as

$$C_u = \frac{D_{10}}{D_{60}}, \tag{2}$$

Fig. 9. Main stages of the field test

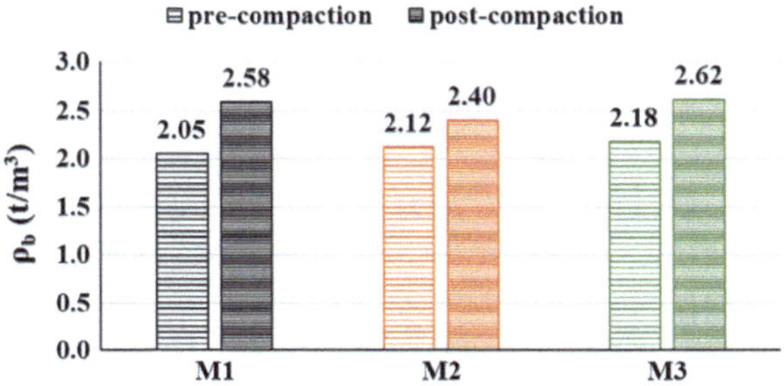

Fig. 10. Bulk density of the unbound layer before and after roller compaction

$$C_c = \frac{D_{30}^2}{D_{10}D_{60}}, \tag{3}$$

$$B_{10} = 1 - \frac{D_{10,f}}{D_{10,i}}; \tag{4}$$

where D_{10}, D_{30}, D_{60} are respectively the grain diameter at 10, 30, 60% passing, subscript *f* and *i* respectively stand for final and initial gradation.

2.5 Road Service Life Phase

According to the pavement design manual N200, traffic class F entails at least 10 million repetitions of 10-ton standard axle load (NPRA 2014). The RTLT equipment is used to reproduce the average stress state in the subbase: 20 and 120 kPa are the pressures respectively set in the horizontal and vertical directions in the triaxial cell. These values derive from the modelling outcomes referring to service life scenario (Fig. 12d–f).

Two specimens are tested for each material M1, M2, M3. All the samples have a diameter of 150 mm and height of 240 mm.

The lower size of the subbase investigated in situ is 20 mm; moreover, it is recommended the maximum testing particle size to be smaller than one fifth of the specimen diameter (CEN 2004): the specimens are made of crushed rocks 20/30 mm in size. All the materials are washed and dried before the testing in the triaxial cell. The specimens are subjected to one million load repetitions, since this figure is sufficient enough to highlight discrepancies among the three materials.

3 Test Results and Discussion

3.1 Standard Acceptance Tests Outcomes

Los Angeles and micro-Deval are the standard tests when it comes to accept crushed rocks as building materials in the pavement unbound layers. Three and two specimens for each material are respectively tested, Fig. 11 displays the average results with standard deviation.

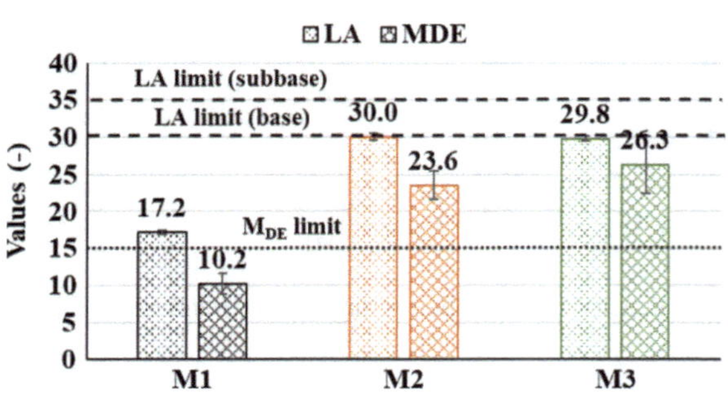

Fig. 11. Los Angeles and micro-Deval values

Material M1 fulfils the code requirements, therefore it can be directly used for road construction. Both material M2 and M3 have LA values lying close to the limit, but both of them exceed the threshold when it comes to MDE values.

3.2 Stress Magnitudes During Construction Phase and Service Life Phase

The contact length between the drum and the soil is the remaining parameter to be defined for modelling purposes. Figure 12 displays the vertical stress in the symmetry plane for respectively 25 mm (a), 75 mm (b) and 125 mm (c).

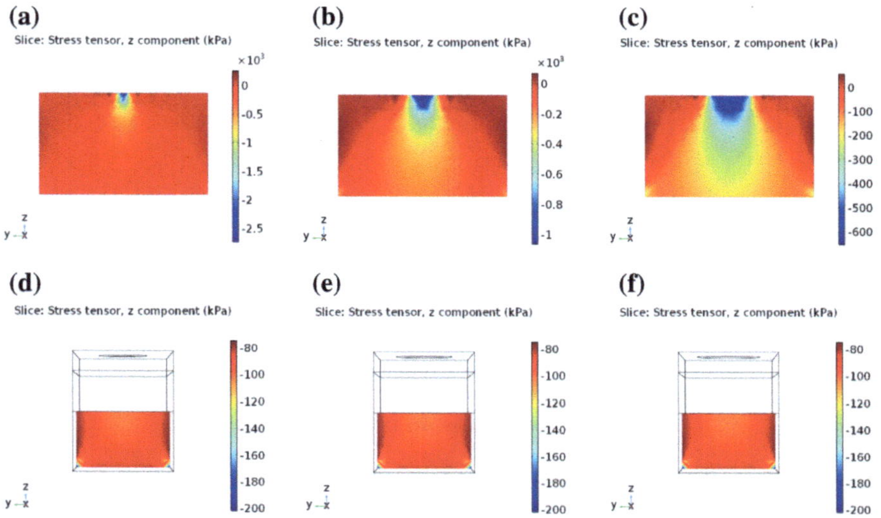

Fig. 12. Vertical stress in the subbase (kPa); for different values of drum contact length: 25 mm (**a**), 75 mm (**b**), 125 mm (**c**); for different values of tyre contact radius: 130 mm (**d**), 150 mm (**e**), 170 mm (**f**)

The vehicle load is the 10-ton standard axle load: each wheel exerts 5 ton. Figure 12 shows the vertical stress in the subbase plane under the wheel for different lengths of tyre contact radius: 130 mm (d), 150 mm (e) and 170 mm (f).

The stress distributions depicted in Fig. 12 refer to M1 resilient curve, as the results connected to M2, M3 resilient curves are highly similar.

The subbase experiences the largest stresses during the compaction phase, moreover, the roller-soil contact length has a large impact on the results regarding stress distribution.

The stress values connected to the service life phase are one order of magnitude smaller; in addition, they are not particularly sensible to the tyre contact radius: major crushing is expected to take place during the former stage than during the latter stage (Gupta 2016).

3.3 Crushed Rocks Performance During Construction Phase

The crushed rocks coming from the twelve field portions are sieved. The steel drum roller exerts its action just along a stripe which comprises six portions. It is assumed that, before the compaction takes place, the grain size distribution referring to the not compacted stripe is the same of the compacted stripe. In total twelve sieving curves are assessed. Therefore for each material type there is a set of two portions subjected to the same treatment (compacted or not compacted), for each set the average curve is assessed: Fig. 13 depicts six sieving curves. The amount of particle breakage during compaction can be visualized thanks to the particle size distribution curves measured before and after the external action.

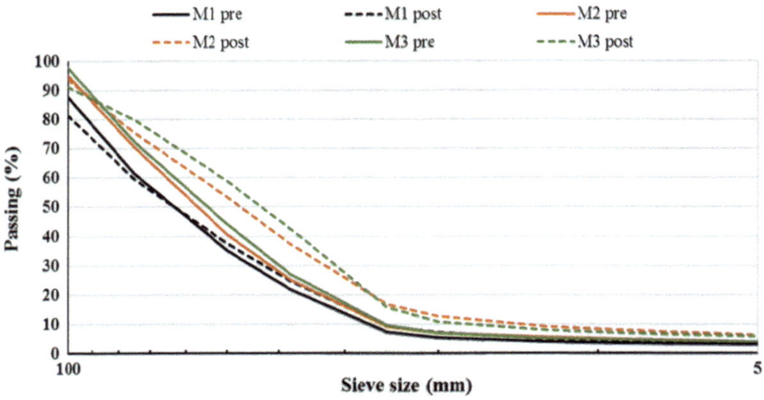

Fig. 13. Sieving curves referring to pre-compaction and post-compaction action

From Fig. 13 it is apparent that there is a slight change in the grain size for material M1, while both materials M2 and M3 undergo substantial crushing. There is an unexpected crossing of lines between 100 and 75 mm regarding pre- and post-compaction: the assumption that the grading curve of the not compacted stripe is the same of the compacted stripe before the roller action turns out to be not completely exact; although this hypothesis is necessary in order not to sieve all the materials placed in situ (about 45 t).

Figure 14 displays the increment in passing for each sieve size used in the sieve analyses; the increment is assessed as the difference in passing values between post-compaction curve and pre-compaction curve, divided by the passing value referring to pre-compaction curve. The maximum increment for material M1 is 31.6% referring to sieve size 20 mm, both materials M2 and M3 have much higher maximum increment, respectively 83.1 and 64.3% referring to sieve size 25 mm. Material M1 crushes less than the other two materials M2 and M3 under compaction load. Each sieving curve is fitted by a third-order equation

$$ax^3 + bx^2 + cx + d = 0, \qquad (5)$$

six polynomial expressions are calculated based on the sieving curves. Considering these new calculated particle size distribution curves, it is possible to have a better estimation of the coefficient of uniformity C_u, coefficient of curvature C_c and particle breakage factor B_{10}.

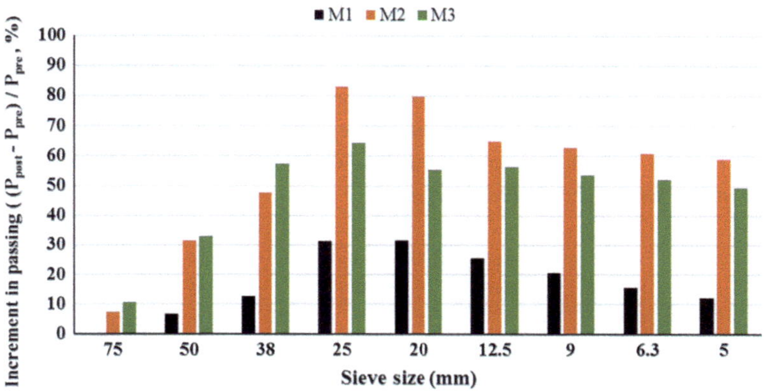

Fig. 14. Increment in passing for each sieve size after compaction

In addition, the optimum vertical translation of the pre-compaction curve towards the post-compaction curve is estimated with the least square methods; the value ζ quantifies this translation and also contributes to evaluating the overall difference in the curve shapes.

Figure 15 displays the coefficient of uniformity, coefficient of curvature and particle breakage factor for each material. Material M1 experiences the lesser modifications in coefficient of uniformity and curvature, its particle breakage factor is smaller than the values referring to M2, M3.

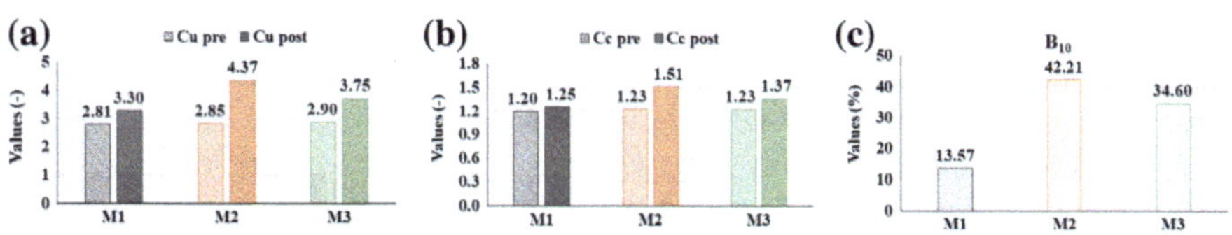

Fig. 15. Coefficient of uniformity (**a**), coefficient of curvature (**b**) and particle breakage factor (**c**) of the calculated sieving curves before and after compaction

It is possible to establish a relationship between the standard tests and the particle breakage factor as follows

$$LA = \alpha B_{10}, \tag{6}$$

$$M_{DE} = \beta B_{10}, \tag{7}$$

the best fitting α, β parameters values in this case are $\alpha = 0.95$ and $\beta = 0.69$. The displacement along the y-axis of the pre-compaction curve assessed with the least square method is $\zeta = -0.14$ for M1, $\zeta = 6.68$ for M2 and $\zeta = 7.3$ for M3. The pre- and post-compaction curves for material M1 are much closer compared to the ones concerning materials M2, M3.

3.4 Crushed Rocks Performance During Service Life Phase

Specimens made of crushed rocks M1, M2, M3 in size 20/30 mm are tested in the triaxial cell. A set of two specimens for each type is tested and sieved both before and after the load action, therefore twelve sieving curves are assessed. Average curves are estimated: Fig. 16 depicts six sieving curves. The trends of pre- and post-load curves are very similar for all the material types.

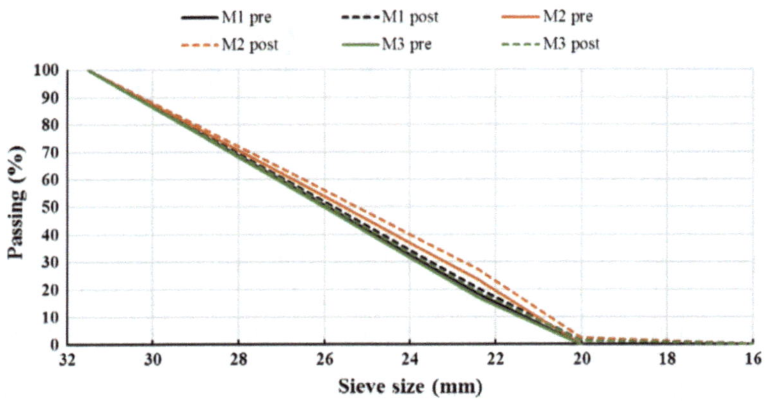

Fig. 16. Sieving curves referring to pre-load and post-load action

Figure 17 shows the passing values after the loading action. It generates new grain sizes (smaller than 20 mm) with approximately the same percentage for all the materials, even if M1 crushes less than M2, M3. Figure 18 shows the coefficient of uniformity, coefficient of curvature and particle breakage factor. The first two parameters do not undergo remarkable changes due to the loading action. The particle breakage factors are close to the lower limit zero, which would imply no particle breakage. The curves are considerably close to each other for all the materials and the loading action does not bring to significative crushing.

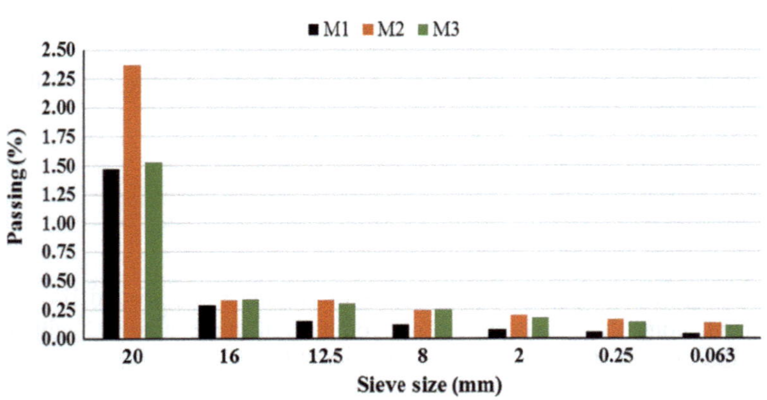

Fig. 17. Passing of the grain sizes generated during the load action

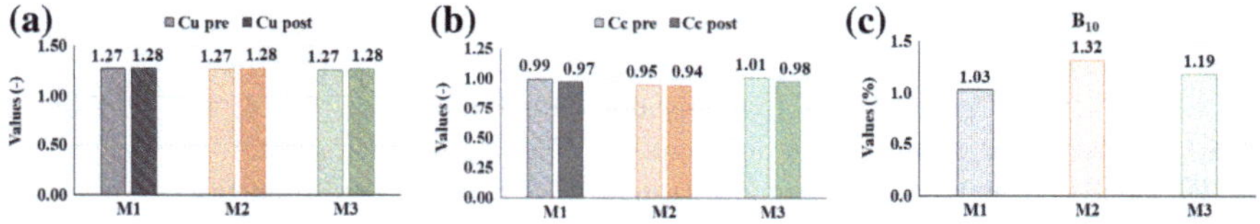

Fig. 18. Coefficient of uniformity (**a**), coefficient of curvature (**b**) and particle breakage factor (**c**) of the sieving curves before and after load

4 Conclusions

The research has focused on the mechanical assessment of crushed rocks derived from blasted rock surplus as a result of tunnel installations along E39 highway. Possible use in the road unbound layers close to the place of production would provide a sustainable cost-benefit application. Three material types M1, M2, M3 have been chosen to represent adequately the bedrock geologies encountered along the entire highway alignment. Material M1 complies with the requirements set by the pavement design manual N200 in terms of Los Angeles and micro-Deval limit values; vice versa materials M2 and M3 do not fulfil them. Material M1 stands for the major part of the geology encountered along the highway, therefore there is a huge potential for the excavated rocks to be directly used in pavement unbound layers.

The research has also investigated the ability of the three materials to withstand compaction and trafficking while used in the unbound layers of a road structure. The crushing process of the materials is investigated during the construction phase, achieved by a full scale testing to assess aggregate soundness after rolling, and during service life phase, simulated with the triaxial cell apparatus. The modelling connected to these two scenarios highlights a remarkable difference in the stress state. The crushing is substantially more significant in the former stage than in the latter stage for all the materials.

Coefficient of uniformity C_u, coefficient of curvature C_c, particle breakage factor B_{10} and optimum vertical translation ζ are the parameters selected to describe the grain size distribution curves before and after drum roller compaction. There is a good correlation between the crushing standard tests and the crushing taking place in situ. Moreover, two relationships involving Los Angeles value, micro-Deval value and particle breakage factor have been proposed. These parameters clearly show material M1 performs better than materials M2, M3.

No significant differences in grain size have been observed due to the service life phase loading: all the materials perform similarly in this scenario.

References

Aatheesan, T., et al.: Beneficial use of brick rubble as pavement subbase material. Adv. Transp. Geotech. (2008). https://doi.org/10.1201/9780203885949.pt10

Arulrajah, A., et al.: Geotechnical and geoenvironmental properties of recycled construction and demolition materials in pavement subbase applications. J. Mater. Civ. Eng. (2013). https://doi.org/10.1061/(ASCE)MT.1943-5533.0000652

Barbieri, D.M., et al.: Laboratory investigation on unbound materials used in a highway with premature damage. In: Bearing capacity of Roads, Railways and Airfields (2017)
CEN: EN 13286-4. Unbound and hydraulically bound mixtures. Part 4: test methods for laboratory reference density and water content. Vibrating hammer (2003)
CEN: Unbound and hydraulically bound mixtures. Part 7: cyclic load triaxial test for unbound mixtures (2004)
CEN: EN 1097-2. Tests for mechanical and physical properties of aggregates. Part 2: methods for the determination of resistance to fragmentation (2010)
CEN: EN 1097-1. Tests for mechanical and physical properties of aggregates. Part 1: determination of the resistance to wear (micro-Deval) (2011)
CEN: EN 933-1. Tests for geometrical properties of aggregates. Part 1: determination of particle size distribution. Sieving method (2012)
Dunham, K.K.: Coastal highway route E39—extreme crossings. Transp. Res. Proc. (2016). https://doi.org/10.1016/j.trpro.2016.05.102
Erichsen, E., Aasly, A.: Mineralressurser i Norge 2015. Mineralstatistikk og bergindustriberetning (2016) (in Norwegian)
Fladvad, M., et al.: Comparison of practice for aggregate use in road construction—results from an international survey. In: Bearing Capacity of Roads, Railways and Airfields
Gorman, P.B., Mooney, M.A.: Monitoring roller vibration during compaction of crushed rock. In: 20th International Association for Automation and Robotic in Construction (2003)
Gupta, A.K.: Effects of particle size and confining pressure on breakage factor of rockfill materials using medium triaxial test. J. Rock Mech. Geotech. Eng. (2016). https://doi.org/10.1016/j.jrmge.2015.12.005
Hardin, B.O.: Crushing of soil particles. J. Geotech. Eng. (1985). https://doi.org/10.1061/(ASCE)0733-9410(1985)111:10(1177)
Hicks, R.G., Monismith, C.L.: Factors influencing the resilient response of granular materials. Highway Research Record (1971)
Kwon, J., et al.: Aggregate base residual stresses affecting geogrid reinforced flexible pavement response. Int. J. Pavement Eng. (2008). https://doi.org/10.1080/10298430701582347
Lade, P.V., et al.: Significance of particle crushing in granular materials. J. Geotech. Eng. (1996). https://doi.org/10.1061/(ASCE)0733-9410(1996)122:4(309)
Lekarp, F., et al.: State of the art. I: resilient response of unbound aggregates. J. Transp. Eng. (2000a). https://doi.org/10.1061/(ASCE)0733-947X(2000)126:1(66)
Lekarp, F., et al.: State of the art. II: permanent strain response of unbound aggregates. Journal of Transportation Engineering (2000b). https://doi.org/10.1061/(ASCE)0733-947X(2000)126:1(76)
Li, D., Selig, E.T.: Resilient modulus for fine-grained subgrade soils. J. Geotech. Eng. (1994). https://doi.org/10.1061/(ASCE)0733-9410(1994)120:6(939)
Mooney, M., Dietmar, A.: Vibratory roller integrated measurement of earthwork compaction: an overview. In: Seventh International Symposium on Field Measurements in Geomechanics (2007). https://doi.org/10.1061/40940(307)80
Neeb, P.R.: Byggeråstoffer (1992) (in Norwegian)
NGU (Norges Geologiske Undersøkelse, Geological Survey of Norway). Accessed 1 Oct 2017. http://www.ngu.no/
NPRA, Norwegian Public Roads Administration: Håndbok N200 Vegbygging (2014) (in Norwegian)
NPRA, Norwegian Public Roads Administration: The E39 Coastal Highway Route. 10 May 2017, Accessed 1 Oct 2017 (2017). http://www.vegvesen.no/vegprosjekter/ferjefriE39/English
Núñez, W.P., et al.: Using unbound aggregates resulting from amethyst mining in low volume roads. Adv. Transp. Geotech. (2008). https://doi.org/10.1201/9780203885949.ch28

Onyango, M., Busch, C.: Use of naturally occurring pozzolans for road construction in Tanzania. Transp. Res. Rec. J. Transp. Res. Board (2007). https://doi.org/10.3141/1989-61

Otto, A., et al.: Method for increasing the use of locally available materials for road construction in Ethiopia by allowing for climatic variations. Transp. Res. Rec. J. Transp. Res. Board (2015). https://doi.org/10.3141/2474-13

Ramberg, I.B., et al.: Landet blir til (2013) (in Norwegian)

Yoo, T.S., Selig, E.T.: Dynamics of vibratory-roller compaction. Int. J. Rock Mech. Min. Sci. Geomech. (1980)

Coupled Eulerian-Lagrangian Modeling to Study the Long-Runout Landslide: A Case Study

Sheng-yang Feng[1,2(✉)], Hong-quan Li[3], Xiang-yang Li[1], Yong Liu[1,2], and Zhi Chen[1]

[1] School of Environment and Safety Engineering, University of South China, Hengyang, China
fengshengyang@hotmail.com
[2] Hunan Province Engineering Technology, Research Center of Uranium Tailings Treatment, Hengyang, China
[3] School of Civil Engineering, Central South University, Changsha, China

Abstract. The coupled Eulerian-Lagrangian (CEL) method is a latest technology for simulating the large deformation and discrete geotechnical problem. In this paper, we use this new method to simulate the long-runout landslide which occurred at the Hong'ao landfill on December 20th, 2015, in Shenzhen, China. This landslide killed 77 people and destroyed 33 houses, regarded as one of the largest landfill landslide in the world. The field survey and previous research results after the accident are applied to validate the numerical model of the landslide. The long-runout behavior of the landslide is studied in terms of runout area, velocity, kinematic energy. The runout area of the simulation landslide is close to that of the actual. The maximum simulated velocity of the landslide is up to 35.41 m/s at 68 s which is close to the result simulated by the LS-RAPID. The maximum kinetic energy of the landslide is 2230.4 GJ at 68 s, and the frontier buildings might be impacted at 78.3 s, when the total energy of the landslide can be 2097.4 GJ, roughly equivalent to the explosive energy of 5 t TNT.

Keywords: Landslide · Coupled Eulerian-Lagrangian method
Digital elevation model · Numerical simulation

1 Introduction

Long-runout landslide is a kind of geological disaster phenomenon with extremely destructive force. Under the control of topography, the landslide moves along a certain trajectory with extraordinary fluidity and velocity. It often causes enormous economic losses and casualties. The study on the movement mechanism of long-runout landslide and the calculation of sliding velocity, distance and coverage area can provide scientific reference for the risk prediction and mitigation of this kind of disaster (Yin et al. 2017). Researches on landslide and debris flow movement mainly include 4 aspects, such as empirical statistics (Saha et al. 2005; Lee et al. 2008; Nandi and Shakoor 2010; Turner et al. 2015), mechanical model (Kent 1966; Blasio 2008; Quinnp et al. 2011;

W.-C. Cheng et al. (eds.), *Tunneling in Soft Ground, Ground Conditioning and Modification Techniques*, Sustainable Civil Infrastructures,
https://doi.org/10.1007/978-3-319-95783-8_20

Barbero and Barpi 2012; Davies et al. 2014), physical model test (Fioravante 2007; Kong et al. 2013; Chi et al. 2016; Zhang et al. 2017) and numerical simulation, in which the numerical simulation is one of the most commonly used methods. At present, the numerical simulation used in long-runout landslide prediction can be divided into 3 categories, which are discrete medium model, continuum model and coupling model. In the discrete model, Tang et al. (2009), Lo et al. (2011), Zhou et al. (2013), Yuan et al. (2014) used the distinct element method (DEM) to study the post-failure kinematic and motion behavior of the landslide runout process. The discontinuous deformation analysis (DDA) method is another discrete medium model used by Wu and Chen (2011), Zhang et al. (2014). In the continuous model, the smoothed particle hydrodynamics (SPH) (Bui et al. 2011; Huang et al. 2012; Dai et al. 2014; Pastor et al. 2014), the material point method (MPM) (Sun et al. 2015; Li et al. 2016), and the computational fluid dynamics (CFD) methods (Degirolamo et al. 2011; Montagna et al. 2011; Horrillo et al. 2013) were applied to simulate the extremely rapid, massive, flow-like motion of the landslide runout process. Recognizing the fact that the long-runout landslide process involves the large deformation. Recently, Coupled Eulerian–Lagrangian (CEL) method has been developed to solve the difficulties in the large deformation problems. The CEL method can solve the problem of large deformation. It combines the advantages of Euler and Lagrangian mesh and adopts the Laplace mesh for topography and the Euclidean mesh for sliding land, which overcomes the shortcoming of the significant mesh distortion in the sliding zone. Recently, Qiu et al. (2011), Tho et al. (2012) used the CEL method to simulate large deformation problems with good results.

In this paper, we use the CEL method to investigate a long-runout landslide in south China (113° 56′ 22″E, 22° 42′ 38″N). The field survey and previous research results after the accident are applied to validate the numerical model of the landslide. The verification results show that the CEL method is a practical numerical tool for risk assessment of the landslide. This landslide occurred in the Hong'ao landfill on December 20th, 2015 in Shenzhen (Fig. 1). 2.75 million cubic meters of construction solid waste (CSW) slided down to the Hengtaiyu industrial park, where 77 people were killed and 33 houses were buried. This huge landslide is regarded as one of the largest landfill slope failure in the world (Yin et al. 2016).

2 Dem of the Landfill

Calibrated digital elevation model (DEM) derived from shuttle radar topographic mission (SRTM) is applied to establish the CEL geometry model. The SRTM is an international research program that obtained digital elevation models to generate the most complete high-resolution digital topographic database of Earth (Ramakrishnan et al. 2013). Firstly, we use the SRTM database to obtain the topographic data of the Hong'ao landfill and its surrounding topography in the SRTM Official website. Secondly, the Global Mapper software was used to convert the topographic data into the DEM shown in Fig. 2.

(a) Satellite image before the landslide

(b) The UAV aerial map of landslide

Fig. 1. The Hong'ao landfill landslide in Shenzhen City

3 Finite Element Numerical Simulation Analysis

Obviously, the landslide runout process involves the large deformation problem. The conventional FE model is based on the Lagrangian framework. The main disadvantage of this type of model is that a significant mesh distortion occurs in the sliding zone, causing non-convergence of the solution. The CEL method is able to overcome extreme mesh distortion problems. It combines the advantages of the Lagrangian formation with that of the Eulerian formation (Pucker and Grabe 2012). In CEL method, the Eulerian material is traced when it flows through the mesh by computing its Eulerian volume fraction (EVF). Each Eulerian element has a percentage, representing the portion of the element filled with a material. If an Eulerian element is wholly occupied by a material, its EVF is 1; if no material in the element, its EVF is 0.

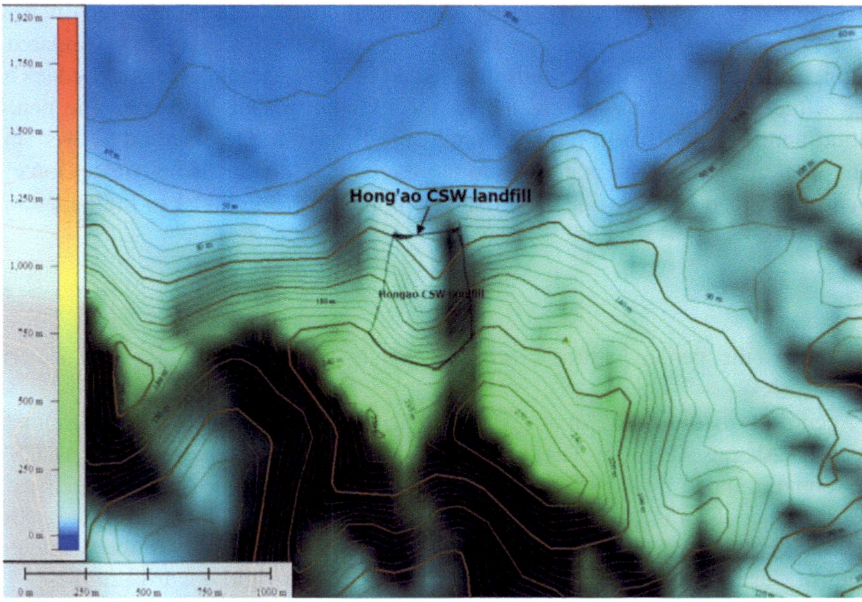

Fig. 2. The DEM of the Hong'ao landfill

The general contact based on a penalty contact method is used for the contact between the Eulerian materials and the Lagrangian. The Lagrangian elements moves through the Eulerian mesh with no resistance until encountering an Eulerian element filled with material (EVF $\neq 0$).

The CEL method in Abaqus/Explicit uses an explicit time integration scheme. The central difference rule is employed for the solution of the non-linear system of differential equations. The unknown solution for the next step can be found directly from the solution of the previous time step, so one advantage of using the explicit time integration is that no iteration is needed in the next step. Another advantage is that the robustness regarding difficult contact conditions. Explicit calculations are not strictly stable. Numerical stability is guaranteed by introduction of the critical time step size Δt_{crit} which depends on the characteristic element length L_e and the dilatory wave speed C_d. We can calculate the critical time step size in every time step as follow

$$\Delta t_{crit} = \frac{L_e}{C_d} \tag{1}$$

The above formula reveals that the waves in the model can propagate through one element per time step at most. It has been stated that non-linear simulations are stable if the critical time step size is kept equal in all steps. Attribute to different stiffnesses during loading and unloading, at least one wrong critical time step size is needed to be calculated in every time step. During the CEL simulations it became evident that the critical time step size has to be reduced by a factor of about 0.1 to receive a stable solution. By using the CEL method the contact between Eulerian domain and

Lagrangian domain is discretized using the general contact algorithm, which is based on the penalty contact method. The penalty contact method is less strict compared to the kinematic contact method. Seeds are created on the Lagrangian element edges and faces while anchor points are created on the Eulerian material surface. The penalty method approximates hard pressure-overclosure behavior. This method allows small penetration of the Eulerian material into the Lagrangian domain. The contact force F_p which is enforced between seeds and anchor points is proportional to the penetration distance d_p.

$$F_p = k_p d_p \tag{2}$$

where k_p is the penalty stiffness which depends on the Lagrangian and Eulerian material properties.

According to the DEM of the Hong'ao landfill, we establish the FE geometric model in Abaqus (Fig. 3). The FE model is made up of three parts: (a) CSW, (b) mountain and (c) void space. The CSW and the void space are modeled as an Eulerian material using the EC3D8R element (8-node linear brick, reduced integration, hourglass control element). The mountain is modeled as a rigid body in the Lagrangian framework using the R3D4 element (4-node 3-D bilinear rigid quadrilateral).

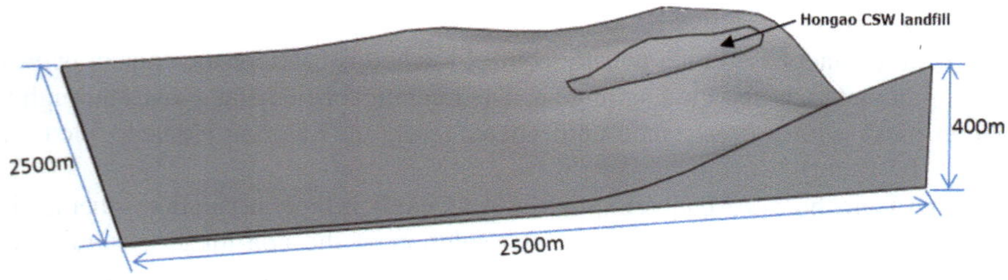

Fig. 3. The FE geometric model of the Hong'ao landfill in Abaqus

The Mohr-Coulomb model is applied to simulate the behavior of the CSW. The parameters of the Hong'ao landfill are given in Table 1, based on survey results taken after the landslide (Yin et al. 2016).

Table 1. Parameters of the Hong'ao landfill

CSW parameters	Value
Density (kg/m^3)	1800
Young's modulus (MPa)	20
Poisson's ratio (–)	0.3
Cohesion (kPa)	10
Friction angle (°)	26
Dilatancy angle (°)	0

The CSW and void space in the initial condition are created by the volume fraction tool in Abaqus. EVF = 1 for the CSW means that the elements are filled with Eulerian material, whereas EVF = 0 for the void space. Zero vertical velocity boundary is applied to the bottom of the model, while zero horizontal velocity boundary to the lateral side of the model. The soil-void interface which uses a general contact has no velocity boundary so that the CSW is able to slide into the void space if necessary. A 1 m × 1 m mesh is used, and 6.64 million eight-node hexahedron elements are generated in this FE model. The meshed model can be seen in Fig. 4.

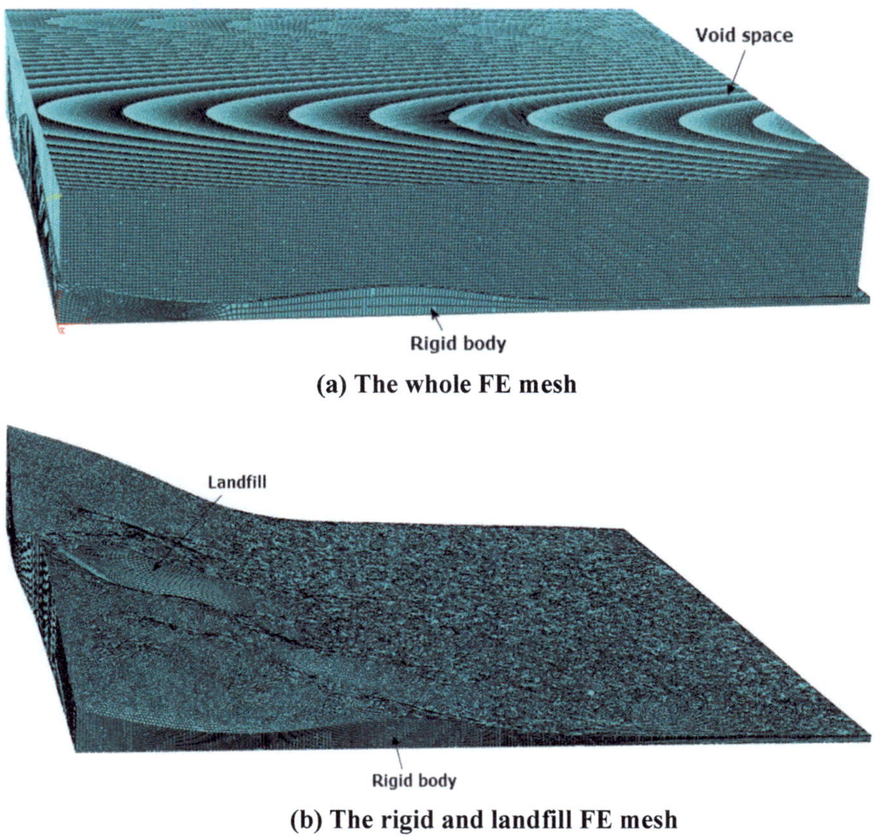

Fig. 4. The FE meshes of the Hong'ao landfill in Abaqus

The FE model consists of two steps. In the first step, the geostatic load is used to bring the landfill to an in situ stress condition under $K_0 = 1$ by keeping the landfill fixed. In the second step, the landslide is going to occur while the landfill unfixed for 120 s under the condition of the geostatic load and the pore water pressure which is generally 200 kPa, with a maximum at the rear of the landfill of about 600 kPa. A computer with 32 CPUs and 384G memory is used to run the FE model for 27.2 h by the parallel computation technology.

4 Results and Discussions

Figure 5 shows the comparison of the FE and satellite zoning map of the Hong'ao landfill landslide. From the FE simulation results, it can be seen that the sliding distance is about 1315 m, and the landslide volume is about 317.6 million cubic metres. The FE zoning map of the Hong'ao landfill landslide is close to that of the satellite. The sliding distance of the former is 19.5% larger than that of the later, and the width of the former is 4.1% larger than that of the later. The possible reason is that the FE model in this paper has been simplified without considering buildings on the runout path so that the landfill can slide farther. It is known that the buildings on the runout path will withstand the impact energy of the landslide and reduce the kinetic energy of the landslide, thus reducing the sliding distance of the landslide. This also shows that the numerical simulation results in this paper are in line with the actual situation.

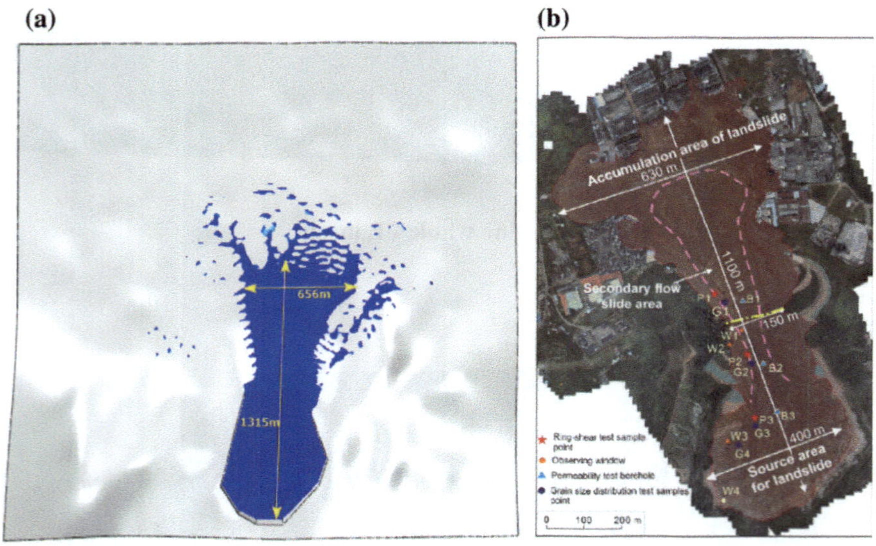

Fig. 5. Zoning map of the Hong'ao landfill landslide. **a** FE image. **b** Satellite image (Yin et al. 2016)

Figure 6 shows the sliding speed of the Hong'ao landfill at different times. From 5 to 30 s, the landslide is at the stage of accelerated sliding with a speed of 4.54 up to 12.13 m/s. At 68 s, the landslide possesses the highest potential energy with the maximum speed of 35.41 m/s in the whole long-runout process. At 90 s, the landslide goes into the slowdown motion stage with a maximum speed of 18.18 m/s. At 120 s, the landslide mass disperses over flat ground with low potential energy, and the landslide is going to slow down with a speed of just only 1.21 m/s due to ground friction.

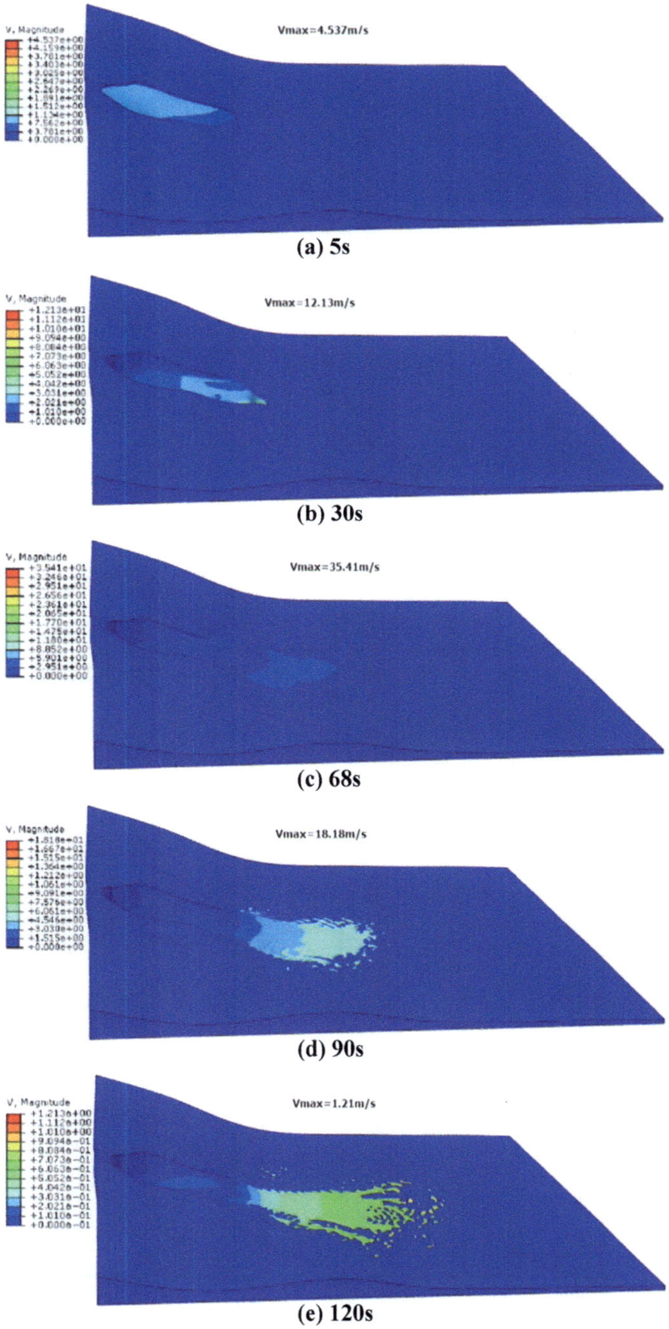

Fig. 6. The speed of the Hong'ao landfill landslide at different times

Figure 7 shows the comparison of the runout velocity of the Hong'ao landfill landslide simulated by the CEL method and the LS-RAPID (Yin et al. 2016). It can be seen that the runout velocity of the landslide simulated by the CEL method in this paper is close to that by the LS-RAPID from 40 s, and the maximum speed of the landslide in this paper is 16.3% larger than that of the later.

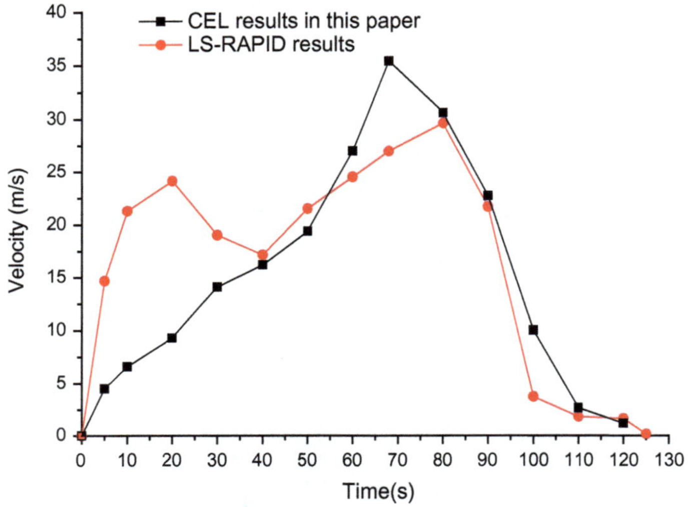

Fig. 7. Comparison of the runout velocity of the Hong'ao landfill landslide simulated by the CEL method and the LS-RAPID

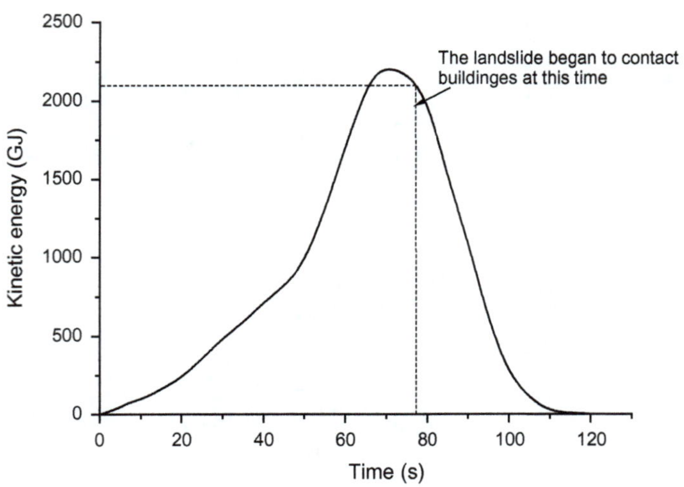

Fig. 8. The kinematic energy evolution of the Hong'ao landfill landslide

The kinetic energy of the landslide is one of important parameters obtained by the simulation. Figure 8 shows the kinematic energy evolution during the Hong'ao landfill landslide. The mechanics of the energy consumption in a large-scale landslide is very complex and no rigorous study is conducted in this paper. The kinematic energy is obtained by the simple equation $E = 1/2(mv^2)$. The maximum energy of the landslide reaches up to 2230.4 GJ at 68 s from the beginning. The landslide will impact the frontier buildings at 78.3 s, when the total energy of the landslide can be 2097.4 GJ, roughly equivalent to the explosive energy of 5 t TNT, which is sufficient to destroy buildings on the runout path of the landslide.

5 Conclusions

In this study, we successfully used the CEL method to simulate the long-runout process of the Hong'ao CSW landslide in Shenzhen City on December 20th, 2015. The long-runout behavior of the landslide is studied on the basis of runout area, velocity, kinematic energy. Limited by the computing capability of the computer, we did not consider the effect of buildings on the runout path of the landslide, so that the runout area of the simulation is slightly larger than that of the actual landslide, but they are still close. The maximum simulated velocity of the landslide in this paper is up to 35.41 m/s at 68 s, which is close to the result simulated by the LS-RAPID. The maximum kinetic energy of the landslide reaches up to 2230.4 GJ at 68 s, and the frontier buildings might be impacted at 78.3 s, when the total energy of the landslide can be 2097.4 GJ, roughly equivalent to the explosive energy of 5 t TNT.

Generally, this study demonstrates that the CEL method is a practical numerical tool for simulating the runout process of large scale landslides. Even though further calibration and validation will be required, this method is sufficient to obtain preliminary analysis results which will be of assistance in risk assessment of the landslide.

Acknowledgements. This work was supported by national natural science foundation of China (No. 11705083), the state key laboratory of safety and health for metal mines (No. 2016-JSKSSYS-08), the PHD initial foundation program of the University of South China (No. 2014XQD11), and the natural science foundation of Hunan Province (No. 2017JJ4009).

References

Barbero, M., Barpi, F.: Geomechanical modeling to study the effects of slope instability on buildings: a case study in northern Italy. Icarus **126**(2), 373–394 (2012)

Blasio, F.: Production of frictional heat and hot vapour in a model of self-lubricating landslides. Rock Mech. Rock Eng. **41**(1), 219–226 (2008)

Bui, H.H., Fukagawa, R., Sako, K., et al.: Slope stability analysis and discontinuous slope failure simulation by elasto-plastic smoothed particle hydrodynamics (SPH). Géotechnique **61**(7), 565–574 (2011)

Chi, L., De, Y., Zhong, W., et al.: Model test on rainfall-induced loess–mudstone interfacial landslides in Qingshuihe, China. Environ. Earth Sci. **75**(9), 1–18 (2016)

Dai, Z., Huang, Y., Cheng, H., et al.: 3D numerical modeling using smoothed particle hydrodynamics of flow-like landslide propagation triggered by the 2008 Wenchuan earthquake. Eng. Geol. **180**, 21–33 (2014)

Davies, O., Rouainia, M., Glendinning, S., et al.: Investigation of a pore pressure driven slope failure using a coupled hydro-mechanical model. Eng. Geol. **178**(8), 70–81 (2014)

Degirolamo, P., Cecioni, C., Montagna, F., et al.: Numerical modeling of landslide generated tsunamis around a conical Island. Nat. Hazards **58**(1), 591–608 (2011)

Fioravante, V.: Physical modeling of landslide stabilization methods in an overconsolidated clay. Geotech. Test. J. **31**(2), 175–191 (2007)

Horrillo, J., Wood, A., Kim, G.B., et al.: A simplified 3-D Navier-Stokes numerical model for landslide-tsunami: application to the Gulf of Mexico. J. Geophys. Res. Oceans **118**(12), 6934–6950 (2013)

Huang, Y., Zhang, W., Xu, Q., et.al.: Runout analysis of flow-like landslides triggered by the Ms 8.0 2008 Wenchuan earthquake using smoothed particle hydrodynamics. Landslides **9**(2), 275–283 (2012)

Kent, P.E.: The transport mechanism in catastrophic rock falls. J. Geol. **74**(1), 79–83 (1966)

Kong, J., Cai, Q., Zhang, Y., et al.: Physical model test of debris landslide reinforcement with single row micro-pile. J. Mt. Sci. **31**(4), 399–405 (2013)

Lee, C.T., Huang, C.C., Lee, J.F., et al.: Statistical approach to earthquake-induced landslide susceptibility. Eng. Geol. **100**(1), 43–58 (2008)

Li, X., Wu, Y., He, S., et al.: Application of the material point method to simulate the post-failure runout processes of the Wangjiayan landslide. Eng. Geol. **212**, 1–9 (2016)

Lo, C.M., Lin, M.L., Tang, C.L., et al.: A kinematic model of the Hsiaolin landslide calibrated to the morphology of the landslide deposit. Eng. Geol. **123**(1), 22–39 (2011)

Montagna, F., Bellotti, G., Risio, M.D.: 3D numerical modeling of landslide-generated tsunamis around a conical island. Nat. Hazards **58**(1), 591–608 (2011)

Nandi, A., Shakoor, A.: A GIS-based landslide susceptibility evaluation using bivariate and multivariate statistical analyses. Eng. Geol. **110**(1), 11–20 (2010)

Pastor, M., Blanc, T., Haddad, B., et al.: Application of a SPH depth-integrated model to landslide runout analysis. Landslides **11**(5), 793–812 (2014)

Pucker, T., Grabe, J.: Numerical simulation of the installation process of full displacement piles. Comput. Geotech. **45**(9), 93–106 (2012)

Qiu, G., Henke, S.: Controlled installation of spudcan foundations on loose sand overlying weak clay. Mar. Struct. **24**(4), 528–550 (2011)

Qiu, G., Henke, S., Grabe, J.: Application of a Coupled Eulerian-Lagrangian approach on geomechanical problems involving large deformations. Comput. Geotech. **38**(1), 30–39 (2011)

Quinnp, E., Diederichsm, S., Rower, K., et al.: A new model for large landslides in sensitive clay using a fracture me. Can. Geotech. J. **48**(8), 1151–1162 (2011)

Ramakrishnan, D., Singh, T.N., Verma, A.K., et al.: Soft computing and GIS for landslide susceptibility assessment in Tawaghat area, Kumaon Himalaya, India. Natural Hazards **65**(1), 315–330 (2013)

Saha, A.K., Gupta, R.P., Sarkar, I., et al.: An approach for GIS-based statistical landslide susceptibility zonation—with a case study in the Himalayas. Landslides **2**(1), 61–69 (2005)

Shahabi, H., Hashim, M.: Landslide susceptibility mapping using GIS-based statistical models and remote sensing data in tropical environment. Sci. Rep. **5**(3), 9899 (2015)

Sun, Y., Yang, J., Song, E.: Runout analysis of landslides using material point method. In: Iop Conference Series: Earth and Environmental Science (2015)

Tang, C.L., Hu, J.C., Lin, M.L., et al.: The Tsaoling landslide triggered by the Chi-Chi earthquake, Taiwan: insights from a discrete element simulation. Eng. Geol. **106**(1–2), 1–19 (2009)

Tho, K.K., Leung, C.F., Chow, Y.K., et al.: Eulerian Finite-Element technique for analysis of jack-up spudcan penetration. Int. J. Geomech. **12**(1), 64–73 (2012)

Turner, D., Lucieer, A., De Jong, S.: Time series analysis of landslide dynamics using an unmanned aerial vehicle (UAV). Remote Sens. **7**(2), 1736–1757 (2015)

Wu, J.H., Chen, C.H.: Application of DDA to simulate characteristics of the Tsaoling landslide. Comput. Geotech. **38**(5), 741–750 (2011)

Yin, Y., Li, B., Wang, W., et al.: Mechanism of the December 2015 catastrophic landslide at the shenzhen landfill and controlling geotechnical risks of urbanization. Engineering **2**(2), 230–249 (2016)

Yin, Y., Xing, A., Wang, G., et al.: Experimental and numerical investigations of a catastrophic long-runout landslide in Zhenxiong, Yunnan, southwestern China. Landslides **14**(2), 1–11 (2017)

Yuan, R.M., Tang, C.L., Hu, J.C., et al.: Mechanism of the Donghekou landslide triggered by the 2008 Wenchuan earthquake revealed by discrete element modeling. Nat. Hazards Earth Syst. Sci. **14**(5), 1195–1205 (2014)

Zhang, Y., Wang, J., Xu, Q., et al.: DDA validation of the mobility of earthquake-induced landslides. Eng. Geol. **194**, 38–51 (2014)

Zhang, Z., Wang, T., Wu, S., et al.: Investigation of dormant landslides in earthquake conditions using a physical model. Landslides **1**, 1–13 (2017)

Zhou, J., Cui, P., Yang, X.: Dynamic process analysis for the initiation and movement of the Donghekou landslide-debris flow triggered by the Wenchuan earthquake. J. Asian Earth Sci. **76**, 70–84 (2013)

Optimizing Arterial Signal with Delay and Queue

Jin Wang(✉), Zou Zhiyun, and Gao Jianzhi

School of Civil Engineering & Mechanics, Huazhong University of Science & Technology, Wuhan 430074, China
wj7040@163.com

Abstract. Signal synchronization is an important countermeasure to improve efficiency and relieve congestion for urban arterials. Scientifically evaluating the effects is the foundation for signal synchronization optimization. A signal synchronization evaluation and optimization model is proposed based on delay and queue length assessment at arterial intersections. With traffic wave theory, taking intersections' phase and sequence, split, movement, link length and offset into considerations, the model can simulate the evolution of queue endpoint phase by phase, and get a polygon which can be used to calculate the control delay. The case study indicated that the model can accurately designate the influence of variable signal coordination parameters, and can be used to arterial signal synchronization appraisal and optimization.

Keywords: Traffic engineering · Arterial signal optimization · Traffic wave Traffic delay · Queue length

1 Introduction

Arterial signal synchronization is an important countermeasure to relieve traffic congestion. Most studies on arterial signal optimization fall into two categories: the mathematical programming method and the simulation-based method [1, 2].

The mathematical programming method formulates a model to maximizing the arterial progression bandwidth. Little (1966) [3] proposed the bi-direction arterial progression bandwidth maximizing model with mixed-integral linear programming model which named MAXBAND, and solved the model by branch and branch algorithm. Based on the MAXBAND model. Focus on the variations of volumes and bandwidth requirements through sections of an arterial, Gartner [4, 5] proposed alterable progression bandwidth model–MULTIBAND for bi-direction arterial and urban grid network. Generally, the mathematical programming method assumes that the travel time between two adjacent intersections is constant disregarding the influence of queue at the downstream intersection [6], so it can't adapt the complicated environment such as approach block or link block caused by long queue at intersection [7], and it can't be used in over-saturated state. Otherwise the mathematical programming method is regardless of turning volume at intersection [2], so it can't work well when there is magnitude of left-turn movements.

W.-C. Cheng et al. (eds.), *Tunneling in Soft Ground, Ground Conditioning and Modification Techniques*, Sustainable Civil Infrastructures,
https://doi.org/10.1007/978-3-319-95783-8_21

The simulation-based methods describe the complicated traffic phenomenon by a series of mathematic models, which can identify the movements, calculate various evaluation indexes for multi-phase signals in a network, so it is practically used extensively. The most famous simulation-based model is TRANSYT, which optimized the network signals with macroscopic traffic-flow model. The model is used in SCOOT (1981) [8] and succeed in worldwide. TRANSYT composed with traffic simulation model and signal optimization model simulates traffic flow progression in a network by mathematical method, calculates the traffic evaluation indexes, and circulates the traffic simulation model and signal optimization model to generate a coordinated signal plan. In TRANSYT the traffic fleet discrete model is used, which treats the vehicles' speeds with a certain distribution, so as to get the vehicle arrival model at the downstream intersection. But in the real world with the magnitude of turning movements, the adjacent vehicles are mutually influenced and restricted, so it is difficult to satisfy a certain stochastic distribution model. Otherwise the model assumes a point queue which doesn't count for queue length, so it limits the applicability for over-saturated traffic.

With different mathematic formations many simulation-based method were developed, such as store-forward models [9], queue-dispersion model [10], and traffic wave model, etc., among which traffic wave model was extensively used and has gotten deeply developed recently [11].

With traffic wave model, this paper described the evolution of traffic flow in an arterial, calculated queue length and delay considering up-stream and down-stream signal and movements, and built a model to optimize signal coordination for the arterial.

2 Basic Principles

In the model, the traffic flow from up-stream intersection is discretized to streaks of homogeneous traffic waves by phase and phase. The evolution of the tail of queue at down-stream intersection is characterized by stop wave in red light and proceeding wave in green light. Then the maximum queue length and delay can be calculated out with the evolution trail of the queue end. Let the position of the tail of queue at downstream intersection at time t_k is L_k, and the spatiotemporal point of the tail of queue is notated as (t_k, L_k). Considering emit flow from upstream-intersection and signal change at downstream intersection phase by phase, the evolution of spatiotemporal point of the tail of queue at downstream intersection is **depicted** as Fig. 1.

In the Fig. 1, the left bar represents phases arrangement at upstream intersection, the right bar represents the signal of the lane group at downstream intersection. The oblique line densities represent the flow rates to the lane group from upstream intersection. The round points represent the spatiotemporal points of the tail of queue, which evolution trail is : $(t_{q0}, L_{q0}) \rightarrow (t_1, L_1) \rightarrow (t_2, L_2) \rightarrow (t_k, L_k) \rightarrow (t_j, L_{max})$, where (t_{q0}, L_{q0}) is the initial queue end, $(t_k, L_k)$is the kth queue end, (t_j, L_{max}) is the maximal queue end. The area of polygon composed by the queue ends and the bar can be used to calculate delay for the downstream intersection. The maximum queue length and delay at the opposite direction at the arterial can be calculated by the same way.

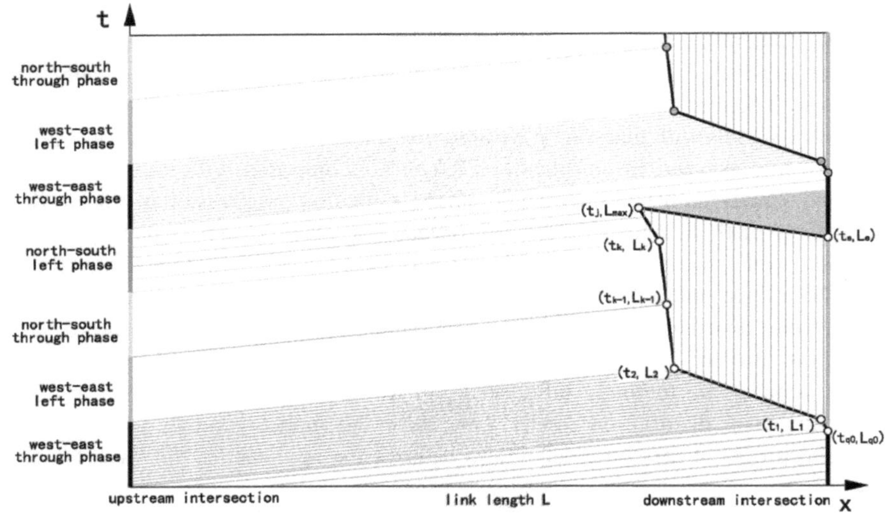

Fig. 1. Evolution of queue length

The arterial signal coordination optimization can be processed by the below steps:

(1) Calculate the flow rate. The flow rates to the downstream lane group can be calculated based on the phase arrangement at upstream intersection.
(2) Decide the initial point of evolution for the queue end.
(3) Evolution calculation of queue end. With the phase arrangement and flow rate at upstream intersection, the queue end can be calculated phase by phase.
(4) The maximum queue length calculation. When proceeding wave catches the stop wave, the maximum queue occurs.
(5) Delay calculation. The delay at downstream intersection can be decided by the area of the polygon composed by the queue ends.
(6) Build the queue and delay model in the opposite direction of the arterial by the same way.
(7) Optimize arterial signals by minimizing the delay and queue.

3 Models

3.1 Flow Rate in Phases

Let the left, through and right movement volumes of the *i*th approach at upstream intersection are F_{iL}, F_{iT} and F_{iR}. The 0–1 variables which represent if the movements drive to the object approach of downstream intersection are ξ_{iL}, ξ_{iT} and ξ_{iR}. The 0–1 variable which represent if the movements are permitted in phase p are θ_{iLp}, θ_{iTp} and θ_{iRp}. The green time in phase p is g_p. The circle length of the intersection is *C*. The proportion of object lane group volume **to** the approach is η. The flow rate to the lane group of downstream intersection from the phase p of upstream intersection can be calculated as below:

$$f_p = F_{iL} * \xi_{iL} * \theta_{iLp} * C/\sum_p (g_{p*}\theta_{iLp}) + F_{iT} * \xi_{iT} * \theta_{iTp} * C/\sum_p (g_{p*}\theta_{iTp}) + F_{iR} * \xi_{iR} * \theta_{iRp} * C/\sum_p (g_{p*}\theta_{iRp}) \quad (1)$$

3.2 The Initial Point of Evolution

Generally, in under-saturate status the queue length when the green light end is zero, so the queue end (t_0, L_0) can be gotten as below:

$$t_0 = T + g \quad (2)$$

In which:
T—offset, s;
g—the green time for the object lane group at downstream intersection, s;

When it is over-saturate, the initial queue length is the remainder vehicles length at the green time end, and the initial point of the queue end is: (t_0, L_0) = (T + g,L_0).

The time when the vehicle of the initial queue end sets out from the upstream-intersection can be calculated as below:

$$t_s = t_0 - \frac{L - L_{q0}}{v_L} \quad (3)$$

In which:
L—link length between the two adjacent intersections, m;
V_L—speed on the link, m/s.

The phase at time t_s in the upstream intersection can begotten as below:

(1) Let p = 1, (2) if $t_s < g_p$, then $f = f_p$, (3) or else $p = p + 1$, go to step (2).

3.3 Evolution of the Queue End

In phase g_p, the last vehicle will arrive at the queue end at the spatiotemporal point (t_k, L_k), which is figured as Fig. 2.

Considering the evolution of the horizontal ordinates from the point (t_{k-1}, L_{k-1}) to the point (t_k, L_k), the equation can be gotten as below:

$$t_{k-1} + \frac{L'_q}{v_{pk}} = g_p + \frac{L - L_{k-1} - L'_q}{v_L} \quad (4)$$

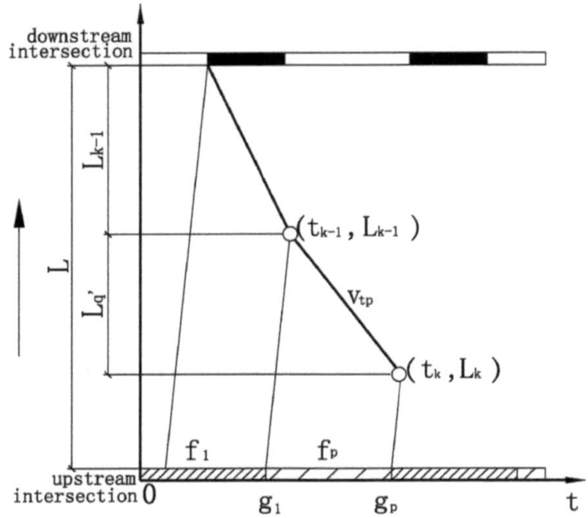

Fig. 2. Evolution of spatiotemporal coordinates of the queue end

Solve the equation, as below:

$$L_q' = \left(g_p - t_{k-1} + \frac{L - L_{k-1}}{v_L}\right) * \frac{v_L * v_{pk}}{v_L + v_{pk}} \tag{5}$$

In which:
L_q'—the part of queue addition in current phase, m;
v_{pk}—The stop wave speed at downstream intersection generated by the vehicles in phase p, m/s;

So:

$$t_k = t_{k-1} + \frac{L_q'}{v_{pk}} = t_{k-1} + \left(g_p - t_{k-1} + \frac{L - L_{k-1}}{v_L}\right) * \frac{v_L}{v_L + v_{pk}} \tag{6}$$

$$L_k = L_{k-1} + L_q' = L_{k-1} + \left(g_p - t_{k-1} + \frac{L - L_{k-1}}{v_L}\right) * \frac{v_L * v_{pk}}{v_L + v_{pk}} \tag{7}$$

When the last vehicle in the upstream intersection arrives the queue end at the red signal time, the model goes to the next phase of upstream intersection and continue to do the evolution calculation. The formula to judge the downstream signal status when the last vehicle arrive the queue end is as below:

If $t_k > T + C$, it's green, or else, it's red.

3.4 The Maximum Queue Length

(1) Calculate the position of proceeding wave

When the green light of the downstream intersection turns on, the proceeding wave start drawing back toward the upstream intersection. At the time t_k, the proceeding wave will arrive at the position as bellow:

$$L_i = (t_k - T - C) * v_q \tag{8}$$

In which:
v_q—the speed of proceeding wave, m/s.

(2) Judge if the queue is dissipating

When the downstream signal is green, the proceeding wave will gradually proceed to the upstream intersection. If the proceeding wave has arrived to the plausible queue end, the maximum queue will occur, which is as below:

If $L_i < L_k$, then $p = p + 1$, and go on the evolution calculation, or else maximum queue occurs, and go to calculate it.

(3) Calculate the maximum queue length

If $L_i \geq L_k$, it denotes that the queue has dissipated before the last vehicle in the current phase arrives at the possible maximum queue end. The possible maximum queue end in the current phase is the point (t_k, L_k). The real maximum queue end is the intersection point of the proceeding wave and the queue wave, which is figured as Fig. 3.

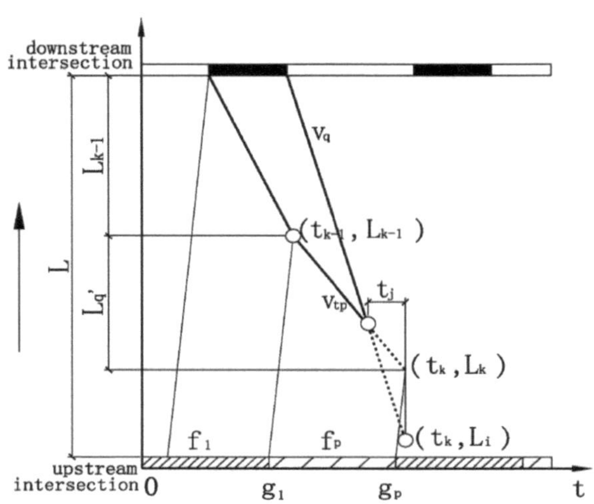

Fig. 3. The maximal queue length calculation

The time span from the real maximum queue end to the possible maximum queue end is as below:

$$t_j = \frac{L_i - L_k}{v_q - v_{pk}} \tag{9}$$

The distance from the real maximum queue end to the stop line is the maximum queue length, which can be calculated as bellow:

$$L_{\max} = L_k - v_{pk} * t_j \tag{10}$$

3.5 Delay Calculation

The polygon enclosed with the queue end points can be used to calculate the delay at the downstream intersection. To divide the polygon's area by average stopping space and the number of arriving vehicles in a circle, the delay can be calculated as below:

$$d_t = \frac{S * \sum_{p=1}^{n} f_p * \frac{3600}{C}}{D_t} \tag{11}$$

In which:
S—The area of the polygon enclosed with the queue end points;
D_t—verage stopping space;
n—the number of phase.

3.6 Model of the Opposite Direction of the Arterial and Coordination Optimization

To model the opposite direction of the arterial, it should renew the designation of the upstream and downstream intersections, and set the corresponding movements and signal parameters. Then with the same method in the formula 3.1–3.5, the maximum queue length and delay can be gotten. With the bi-direction model of queue and delay models, we can optimize the arterial coordination by adjusting the signal parameters such as offset, circle length, split and etc.

4 Case Study

Set an arterial, in which the distance of adjacent intersections is 300 m. The number of lanes in the approaches are all three with one for left, one for through and one for right. The volume in each direction is 500 pcu/h and the left turn rate and right turn rate is same as 15%. The saturate volume is 1800 pcu/h, the average stopping distance is

7.5 m. The speed in the link is 10 m/s. The signal circle length is 120 s. The intersections are all signalized by four phases and each phase has 27 s green light and 3 s yellow light. Calculate the delay in different offsets by SYNCHRO software and the proposed model, which results are as Table 1.

Table 1. The delay in the arterial

Offset (s)	Delay with the proposed model (s)			Delay with SYNCHRO software (s)			The difference of delay in the arterial (%)
	Delay in the forward direction	Delay in the opposite direction	Delay in the arterial	Delay in the forward direction	Delay in the opposite direction	Delay in the arterial	
0	66.6	66.6	66.6	70.6	69.7	70.2	5.1
10	48.1	64.1	56.1	54.1	62.6	58.3	3.9
20	28.7	61.6	45.1	27.7	57.2	42.5	−6.3
30	9.2	58.9	34.1	12.1	63.0	37.5	9.3
40	16.4	50.4	33.4	20.6	54.5	37.5	10.9
50	24.9	41.9	33.4	29.1	46.0	37.5	10.9
60	33.4	33.4	33.4	37.5	37.5	37.5	10.9
70	41.9	24.9	33.4	46.0	29.1	37.5	10.9
80	50.4	16.4	33.4	54.5	20.6	37.5	10.9
90	58.9	9.2	34.1	63.0	12.1	37.5	9.3
100	61.6	28.7	45.1	57.2	27.3	42.2	−6.9
110	64.1	48.1	56.1	63.5	54.1	58.8	4.6
120	66.6	66.6	66.6	70.6	69.7	70.2	5.1

From the above table we found that the proposed model has almost the same resolution as the SYNCHRO software, and it can sensitively reflect the influence of signal parameters.

In the above case, because the volumes are same in the forward and opposite directions, it exits a region in which the delay in the arterial is fixed regardless of the signal parameters.

To reflected the impacts of other parameters, we varied the adjacent intersection distances in 150, 250, and 350 m, increase the forward volume to 530 pcu/h, and decrease the opposite volume to 470 pcu/h. Applied the proposed model to calculate the arterial delay in different coordination parameters, as Fig. 4.

In the Fig. 4, the three bigger round nodes designate the optimal coordination states under different intersection distances.

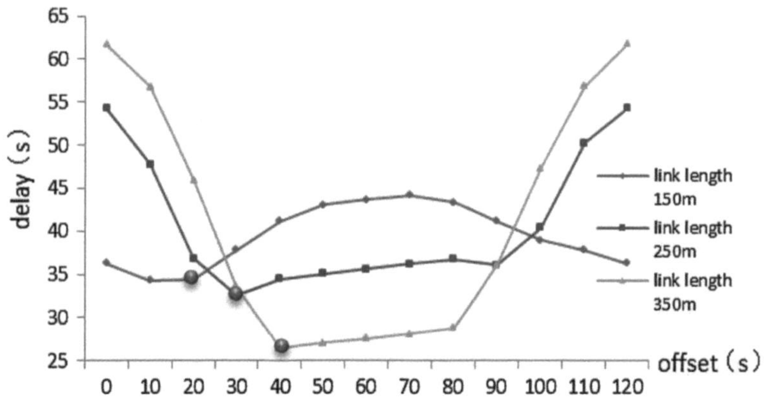

Fig. 4. Variation of arterial delay with different coordination parameters

5 Conclusions

By the case study, it was found that the optimizing model of arterial coordination can sensitively reflect the impacts of delay and maximum queue length for the signal and road parameters. The model is suitable for arterial coordination in various conditions. If considering the delay and queue at the crossing road simultaneously, the model will be applied to network signal coordination.

Acknowledgements. This paper is supported by the independent innovation fund of Huazhong university of science & technology (No: 2016YXMS099).

References

1. Lan, C.L., Chang, G.L.: Optimizing signals for arterials experiencing heavy mixed scooter-vehicle flows. Transp. Res. Part C-Emerg. Technol. **72**, 182–201 (2016)
2. Chen, Y.Y., Chang, G.L.: A macroscopic signal optimization model for arterials under heavy mixed traffic flows. IEEE Trans. Intell. Transp. Syst. **15**, 805–817 (2014)
3. Little, J.D.C.: The synchronization of traffic signals by mixed integer linear programming. Oper. Res. **14**, 568–594 (1966)
4. Gartner, N.H., Assman, S.F., Lasaga, F., Hou, D.L.: A multi-band approach to arterial traffic signal optimization. Transp. Res. Part B: Methodol. **25**(1), 55–74 (1991)
5. Gartner, N.H., Stamatiadis, C.: Arterial-based control of traffic flow in urban grid networks. Math. Comput. Model. **35**, 657–671 (2002)
6. Li, Jing-Quan: Bandwidth synchronization under progression time uncertainty. IEEE Trans. Intell. Transp. Syst. **15**(2), 749–759 (2014)
7. Liu, Y., Chang, G.L.: An arterial signal optimization model for intersections experiencing queue spillback and lane blockage. Transp. Res. Part C-Emerg. Technol. **19**, 130–144 (2011)
8. Robertson, D.I., Bretherton, R.D.: Optimizing networks of traffic signals in real-time: the SCOOT method. IEEE Trans. Veh. Technol. **40**, 11–15 (1991)

9. Papageorgiou, M.: An integrated control approach for traffic corridors. Transp. Res. C-Emerg. Technol. **3**(1), 19–30 (1995)
10. Wu, J., Chang, G.L.: An integrated optimal control and algorithm for commuting corridors. Int. Trans. Oper. Res. **6**(1), 39–55 (1999)
11. Sun, Z., Jin, W.L., Ritchie, S.G.: Simultaneous estimation of states and parameters in Newell's simplified kinematic wave model with Eulerian and Lagrangian traffic data. Transp. Res. Part B: Methodol. **104**, 106–122 (2017)

Author Index

W.-C. Cheng et al. (eds.), *Tunneling in Soft Ground, Ground Conditioning and Modification Techniques*, Sustainable Civil Infrastructures,
https://doi.org/10.1007/978-3-319-95783-8

MIX
Papier aus verantwortungsvollen Quellen
Paper from responsible sources
FSC® C105338

If you have any concerns about our products,
you can contact us on
ProductSafety@springernature.com

In case Publisher is established outside the EU,
the EU authorized representative is:
Springer Nature Customer Service Center GmbH
Europaplatz 3, 69115 Heidelberg, Germany

Printed by Libri Plureos GmbH
in Hamburg, Germany